Immunochemical Technology
for Environmental Applications

ACS SYMPOSIUM SERIES **657**

Immunochemical Technology for Environmental Applications

Diana S. Aga, EDITOR
U.S. Geological Survey

E. M. Thurman, EDITOR
U.S. Geological Survey

Developed from a symposium sponsored
by the Division of Environmental Chemistry, Inc.

American Chemical Society, Washington, DC

Library of Congress Cataloging-in-Publication Data

Immunochemical technology for environmental applications / Diana S. Aga, editor, E. M. Thurman, editor.

    p.   cm.—(ACS symposium series, ISSN 0097–6156; 657)

"Developed from a symposium sponsored by the Division of Environmental Chemistry, Inc., at the 211th National Meeting of the American Chemical Society, New Orleans, Louisiana, March 24–29, 1996."

Includes bibliographical references and indexes.

ISBN 0–8412–3487–6

1. Pollution—Measurement—Technological innovations—Congresses.
2. Immunoassay—Congresses.

I. Aga, Diana S., 1967– . II. Thurman, E. M. (Earl Michael), 1946– . III. American Chemical Society. Division of Environmental Chemistry. IV. American Chemical Society. Meeting (211th: 1996: New Orleans, La.) V. Series.

TD193.I484   1996
628.5′028′7—dc21
                                96–47943
                                      CIP

This book is printed on acid-free, recycled paper.

Advisory Board

ACS Symposium Series

Foreword

THE ACS SYMPOSIUM SERIES was first published in 1974 to provide a mechanism for publishing symposia quickly in book form. The purpose of this series is to publish comprehensive books developed from symposia, which are usually "snapshots in time" of the current research being done on a topic, plus some review material on the topic. For this reason, it is necessary that the papers be published as quickly as possible.

Before a symposium-based book is put under contract, the proposed table of contents is reviewed for appropriateness to the topic and for comprehensiveness of the collection. Some papers are excluded at this point, and others are added to round out the scope of the volume. In addition, a draft of each paper is peer-reviewed prior to final acceptance or rejection. This anonymous review process is supervised by the organizer(s) of the symposium, who become the editor(s) of the book. The authors then revise their papers according to the recommendations of both the reviewers and the editors, prepare camera-ready copy, and submit the final papers to the editors, who check that all necessary revisions have been made.

As a rule, only original research papers and original review papers are included in the volumes. Verbatim reproductions of previously published papers are not accepted.

ACS BOOKS DEPARTMENT

Contents

NEW FORMATS OF IMMUNOCHEMICAL
TECHNIQUES

IMMUNOASSAYS IN ENVIRONMENTAL
STUDIES AND MONITORING

Preface

THE INCREASING CONTAMINATION OF THE ENVIRONMENT has prompted regulatory agencies to mandate extensive monitoring of soil, air, water, and food resources, resulting in unprecedented demands on analytical laboratories. Accordingly, immunochemical methods have emerged from screening tools to quantitative analytical techniques in order to meet these demands in more timely and cost-effective ways. In the past few years, the development of sensitive immunoassays and innovative assay formats for environmental applications has been considerable. However, most of these techniques are not well known to many environmental scientists, and their availability is limited primarily to the research laboratories that developed them. Dissemination of this information is necessary to expedite the practical applications of these techniques so that environmental studies are not cost prohibitive.

The symposium upon which this book is based, "Development and Applications of Immunoassays for Environmental Analysis", was held at the 211th National Meeting of the American Chemical Society in New Orleans, Louisiana, March 24–29, 1996. This symposium was organized to facilitate information exchange of recent technological advances in the field of environmental immunoassays between researchers, assay developers, and environmental scientists.

Immunochemical techniques have received increased acceptance from regulatory agencies and consumers because of the number of studies demonstrating the reliability of immunoassays in environmental analysis, such as in large water-quality surveys, characterization of hazardous waste sites, and post-remediation monitoring. Unique applications of immunoassays in the field of environmental chemistry are illustrated in this book with the aim of enhancing the confidence of analysts to employ this technology where it can be a suitable analytical method. The information provided in many chapters of this book will serve as an avenue for users to discuss the limitations of an assay so that developers may look for solutions to circumvent these obstacles.

The previous limitations of immunochemical assays, such as heterogeneity of polyclonal antibodies, cross-reactivity, or single-assay capability, which hindered their wide acceptance by the scientific and regulatory community, are gradually being resolved. For example, the production of homogeneous monoclonal antibodies against environmentally important compounds has advanced immunosensor research and the development of highly specific assays. In addition, recombinant antibodies and molecularly imprinted polymers are promising and attractive alternatives to conventional antibodies because of their lower cost of production and their potentially better detection capabilities. The problem of cross-reactivity may eventually be eliminated, or even turned into a benefit, through statistical manipulations and neural networks that will deconvolute detection signals to identify constituent analytes. New detection systems have also been employed to increase assay sensitivity and improve detection limits. Finally, as multianalyte immunoassays become available, wider applications in studies on the fate and transport of contaminants in the environment will be realized. Overview discussions and experimental results from these areas are provided in this book with the aim of provoking scientific ideas that will advance these technologies from being prototypes to routine, practical tools.

Great efforts by immunochemists are paving the way for the miniaturization and automation of immunoassay techniques. Collaboration among scientists from various disciplines has resulted in the realization of usable immunosensors and field-portable assays, automated flow-injection immunoassays, and high-performance immunoaffinity chromatography for the analysis of pesticides and contaminants from military and industrial sites. The development and optimization of these techniques are discussed in this book. Although the principal focus of this book is to present advances in environmental immunoassays, recent innovations in clinical immunoassays have also been included with the anticipation that these novel techniques will be adaptable for environmental applications. This book will be a valuable reference for scientists in industry, government agencies, and academia who are involved in immunoassay development, and those who are interested in employing immunoassays for environmental studies.

Acknowledgments

We thank the authors and the peer reviewers for their time and willingness to contribute. We express our appreciation for the generous support provided by the American Chemical Society, Analytical Environmental Immunochemical Consortium, Millipore Corporation, and Ohmicron Corporation. Special thanks are due to Lisa Zimmerman, U.S. Geological Survey, Kansas, for co-organizing the symposium.

DIANA S. AGA
E. M. THURMAN
U.S. Geological Survey
4821 Quail Crest Place
Lawrence, Kansas 66049

Chapter 1

Environmental Immunoassays: Alternative Techniques for Soil and Water Analysis

Diana S. Aga[1,2] and E. M. Thurman[1]

[1]U.S. Geological Survey, 4821 Quail Crest Place, Lawrence, KS 66049

Analysis of soil and water samples for environmental studies and compliance testing can be formidable, time consuming, and costly. As a consequence, immunochemical techniques have become popular for environmental analysis because they are reliable, rapid, and cost effective. During the past 5 years, the use of immunoassays for environmental monitoring has increased substantially, and their use as an integral analytical tool in many environmental laboratories is now commonplace. This chapter will present the basic concept of immunoassays, recent advances in the development of immunochemical methods, and examples of successful applications of immunoassays in environmental analysis.

Immunoassay (IA) is a method of analysis that relies on specific interactions between antibodies and antigens to measure a variety of substances, ranging from complex viruses and microorganisms to simple pesticide molecules and industrial pollutants. IA techniques can be qualitative, semiquantitative, or quantitative. The key reagent in IAs are antibodies, which are soluble proteins produced by the immune system in response to infection by foreign substances (called antigen).

IAs are based on the fundamental concept that antibodies prepared in animals can recognize and bind with exquisite specificity to the antigen that stimulated their production. The binding forces involved in specific interactions between antibodies (Ab) and antigens (Ag) are of a noncovalent, purely physicochemical nature. The noncovalent interactions that form the basis of Ab-Ag binding include hydrogen bonds, ionic bonds, hydrophobic bonds, and van der Waals interactions. Each of these interactions is weaker than the covalent bonds. Hence, for an effective Ab-Ag interaction, a large number of such interactions need to be present, and there should be a very close fit between the Ab and the Ag for these noncovalent forces to operate (less than 1 angstrom or 1×10^{-7} mm) (1). This is reflected in the high degree of specificity characteristic of Ab-Ag interactions. The strength of the sum total of noncovalent interactions between a single binding site on an Ab and a single epitope (the part of Ag

[2]Current address: EAWAG, Swiss Federal Institute for Environmental Science and Technology, Ueberlandstrasse 133, CH–8600 Dübendorf, Switzerland

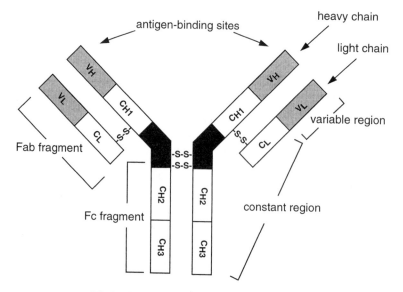

Figure 1. General structure of an antibody.

that directly interacts with the Ab binding site) is the affinity of the Ab for the epitope. Antibodies may posses affinity constants for individual antigens on the order of 10^4-10^{12} M^{-1}. Thus, antibodies make a suitable tool for environmental analysis because of their ability to distinguish between the homologous antigen (target compound) and the myriad of other compounds of widely diverse structure that are found in environmental samples.

The use of antibodies as analytical reagents was first reported in 1959 when Berson and Yalow successfully demonstrated the measurement of picogram levels of human insulin in samples of body fluids using radioimmunoassay (RIA) (2). Since then, various IAs for detecting hundreds of molecules of endogenous and exogenous origin have been described. This technique proved to be reliable, fast, and very sensitive; hence, many other RIAs have been developed for clinical and medical tests since then. Radiolabels were gradually replaced with enzyme labels because of the hazards associated with using radioactive materials. Enzyme-linked immunosorbent assay (ELISA), which was first introduced by Engvall and Perlman in 1971 (3), has become perhaps the most popular IA format used in laboratories today.

The use of immunochemical techniques in the environmental field was first proposed in 1971 by Ercegovich (4), who suggested the use of immunological screening methods for the rapid detection of pesticide residues and for confirming results of conventional analyses. An RIA for the insecticides aldrin and dieldrin was the first reported IA for an environmental contaminant (5). Although a few RIAs still exist in the medical field, they are seldom used in environmental and food analysis because of the need for special handling and disposal of the radioactive materials.

The development of IAs for environmental applications was rather slow. It was only during the early 1990's that the scientific community and regulatory agencies recognized the strength and advantages that IAs can offer (6-7). To date, there are at least 11 commercially available IA kits that have been validated by the U.S. EPA to be included in SW-846 methods, and this list will continue to grow as more kits prove useful and reliable for the particular matrix of concern. Several chapters in this book demonstrate the effectiveness of IAs for the analysis of pesticides (see chapters 18-21, 24-25), metals (see chapters 3, 4, and 23), and other industrial contaminants (see chapters 6 and 22) in soil and water, both in small- and large-scale studies in the United States, Europe, and Asia.

Antibodies and their Production

Antibodies are glycoproteins produced by lymphocyte B cells, usually in conjunction with T-helper cells, as part of the immune system's response to foreign substances. Antibodies (also known as immunoglobulins) are found in the globulin fractions of serum and in tissue fluids that bind in a highly specific manner to foreign molecules.

There are five classes of immunoglobulins, IgG, IgM, IgA, IgD, and IgE. The predominant immunoglobulin in serum is IgG, which has an approximate molecular weight of 160,000 daltons. It has been shown that all five classes of immunoglobulins share a common basic structure comprised of two light chains and two heavy chains linked by disulfide bonds and noncovalent forces. The antibody molecule usually is represented by a Y-shaped molecule to depict its three-dimensional structure (Figure 1). Porter (8) demonstrated that the two arms of the Y-shaped antibody were identical and were able to combine with the antigen; hence, these were called Fab fragments (fragment antigen binding). On the other hand, the tail of the Y-shaped antibody was found to have the ability to crystallize and was subsequently called the Fc fragment (fragment crystallizable). This fragmentation of IgG is shown in Figure 1. The amino acid sequences of the N-terminal regions of both heavy (V_H) and light chains (V_L) show exceptional degrees of variability (hence termed the hypervariable regions or

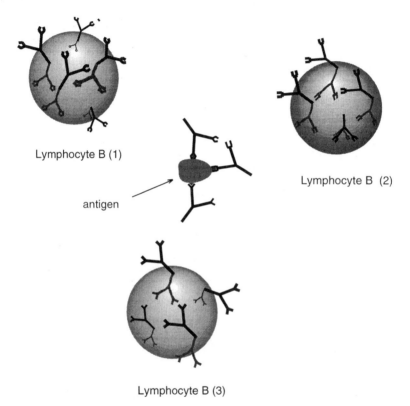

Figure 2. Production of polyclonal antibodies.

complementarity determining regions) and are directly involved in the formation of the binding site. The majority of the chemical structure of each immunoglobulin class is very similar and is referred to as the constant region (C_L and C_H).

Polyclonal antibodies obtained by immunizing experimental animals are the most common type of antibodies used in environmental IA methods. However, monoclonal antibodies, produced using the hybridoma technology developed by Köhler and Milstein (9), have also become popular due to their inherent homogeneity and specificity. Recently, genetically engineered or recombinant antibodies have gained increasing attention due to the lower cost of production and their potentially better analyte detection capabilities. Although the recombinant antibody technology is relatively new, it has been shown that recombinant antibodies may be produced efficiently without resorting to the time-consuming and difficult hybridoma methodologies (10). Chapter 2 (Hall et al.) in this book provides an overview of phage-display technology (one of the most common recombinant antibody technologies) and its applications in the detection of environmental contaminants.

Polyclonal Antibodies. The stimulation of the immune system of an animal can be accomplished by introducing an antigen with sufficient structural complexity, high molecular weight (usually >3,000 daltons), and containing a region that the lymphocyte B cells will recognize as foreign. It is critical that the antigen be foreign to the animal system in which it is being introduced to activate the antibody formation by the immune system. An antigen that can elicit immune response is specifically termed an immunogen (for immune response generating). Small molecules with a molecular weight of <1,000 daltons, such as most environmental contaminants, generally are not immunogenic even though they are foreign to the animal. Therefore, these compounds cannot be used to directly stimulate antibody production. To produce antibodies for IAs of such compounds, the small molecule is physically coupled to a larger, more immunogenic molecule, such as proteins that are foreign to the immunized animal. The small molecule in this arrangement is called a hapten, and the larger molecule to which it is coupled to is referred to as the carrier. Some of the most commonly used carriers include keyhole limpet hemocyanin (KLH), bovine serum albumin (BSA), and other serum albumins that are readily available commercially. A variety of proven methods are available for conjugation of haptens to their carriers, most of them using common commercially available reagents. An example of this is given in chapter 5 in this book (Knopp et al.), which describes a detailed procedure for hapten synthesis and immunogen preparation for the production of polyclonal antibodies against a small molecule, 1-nitropyrene.

Immune response is induced by injecting a complex of the hapten and the carrier (immunoconjugate) into an animal. Because of the multiple epitopes of the immunoconjugate, the antibody-producing cells of the immune system will respond and proliferate differently, giving rise to several antibodies with varying specificities. Each B-cell or lymphocyte will produce a specific antibody that is directed towards certain parts of the immunoconjugate as depicted in Figure 2. These antibodies are called polyclonal antibodies because they are produced from several antibody-producing cells. Polyclonal antibodies are heterogeneous, and thus standardization of antibodies is critical if used in immunochemical techniques. The specificity of the antibodies strongly depends on the chemistry of the conjugation of the hapten to the carrier and the purity of the conjugates used for immunization. The presence of impurities in the reagents used for the synthesis of immunoconjugates will lead to nonspecific antibodies and, thus, to unwanted cross-reactivities of the antisera. Despite the limitations of using polyclonal antibodies, these are the conventional reagent used in assay development because of the low cost and short time required for antibody production. Also, it has been shown that generally polyclonal antibodies are more sensitive than monoclonal antibodies for pesticide analysis (11).

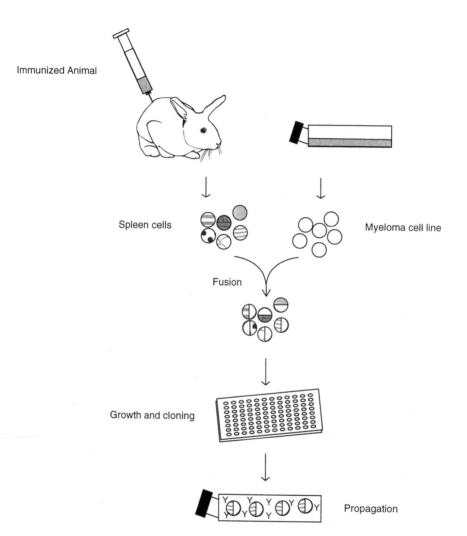

Figure 3. Production of monoclonal antibodies.

Monoclonal Antibodies. Antibodies may be produced by fusing antibody-producing lymphocyte B cells with mutant myeloma cells (tumor cells) using polyethylene glycol treatment (Figure 3). This technology allows the production of antibodies by a single clone of lymphocyte B or plasma cells, hence the term monoclonal antibodies (Mab). The lymphocyte B cell provides the specificity, whereas the myeloma cell confers immortality on the hybridoma clone. The identical copies of the antibody molecules produced contain only one type of heavy chain and one type of light chain. Mabs are homogeneous and are used extensively in clinical IAs, but the high initial cost of developing Mabs has hampered their use for environmental applications. However, because of the necessity of having a homogeneous and unlimited supply of antibodies for IAs that will be used for regulatory purposes, there has been an increasing interest in the use of Mabs for environmental analysis. Lenz et al. (chapter 6 in this book), for instance, review the different monoclonal and polyclonal antibodies used for detecting chemical warfare agents in environmental samples and biological fluids.

Recombinant Antibodies. The limitations encountered in hybridoma technology has triggered the interest of many scientists in using recombinant DNA procedures for antibody production. Although the Mabs have defined specificity and may be produced in virtually unlimited amounts, their production is expensive, time consuming and labor intensive. Recombinant DNA technology is based on isolating genes that encode antibodies from one organism and purifying and reproducing them in another organism. For example, antibody fragments, with the binding properties of the intact antibody, may be expressed in useful amounts by easy-to grow organisms such as *Escherichia coli* (*10*). Using the powerful tools of molecular biology, the binding sites of antibodies can be manipulated to give more desirable characteristics, such as increased specificity and affinity. Antibody fragments may be synthesized from cloned and engineered genes in several nonmammalian hosts. Production of recombinant antibodies for polynuclear aromatic hydrocarbons, phenylureas, and imidazolinone herbicides has been reported (*12*), but the method is still far from being routine due to the complexity of cloning, assembling, and expressing antibody molecules.

Classifications of Immunoassays

Competitive vs. Non-competitive Immunoassays. Immunoassays can be classified as competitive or non-competitive. The main principle of competitive immunoassays is shown in the following immunochemical reaction:

$$2Ab + Ag + Ag^* = AbAg + AbAg^*$$

A free antigen (Ag) competes with a labeled antigen (Ag*) for a fixed and limited number of specific binding sites on the antibody (Ab) molecules. The label can be radioisotopes, enzymes, liposomes, fluorophores, or chemiluminescent compounds. After the reaction is complete, the amount of labeled antigen is determined. The extent of the binding of labeled antigen by the antibody depends on the concentration of unlabeled antigen and this allows the determination of the concentration of the unknown analyte (free antigen). A calibration curve is constructed from known amounts of antigen to determine the concentration of the unknown analyte.

Non-competitive assays are typically used for the assay of large molecules with more than one epitope. An example of this assay is a sandwich ELISA where antibodies are adsorbed on a solid phase to capture the antigen, and a second labeled antibody is added to bind to the other epitope of the antigen. The amount of the labeled antibody is measured and is directly proportional to the antigen present in the sample. Non-competitive assays are generally more sensitive than competitive methods, and enhanced specificity is oftentimes observed because two sites on the analyte must be bound by the antibodies. However, this type of assay is not suited for small molecules, such as most of the environmental contaminants, because the analyte must have multiple epitopes to allow simultaneous binding of two antibodies. Examples of the successful applications of non-competitive assays are in protein analysis, identification of plant viruses or bacteria, and detection of bacterial pathogens, such as Salmonella in food.

Homogeneous vs. Heterogeneous Assays. Another classification of IAs may be based on whether separation of the bound and free phases is required before addition of the detecting compound for end-point detection. Homogeneous assays, which do not require a separation step, use signal modulation for detection. For example, in the homogeneous assay described by Dzantiev et al. (chapter 7 in this book), the activity of the antigen-enzyme conjugate is inhibited after binding to the antibody. The absence of enzyme activity results in a colorimetric change which can be used for quantification. The mechanism of this enzyme inhibition is steric. Another example is the microtiter particle agglutination inhibition assay for 2,4-D (Lukin et al., chapter 8 in this book) that uses the formation of a colored film of agglutinated particles to show a positive result. Homogeneous IA offers the advantage of shorter analysis time relative to its equivalent heterogeneous assay, but it is less commonly used due to its inherent limitations, such as matrix effects and lower sensitivity. In addition, not all antigens are amenable to a system that involves their conjugation to enzymes, substrates, or co-factors, the reactions of which they must modify to produce a functional assay.

In heterogeneous assays, physical separation of the bound and free reagent is required. Separation is made possible by attaching either the antigen or the antibody to a solid phase and removing the excess reagent before addition of the detecting compound. Solid supports are varied and include test tubes, microtiter plates, magnetic beads, latex emulsions, papers, and glass surfaces. Although heterogeneous assays involve more steps due to the sequential rinsings required, removal of the unreacted components results in a less complex matrix for signal detection. Thus, sensitivity and detection limits are usually improved. An example of this type is ELISA, the most common format in environmental IA.

Enzyme-Linked Immunosorbent Assay (ELISA)

Competitive heterogeneous ELISA is the most common format for environmental analysis. Typically, it is the detecting antibody that is attached to the solid support (direct ELISA), although haptens may also be modified so that they can be immobilized on the solid surface (indirect ELISA). In the direct ELISA, the target analyte and an enzyme-labeled hapten (enzyme conjugate) compete for the binding sites of the immobilized antibody. The ratio of free analyte to enzyme conjugate determines the amount of the enzyme conjugate that will bind to the antibody. After unbound materials are removed and washed, an enzyme substrate is added together with a color-producing reagent (chromogen). The enzyme reacts with the substrate, which in turn converts the chromogen into a colored product. The color intensity of the product is inversely proportional to the amount of analyte in the sample because the free analyte prevents the enzyme conjugate from binding. Before quantifying the intensity of the colored product, the enzymatic reaction is stopped by adding concentrated acid to prevent further color development. The most popular label in environmental IA is horseradish peroxidase, although alkaline phosphatase and B-galactosidase have been used as well.

Other Immunochemical Techniques

Although quantitation of many fluorescent labels could rival the signal intensity of radioisotopes, the use of fluorescent labels in IAs has been impeded not only by the cost of fluorimeters but also by the tendency of natural organic compounds to interfere with signal detections. To circumvent this limitation, Swamy et al. (Chapter 12 in this book) examined the use of fluorophores which absorb in the near-infrared (NIR) region as labels in microtiter plate IAs and fiber optic probes. Since very few compounds exhibit fluorescence in the NIR region, the signal-to-background ratio is significantly increased. Immuno-electrochemiluminescence is another alternative approach that shows promise in eliminating background signals in IAs. In this technique, the signal is generated as a result of an oxidation-reduction reaction under well-controlled electric potential; hence, exceptional sensitivities may be obtained (Yu, Chapter 13 in this book). A continuous flow immunosensor using a fluorescent dye that operates in the visible spectrum (fluoresces at 667 nm), where there is very minimal fluorescence background, was successfully used for the detection of sub ppb levels of the explosives TNT (2,4,6-trinitrotoluene) and RDX (hexahydro-1,3,5-trinitro-1,3,5-triazine) (see chapters 16-17 in this book) . Highly sensitive fluorescence and luminescence IAs will find applications in environmental analysis as improvements are made to minimize background interferences in the detection signal.

Sample preparation is as important as the main analytical procedure. Many laboratories are trying to minimize the use of organic solvents, not only for cost savings but for safety and environmental reasons. Sample-preparation techniques that use smaller volumes of solvents are beginning to attract attention. Solid-phase extraction (SPE) is one of the most popular approaches because it allows trace enrichment and cleanup in one step. However, sample cleanup is not always easily achieved, especially in complex matrices, because of the many interfering compounds co-extracted and co-eluted with reversed-phase sorbents. Immunoaffinity sorbents could circumvent this limitation. Immunoaffinity chromatography (IAC) is based on analyte-antibody interactions. These phases are much more selective than phases relying on hydrophobic interactions, adsorption, ion-exchange, etc. They do, however, rely on the availability of specific antisera to the analytes of interest and on the successful immobilization of the antisera onto a suitable solid phase.

Hage et al. (Chapter 10 in this book) provide an example of the effective use of a high-performance immunoaffininty column (HPIAC) to selectively isolate atrazine and its major degradation products (hydroxyatrazine, deethylatrazine, deisopropylatrazine) from water samples. The HPIAC is coupled on-line with a reversed-phase liquid chromatographic (RPLC) column, which in turn separates the isolated compounds for identification and quantitation. The immunoaffinity approach followed by chromatography combines the specificity obtainable from IAs with the confirmation that can be assured from chromatographic methods.

Immunoaffinity chromatography (IAC) is particularly valuable for the isolation of polar ionic compounds, such as most metabolites of herbicides. Despite the many attractive features of IAC, it has not found popularity in the environmental field probably because of the large quantities of antibodies required to prepare immunoaffinity columns. With the advent of recombinant antibody technology, commercial quantities of antibodies eventually may be available for low-cost affinity chromatography and concentration systems.

A more promising source of "packing materials" for immunoaffinity cleanup is the use of molecularly imprinted polymers (MIPs) as replacement to antibodies. MIPs are prepared by polymerizing functional monomers around an imprint species, which produces a cavity that is complementary to the chemical structure of the imprint (*13*). The cavity formed in MIPs can bind specifically with the imprint species in a manner that is similar to the binding of an antibody molecule with an antigen. For this reason, MIPs are referred to as artificial antibodies, and just like polyclonal antibodies, these polymers have a range of binding constants due to the heterogeneity of the population. Unlike antibody production, however, the preparation of MIPs is simpler and does not require the use of live animals or any biological systems. Thus, once the polymerization method is optimized, production of MIPs may be achieved inexpensively for applications that require large quantities such as in IAC. For instance, Muldoon and Stanker (Chapter 26 in this book) succesfully applied MIPs for the isolation of atrazine residues from tissue extracts prior to the analysis by high-performance liquid chromatography (HPLC). As inexpensive antibodies or MIPs to a wider range of pesticides and other contaminants become available, the IAC technique will become more popular in the preparation and cleanup of complex environmental samples.

Attempts have been made in the development of immunosensors and multi-analyte IAs. Immunosensors are detection systems that use immobilized antibodies to capture analytes and are equipped with signal-generating labels for quantitation. Recently, much attention has been focused on developing fiber-optic immunosensors for environmental applications. Two main approaches have been explored; one uses evanescent wave sensors (14-15) and the other uses distal end sensing (16), both of which may be designed for continuous monitoring of multiple analytes in the samples. Immunosensors to detect imidazolinone herbicides (14), polychlorinated biphenyls (15), and atrazine (17) in soil and water samples have been reported. Flow-injection immunoanalysis and flow immunosensors also show promise in continous onsite monitoring (see Chapters 11, 16, and 17 in this book). Recent advances in clinical and biomedical IAs indicate that simultaneous detection of two or more analytes is feasible. Chu et al. (Chapter 14 in this book), for instance, developed the "microspot immunoassay" approach to quantify multiple biomolecules and proposed the feasibility of adapting this technology for environmental contaminants. Although immunosensors and multi-analyte assays are ideal systems for environmental testing, these technologies are still far from routine, practical use.

Applications of Immunoassays in Environmental Analysis

Pesticide Residue Analysis. The number of pesticides, (e.g. herbicides, insecticides, and fungicides) and their use have increased substantially during the past three to four decades, revolutionizing agricultural practices and improving crop yields. The increased use of pesticides has been mirrored by an increased public concern about the effect of pesticides on the environment and their detection in drinking-water supplies, which has resulted in increased monitoring programs and water-quality surveys.

Analysis of pesticides has been based primarily on conventional chromatographic techniques such as gas chromatography (GC) and HPLC. Although these techniques are sensitive and reproducible, they are cumbersome, time consuming, and expensive. The potential of immunochemical techniques for pesticide residue analysis in various matrices such as soil, water, fruits, plants, urine, and blood has been examined previously (18-22). IAs have been successfully used in large water-quality surveys of surface and ground water, field-dissipation studies, and investigations of the fate and transport of pesticides in rivers, reservoirs, and the atmosphere (23-26).

One of the first major applications of IAs in the environmental area was conducted by the U.S. Geological Survey (USGS) in its study of the deposition patterns of herbicides in the atmosphere (23). This study investigated the occurrence and distribution of herbicides in precipitation (rain and snow) for a study area encompassing 26 states in the Midwestern and Northeastern United States. Weekly accumulations of precipitation were analyzed by IA for alachlor and atrazine, the two most extensively used herbicides in the United States. The whole study involved the analysis of more than 6,200 precipitation samples by ELISA. About 37% of these samples were confirmed by gas chromatography/mass spectrometry (GC/MS), which correlated well with the ELISA results. Analysis of a large number of samples by ELISA allowed the estimation of the annual mass deposition in precipitation for alachlor and atrazine as shown in Figure 4. In addition, the data generated on the spatial and temporal distribution of alachlor and atrazine in precipitation revealed the significance of the long-range transport of herbicides in the atmosphere.

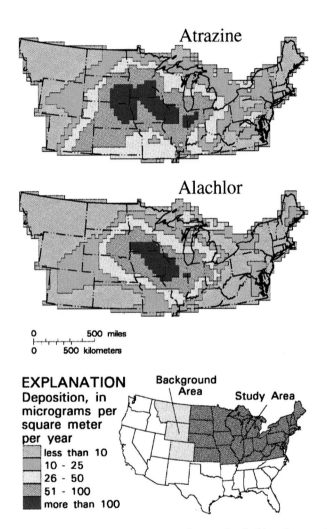

Figure 4. Estimated deposition of atrazine and alachlor in precipitation throughout the Midwestern and Northeastern United States, January through September 1991 (Reproduced from reference 23).

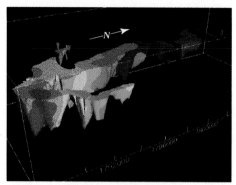

A 1992 pre-application survey,
March 30 to April 10

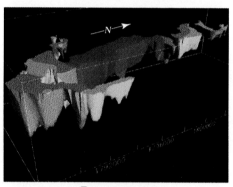

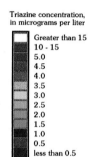

EXPLANATION

Triazine concentration,
in micrograms per liter

Greater than 15
10 - 15
5.0
4.5
4.0
3.5
3.0
2.5
2.0
1.5
1.0
0.5
less than 0.5

B First-flush survey,
June 16-25, 1992

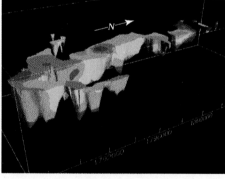

C Summer survey,
August 3 and 4, 1992

Figure 5. Three-dimensional distribution of triazine concentrations observed in Perry Lake during the (A) pre-application survey, March 30-April 10, (B) first-flush survey, June 16-25, and (C) summer survey, August 3-4, 1992 (Reproduced from reference *26*).

Because of the large number of samples analyzed in the above study, an inexpensive analytical method, such as ELISA, was necessary. Furthermore, most of these samples were very small in volume; thus, analysis by GC/MS was not always possible. For small-volume samples, ELISA which required only 300 µL of sample for duplicate analysis was absolutely essential. Many samples were found negative for both herbicides by ELISA and were not analyzed further by GC/MS (except for the selected 10% of the blanks that were used for quality assurance purposes (QA)), resulting in big savings in time and cost of analysis.

ELISA also has been used in many reconnaissance studies of herbicides in surface and ground water. For instance, Goolsby et al. (*24*) used two commercially available IAs of different formats in their study of the temporal and longitudinal variability of dissolved herbicides in the Mississippi River. Results indicate that the two commercial triazine IAs were comparable, but both assays tend to give slightly higher results than GC/MS.

The utility of ELISA in a field-dissipation study of acetanilide herbicides was also successfully demonstrated (*25*). In this study, the use of ELISA, in conjunction with other analytical methods such as HPLC, GC/MS, and fast atom bombardment MS/MS, provided important and new information on the fate and transport of the chloroacetanilide herbicides in soil. In the field study, many soil samples were needed to be analyzed to determine constituent concentration patterns; thus, it could have become a lengthy analysis and an expensive study. However, the low cost of IA allowed the analysis of replicate soil samples, so that the fate and transport of the herbicides and their metabolites were followed more intensively. In addition, the researchers were able to repeat the field study for 2 years; thus, additional information on the effect of field conditions on the persistence and behavior of the herbicides and their metabolites in soil were obtained.

IAs have resulted in many unique field studies including that of Fallon (*27*), where ELISA was used to determine the three-dimensional (3-D) distribution, transport, and relative age of atrazine in a reservoir. Reservoirs are an important component of water supply because they are used for flood control, irrigation, recreation, wildlife habitat, and drinking-water supply. The 3-D images produced in this study modeled the changes in atrazine concentrations during 1 crop year and provided information on the residence time of atrazine in the reservoir. Figures 5A-C show examples of the 3-D distribution of atrazine in Perry Lake, Kansas, determined using ELISA, before application, during the first flush, and during the summer of 1992. The cost of this study would have been prohibitive if IAs were not used.

IAs also are used in wastewater-treatment plants to monitor influent and effluent on a daily basis. For example, diazinon, which is used to control a variety of insects and pests, commonly occurs in municipal effluents in the Southern United States because of its widespread use in industrial, agricultural, and domestic areas. The city of Denton, Texas, used ELISA as a supplement to GC to determine diazinon levels because the standard toxicity tests for diazinon could not track its concentrations well enough (*27*). The results from ELISA allowed the study team to hypothesize that diazinon was entering the sanitary sewer through cracks in pipes due to the runoff from treated lawns after a storm event.

Water-quality monitoring is essential for understanding the condition of water and related resources and for providing a basis for the adoption of decisions that promote the wise use and management of water resources. The use of IAs in monitoring the level of pesticide contamination in the environment could make frequent monitoring more feasible. It has been reported that the U.S. EPA-required quarterly sampling to comply with the water-quality regulations generally underestimates annual mean herbicide concentrations due to the seasonality of herbicide occurrence (*28*). For example, the seasonal nature of herbicide occurrence and transport in Midwestern rivers makes it difficult to calculate accurate estimates of annual mean concentrations using calendar-based sampling strategies with a limited number of samples. A more accurate representation of annual mean concentrations could be obtained by frequent sampling, especially during spring and early summer runoff. For this purpose, IA would be the most cost-effective analytical tool that would give reliable and timely results. IAs could be particularly useful for monitoring pesticides for which environmental and health effects are not fully understood and for those that presently do not have designated MCLs. In addition, Federal water-quality regulations for existing pesticides also may have to be reconsidered if new knowledge on their occurrence and fate in the environment is revealed as a result of more frequent surveys.

Site Assessment and Remediation. IAs have found wide applications in many site-assessment and remediation operations. Because of their field portability and short analysis time, IAs allow the analyst to delineate areas of site contamination more quickly and determine more rapidly the volume of contaminated soil onsite. During site-remediation operations, the real-time information provided by IA can facilitate rapid decision-making and enable the user to manage assessment work processes more effectively. Finally, the cost effectiveness of IAs compared to mobile GC helps consultants and contractors, as well as site owners and facility operators, achieve significant project savings.

One example of a wise usage of IA is in the development of a contamination map of a retail gasoline station site where benzene, toluene, ethylbenzene and xylene (BTEX) were suspected of contaminating a shallow aquifer (29). The problem was to determine which gasoline-storage tanks were leaking and in what direction the BTEX plume was moving. A field-portable IA allowed detailed mapping of the site because of the feasibility to analyze soil samples from the surface and at several depths in a short period of time. IA was used to screen samples to reduce the number of offsite laboratory analyses. Furthermore, as soon as assay results indicated noncontamination of the sample location, sampling was halted, minimizing unnecessary sample collection. This resulted in a $3,000 savings in sample-analysis costs and completion of the project 20 days earlier than if conventional chromatographic analysis was used.

Another example of a successful use of IA is in the characterization of old military sites where soil and ground water may have been contaminated by the long-term storage of explosives. Bart et al. (Chapter 17 in this book) used a field-portable, continuous-flow immunosensor to develop contour maps of the concentrations of TNT and RDX in the saturated zone. The contour maps obtained using the immunosensor were similar to those obtained using the U.S. EPA-approved HPLC method of analysis for these two compounds. Dombrowski et.al (Chapter 18 in this book) also used IAs to determine the extent of ground-water contamination by cyclodiene insecticides and triazine herbicides at the Rocky Mountain Arsenal (RMA), near Denver, Colorado. The RMA is a 70-km^2 tract of land used by the U.S. Army for the production of chemical weaponry, and later leased by private industry to manufacture agricultural pesticides. These studies proved that immunochemical techniques are effective in characterizing military cleanup sites, with less cost, less sampling, and less time involved.

Another obvious area where IA is applicable is in the routine analysis of ground-water discharge following remediation to monitor cleanup progress. The timely results that IA provides is of great value because often times pollution effects are detected only after they have become severe and widespread. Lack of information may result in costly delays in making or revising regulatory decisions.

Advantages and Limitations of Immunoassays

One important aspect in the successful use of IAs is the proper interpretation of the results. A common mistake committed by new users of IA technology is interpreting the results without validating the data or understanding the limitations of the technique. In deciding whether or not the use of IA is practical, it is important to recognize these limitations. First, the purpose of the study should be examined carefully to determine if IA is indeed the appropriate method to use. If the study involves analysis of multiple analytes, IA may not be practical. In this case, the cost and time saved to carry out repeated analysis for each contaminant using a compound-specific assay will be negligible compared to a single chromatographic analysis that can simultaneously quantify several analytes. However, in a case where instrumental analysis involves tedious sample preparation and expensive instrumentation, conducting multiple IAs may still be advantageous over conventional methods. In instances where real-time data are critical and field assays are needed, IAs may be the best method of choice. Furthermore, if a study involves analysis of a large number of samples, most of which are predicted to be negative, screening by IA will clearly benefit the analyst by reducing the sample load for the more expensive conventional method.

One important application of IA is the analysis of polar and ionic contaminants that are difficult to analyze by GC techniques. Most ionic compounds require derivatization before they become amenable to GC analysis; hence, these compounds typically are analyzed by HPLC. A problem may arise if the compounds do not posses functional groups that would allow sensitive detection electrochemically, fluorimetrically, or by absorption in the UV region. In this case, IA should be used if available.

Other advantages offered by using IAs include elimination of large volumes of waste organic solvents and the subsequent cost for their disposal. Moreover, the time and expense required to train personnel to conduct GC or HPLC analyses and the cost of maintenance of the instruments are far greater than that needed to train staff to conduct immunochemical assays. Although the training process for IA is relatively easy compared to chromatographic analysis, it is a common misconception that training is not required to perform IAs. On the contrary, adequate training of the analyst is critical in obtaining consistent and reliable results. In addition to a good understanding of the basic principles of IA, the training should include knowledge of proper laboratory techniques. For instance, pipetting errors, inconsistent timing, and cross contamination, are some of the errors most commonly encountered by untrained analysts. The analyst needs to understand the principles of the assay, and should be able to trace sources of errors in order to troubleshoot potential problems. QA is an important part of using IA for field screening. Harrison (chapter 29 in this book) points out important QA issues and provides recommendations for both users and manufacturers to improve overall IA performance.

Many water-quality surveys have used IAs for the analysis of the major crop herbicides in the United States (e.g. alachlor, atrazine, cyanazine, metolachlor). Most of these surveys used ELISA both as a quantitative tool and as a screening method to reduce sample load for conventional analysis. Because IAs can be used to screen large sample sets at realistic costs, they can facilitate large-scale surveys and monitoring programs that are currently unrealistic. Although immunoassays are intended to complement rather than replace conventional analytical techniques, commercial kits are now validated for many applications and can be used as a stand-alone analytical method, with fewer samples needing verification. For instance, in Figures 6, comparison of immunoassay results for alachlor, atrazine, cyanazine, and metolachlor with GC/MS analysis shows the good correlation between the two methods in the analysis of surface-water samples. The slopes of the graph for atrazine (0.96) and cyanazine (1.10) and the strong correlation between GC/MS and IA, indicate that the response of the IA kits is specific to the target analyte. On the other hand, the slopes of the alachlor (1.37) and metolachlor (1.30) graphs are slightly higher than unity, and the correlation with GC/MS is lower (r^2=0.84), which indicate some interference from the sample matrix. Except for alachlor, where 16% of the samples were false positive near the assay detection limit (between 0.10 to 0.25 μg/L), the ELISA for atrazine, cyanazine, and metolachlor did not show any false-positive or false-negative results.

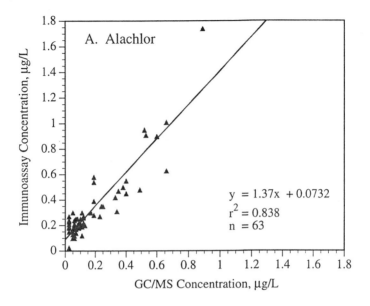

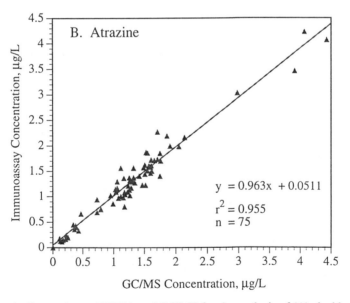

Figure 6. Comparison of ELISA and GC/MS for the analysis of (A) alachlor, (B) atrazine, (C) cyanazine, and (D) metolachlor in water samples from reservoirs in the Midwestern United States (24).

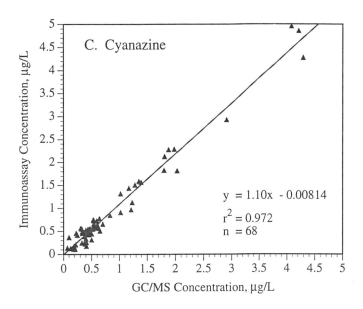

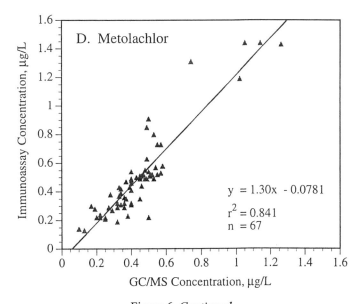

Figure 6. *Continued*

The above example brings out a very important aspect in the interpretation of IA results. Antibodies can cross-react with compounds that closely resemble the target analyte, producing false-positive results or higher readings than conventional methods. Although this limitation may be advantageous in screening for a class of related compounds, this means that IA results need to be verified to obtain data with a high level of confidence, particularly if quantitation of a specific analyte is important. Knowledge of the cross-reactivity pattern of the antibodies used in the assay, therefore, is critical in interpreting results. Information on the type of compounds present, sample pH, organic-matter content, and the type of sample (ground, surface, precipitation, salt water, etc.) are also important factors that need to be considered before drawing conclusions from the IA data. Most of the commercially available ELISA kits for environmental analysis have package inserts that include information on the cross-reactivity of the antibody towards some of the known metabolites and related compounds. However, a complete list of cross-reacting compounds may not be possible because many of these compounds are not known and others are not available to the kit manufacturers.

Cross-reactivity may be turned into a positive characteristic of IA. For instance, Jones et al. (chapter 27 in this book) discuss how an immunoarray of properly selected antibodies and an appropriate statistical analysis can be used to identify and quantitate closely related compounds that are ordinarily hard to differentiate by IA. Triazine herbicides are perfect examples of structurally similar compounds that may occur together in environmental samples. There have been extensive antibody development work done for triazine compounds using different haptens and assay strategies, yet not a single antibody is 100% monospecific to a particular triazine compound. In such cases, mathematical models may be developed to aid in the proper interpretation of results. Wittman et al. (chapter 28 in this book) also present the use of neural networks to correctly identify and quantify cross-reacting analytes and to solve the problem of overestimation due to cross-reactivity. The application of neural network produced promising results in several IA formats, such as the dipstick and eight-well test strip. Immobilization of class-specific antibodies in an array could be used in conjunction with neural network to deconvolute the detection signals into the constituent analytes using pattern-recognition strategies. It appears that, with more collaboration between immunologists, statisticians, and chemists, cross-reactivity problems may be eliminated or turned into an advantage.

Conclusions

Despite many clear demonstrations that immunochemical techniques provide high-quality data in a fraction of time and price required by common chromatographic methods, the acceptance of this technology in environmental analysis has been very slow. However, the rate of acceptance of the technology has increased substantially in the last few years, as manifested by the increase in related literature and participation in many scientific meetings. The supportive action given by regulatory agencies and the private sector has helped in the promotion of IA technology in environmental monitoring. For example, the U.S. EPA Characterization Research Division-Las Vegas sponsors the annual Immunochemistry Summit meeting to facilitate a dialog between IA developers, regulatory agencies, university researchers, instrument manufacturers, and users to discuss current trends and issues concerning regulatory acceptance of IA methods.

There have been many studies to date demonstrating that the advantages offered by environmental IAs outweigh the limitations. As more and more analytical chemists illustrate new applications of IAs, the current limitations of this technology may be addressed and solved accordingly. The applications of immunochemical technology in environmental analysis will continue to grow as more sensitive assays are developed and as automated and miniaturized systems become available.

Acknowledgments

The authors wish to thank the members of the USGS laboratory in Lawrence, Kansas, who have been involved in a number of the IA projects mentioned in this chapter: Angela Buckley, Tonya Dombrowski, Janice McClelland, James Fallon, Michael Meyer, Michael Pomes, Elizabeth Scribner, Michelle Yockel, and Lisa Zimmerman. Special thanks to Donald Goolsby, USGS, who is the project chief of the Mississippi River and precipitation studies. We acknowledge the artwork of William Battaglin, USGS, Denver, Colorado, who is responsible for the colored illustrations. The use of brand, trade, or firm names in this book is for identification purposes only and does not constitute endorsement by the U.S. Geological Survey.

Literature Cited

1. Kuby, Janis. *Immunology*; W.H. Freeman and Co: New York, NY, 1992; pp 1-585.
2. Berson, S.A.; Yalow, R.S. *J. Clin. Invest.* **1959**, *38*, 1196.
3. Engvall, E.; Perlman, P. *Immunochemistry*. **1971**, *8*, 871.
4. Ercegovich, C.D. In *Analysis of Pesticide Residues*; American Chemical Society: Washington, D.C., 1971; pp 162-177.
5. Langone, J.L.; Van Vunakis, H. *Res. Commun. Chem. Pathol. Pharmacol.* **1975**, *10*, 163.
6. Van Vunakis, H. In *Immunochemical Methods for Environmental Analysis*; Van Emon, J.M.; Mumma, R.O., Eds.; ACS Symposium Series; American Chemical Society: Washington, D.C., 1990, Vol. 442; pp 1-12.
7. Vanderlaan, M.; Stanker, L.; Watkins, B. In *Immunoassays for Trace Chemical Analysis*; Vanderlaan, M.; Stanker, L.; Watkins, B.; Roberts, D., Eds.; ACS Symposium Series; American Chemical Society: Washington, D.C., 1990, Vol. 451; pp 1-13.
8. Porter, R.R. *Biochem. J.* **1959**, *73*, 119.
9. Köhler, G.; Milstein, C. *Nature*, **1975**, *256*, 495.
10. Choudary, P.V.; Lee, H.A.; Hammock, B.D.; Morgan, M.R.A. In *New Frontiers in Agrochemical Immunoassay*. Kurtz, D.A.; Skerritt, J.H.; Stanker, L., Eds.; AOAC International: Arlington, VA, 1995, pp 171-185.
11. Hock, B.; Giersch, T.; Kramer, K.; Dankwardt, A. In *New Frontiers in Agrochemical Immunoassay*. Kurtz, D.A.; Skerritt, J.H.; Stanker, L., Eds.; AOAC International: Arlington, VA, 1995, pp 149-162.
12. Bell, C.W.; Roberts, V.A.; Scholthof, K.B.G.; Zhang, G.; Karu, A. In *Immunoanalysis of Agrochemicals: Emerging Technologies*. Nelson, J.O.; Karu, A.E.; Wong, R.B., Eds.; ACS Symposium Series; American Chemical Society: Washington, D.C., 1995, Vol. 586; pp 50-71.
13. Andersson, L.I.; Nicholls, I.A.; Mosbach, K. In *Immunoanalysis of Agrochemicals: Emerging Technologies*. Nelson, J.O.; Karu, A.E.; Wong, R.B. Eds.; ACS Symposium Series; American Chemical Society: Washington, D.C., 1995, Vol. 586; pp. 89-96.
14. Anis, N.A.; Eldefrawi, M.E. *J. Agric. Food Chem.* **1993**, *41*, 843.
15. Zhao, C.Q.; Anis, N.A.; Rogers, K.R.; Kline, R.H.; Wright, J.; Eldefrawi, A.T.; Eldefrawi, M.E. *J. Agric. Food Chem.* **1995**, *43*, 2308.
16. Walt, D.R.; Agayn, V.; Healey, B. In *Immunoanalysis of Agrochemicals: Emerging Technologies;* Nelson, J.O.; Karu, A.E.; Wong, R.B., Eds.; ACS Symposium Series; American Chemical Society: Washington, D.C., 1995, Vol. 586; pp 186-196.

17. Oroszlan, P.; Duveneck, G.L.; Ehrat, M.; Widmer, H.M. *Sens. Actuators.* **1993**, *11*, 301.
18. Brecht, A.; Abuknesha, R. *Trends Anal. Chem.* **1995**, *14*, 361.
19. Vanderlaan, M.; Watkins, B.E.; Stanker, L. *Env. Sci. Technol.* **1988**, *22*, 247.
20. Hammock, B.D.; Gee, S.J.; Harrison, R.O.; Jung, F.; Goodrow, M.H.; Li, Q.X.; Lucas, A.D.; Székács, A.; Sundaram, K.M.S. In *Immunochemical Methods for Environmental Analysis*; Van Emon, J.M.; Mumma, R.O. Eds.; ACS Symposium Series; American Chemical Society: Washington, D.C., 1990, Vol. 442; pp 112-139.
21. Sherry, J.P. *Crit. Rev. Anal. Chem.* **1992**, *23*, 217.
22. Van Emon, J.M.; Lopez-Avila, V. *Anal. Chem.* **1992**, *62*, 79A.
23. Goolsby, D.A.; Thurman, E.M.; Pomes, M.L.; Meyer, M; Battaglin, W.A. In *Selected Papers on Agricultural Chemicals in Water Resources of the Midcontinental United States;* Open-File Report 93-418; U.S. Geological Survey: Denver, CO, 1993; pp 75-89.
24. Goolsby, D.A. In *Chemical Data For Water Samples Collected During Four Upriver Cruises on the Mississippi River between New Orleans, Louisiana, and Minneapolis, Minnesota, May1990-April 1992*; Open-File Report 94-523; U.S. Geological Survey: Denver, CO, 1995; pp 19-87.
25. Aga, D.S. *Analytical Applications of Immunoassays in Environmental and Agricultural Chemistry: Study of the Fate and Transport of Herbicides*; Ph.D. Dissertation; University of Kansas: Lawrence, KS, 1995.
26. Fallon, J. *Determining the Three-Dimensional Distribution, Transport, and Relative Age of Atrazine and Selected Metabolites in Perry Lake, Kansas.* Masters Thesis, University of Kansas, 1994.
27. Dohrman, L. *Water World.* Feb. **1996**, PennWell Publication: USA, v. 12, no.2.
28. Battaglin, W.A.; Hay, L.E. *Env. Sci. Technol.* **1996**, *30*, 889.
29. Lankow, R. *Hazmat World*, March **1994**. Advanstar Publication: USA.

ANTIBODY PRODUCTION
AND ASSAY DEVELOPMENT

Chapter 2

Phage-Display Technology for Environmental Analysis

J. Christopher Hall, Graham M. O'Brien, and Steven R. Webb

Department of Environmental Biology, University of Guelph, Guelph, Ontario N1G 2W1, Canada

With the advent of recombinant antibody technology such as phage-display, it has become possible to produce antibody fragments by cloning, introducing, and expressing antibody genes in simple host systems such as *E. coli*. Theoretically, phage-display could generate highly specific antibodies more cost effectively and efficiently than conventional monoclonal systems. This technology is discussed with regard to its advantages and disadvantages for applications such as detection, quantification, and bioremediation of hazardous chemicals in the environment. Comparison of monoclonal and recombinant phage-displayed antibodies developed for the analysis of trace contaminants in food and the environment are discussed.

Antibodies have been used successfully during the last two decades for detection of environmental contaminants. Advantages of immunoassay technology include speed of analysis, reduced costs, amenability to automation, reduced sample preparation, as well as increased sensitivity and specificity (1). Key disadvantages include matrix interference with antibodies, the potential lack of availability of large amounts of homogeneous antibody, and difficulty in producing immunogens due to the small size of many environmental contaminants.

Many of the successful, commercially available immunoassays for detection of environmental contaminants are polyclonal-based assays. However, there is little or no homogeneity of the desired antibodies among immunized animals of the same species, and it is not possible to produce an unlimited supply of homogeneous antibody. Monoclonal antibody technology has allowed us to produce unlimited quantities of a homogenous antibody with high specificity. However, monoclonal antibodies have some limitations including high cost of production, a requirement for highly trained personnel, rigorous screening procedures, initial lack of stability, and difficulty in characterizing the genes associated with antibody production (1).

With recombinant antibody technology it has become possible to rapidly produce functional antibody fragments designed for specific applications. Antibody genes can be cloned, introduced, and expressed within inexpensive and relatively simple host systems. Although several non-mammalian host systems (yeast, fungi, plant and insect cells) have been used to produce recombinant antibodies, the most common vehicle is *E. coli.* (2). *E. coli* has rapid growth properties that facilitate direct antibody production requiring no proteolytic preparation (2). One of the most common recombinant antibody technologies that utilizes *E. coli* is *phage-display.* With this technique, antibody fragments can be expressed either fused to the N-terminus of bacteriophage coat proteins or as soluble proteins (Figures 1 and 2).

Theoretically, phage-display and other recombinant technologies may be used to generate antibodies for environmental analysis more cost effectively and efficiently than with conventional monoclonal systems. Other potential advantages of this technology include: i) extending the type and number of antibodies normally obtained by conventional methods by several orders of magnitude, ii) by-passing the immunization step, and hence the use of animals to obtain antibodies specific for proteins and haptens (3), iii) relatively easy isolation and expression of cloned genes in a bacterial host because the selection of the desired display protein also co-selects the gene encoding it, and iv) the potential to further improve binding properties of the selected antibody by using protein engineering techniques.

In this chapter we will provide a general overview of phage-display antibody technology with specific examples of its use for detection of environmental contaminants. Advantages and disadvantages of this technology will also be discussed.

Types Of Antibody Fragments Used For Phage-Display

The two most common antibody fragments are ScFv (single chain variable fragment) and Fab (antigen binding fragment) (Figure 1). The ScFv is composed of a variable heavy chain protein (V_H) and a variable light chain protein (V_L), linked by a flexible polypeptide, usually $(Gly_4Ser)_3$ (4). The Fab is composed of an entire light chain (the V_L and the constant light chain region (C_L)) and a section of heavy chain (the V_H and a constant heavy chain protein domains (C_H); also termed the Fd region) that corresponds in size and function to the light chain region (5). Only one of the Fab chains is bound to the phage surface and the other is secreted into the periplasm where the two chains associate through a disulfide bond (6). The inclusion of the constant domains facilitate this association. Fv antibody fragments (V_H and V_L with no peptide linker) have also been produce but only to a limited extent.

Each type of fragment has particular benefits and drawbacks. The Fab fragment is speculated to be more stable at 37°C and has sometimes shown higher antigen affinity than a comparable ScFv fragment (4). However, the heavy and light regions of the Fab fragment do not always associate which can lead to high background levels and isolation difficulties during screening (10). In ScFv fragments V_L and V_H proteins are covalently attached by a linker peptide and

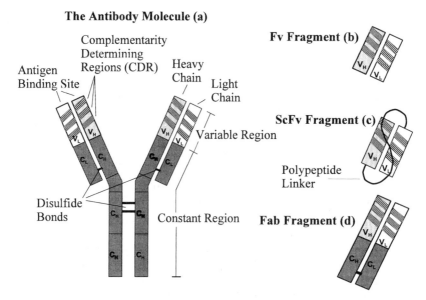

Figure 1. Representation of a functional antibody molecule (a). Phage-display technology has been used to express the Fv (b), ScFv (c) and Fab (d) antigen binding fragments (Adapted from refs. 7 and 8).

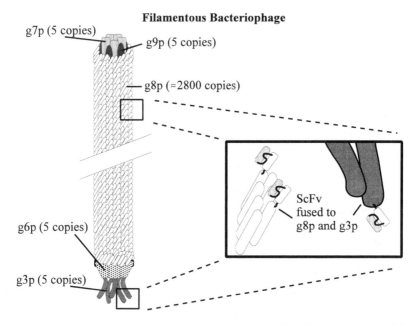

Figure 2. Structure of filamentous bacteriophage M13 showing the location of the g3p and g8p sites used for display of antibody fragments (Adapted from ref. 9).

therefore, they are less likely to dissociate. However, the linker can interfere with antigen binding.

Phage Proteins Used for Expression of Antibody Fragments

Two predominant approaches for the display of antibody fragments on the surface of filamentous phage have been developed. Antibody fragments can be expressed by fusing antibody variable genes with either the bacteriophage major coat or minor coat proteins which are encoded by gene VIII (*g8*) or gene III (*g3*), respectively (Figure 2). Once expressed, antibody fragments can be selected on the basis of affinity, binding kinetics, or avidity (11).

Theoretically, expression systems using *g8* result in the display of over 2000 antibody fragments per phage. However, several studies have shown that the actual number of fragments displayed is usually less than 60 peptides per phage (12, 13, 6) The reason for this is not clear. Antibody fusion to the gene VIII protein (g8p) may result in conformational changes to g8p making them unable to associate to produce the phage coat unless they are sufficiently diluted amongst wild-type (unfused) g8p. Fusion to g8p leads to multivalent display where avidity is the predominant factor driving antibody fragment selection (12, 14, 15). This makes it difficult to separate phage expressing many low affinity antibody fragments from phage displaying a few high affinity antibodies. An advantage of g8p display is that the resulting fusion has little or no effect on the infectivity of the recombinant phage (13).

In contrast to the g8p, there are only five copies of gene III protein (g3p) per phage. The function of g3p during infection is attachment of the phage to the pilus of *E. coli* (16). Several research groups have used the g3p as an expression site for ScFv (17) or Fab (18) antibody fragments. However, since g3p is involved in the process of bacterial infection, phage infectivity can be prevented or impaired if all or most g3p are fused to antibodies (13). Methods such as the use of helper phage, which will be discussed later, have been developed that enable the display of only one g3p antibody fusion protein, effectively minimizing this problem (6, 19). Therefore, antibody fragment selection using g3p fusion display, is based on affinity rather than avidity (20).   These screening properties permit researchers to discriminate between phage-display antibody fusion variants with affinities for the antigen of better than one nanomolar. Since many applications require high affinity antibodies, the *g3* system has been preferred for antibody display (6).

Genetic Vectors For Antibody Expression

To allow expression of the Fab or ScFv binding fragments on bacteriophage, their respective genes must be cloned into an appropriate vector construct. The antibody genes can be obtained from peripheral and spleen lymphocytes, and plasma cells, directly as DNA, or more commonly, by isolating antibody messenger RNA (mRNA). High levels of mRNA can be readily isolated from plasma cells because their cellular levels of antibody mRNA are around 30,000 copies per cell, as opposed to 100 copies per cell in non-stimulated B cells (21). Isolating specific mRNA from monoclonal hybridomas is also relatively easy because they produce mRNA in large quantities. Once the mRNA has been extracted, it is converted into complementary DNA (cDNA) using reverse transcription (4). The specific heavy and light chain DNA sequences are amplified using the polymerase chain reaction (PCR) and cloned into a vector designed to express either Fab or ScFv fragments (Figure 3).

A *phage vector* construct is created by incorporating antibody genes into a bacteriophage genome at the N-terminus encoding region of *g3* or *g8* (4). Vectors are introduced into bacterial cells by either $CaCl_2$ transformation or electroporation (22). The phage DNA replicate inside the transformed bacterial cells and new phage particles are assembled displaying the specific antibody fusion protein.

Antibody genes can also be incorporated into *phagemid vectors* which are plasmids containing a phage origin of replication and either *g3* or *g8* (6). An example of a phagemid system is pCANTAB 5 E developed by Pharmacia (23) and designed for the production of ScFv (Figure 3). This phagemid carries the origin of replication for M13 phage, but it lacks the genes required to replicate and package single-stranded phagemid DNA. When introduced into *E. coli*, phagemids will replicate as plasmids. Phagemids carry an antibiotic resistance marker, (ampicillin in pCANTAB 5 E) so that transformed *E. coli* cells can be selected. To produce M13 expressing the antibody, the transformed *E. coli* must be infected with a helper phage (M13K07 is used for the pCANTAB 5 E system) in a process known as phage rescue (Figure 3). Helper or rescue phage encode all the necessary phage proteins but have a defective origin of replication, and therefore they will preferentially package phagemid DNA (24). Helper phage DNA also contain an antibiotic resistance marker (kanamycin on M13K07) which allows for the selection of infected *E. coli*. After bacterial replication functional phage are produced which express antibody fusion proteins (6).

Phagemids are the most commonly used vectors because they have some significant advantages over phage vectors. Phagemid are 100-fold more efficient than phage vectors at being transformed into bacterial cells (22). This feature is important when trying to create large, diverse libraries containing many different antibody genes. In the presence of helper phage, phagemid systems encode both the fusion protein and the wild type protein; therefore, with *g3*, expression can be manipulated to display approximately one monovalent fragment per phage (24). This circumvents infectivity problems associated with phage vectors and allows selection of high affinity antibodies.

Expression of Phage-Displayed and Soluble Antibodies Using pCANTAB 5 E

The phagemid pCANTAB 5 E used in the Pharmacia Recombinant Phage Antibody system (23) is described below and shown in Figure 3. This system demonstrates the general principles of a phage-display system and has been successfully used it in our laboratory. In the Pharmacia system, an amber translational stop codon lies between the ScFv gene and the *g3* sequence (4). In the transformed suppressor strain of *E. coli* TG1, translation proceeds through the amber stop codon producing a fusion ScFv-*g3*p on the end of the phage. In non-suppressor strains of *E. coli* such as HB2151, translation stops at the amber codon resulting in biosynthesis of only the ScFv which accumulates in the periplasm and can be utilize as soluble antibody.

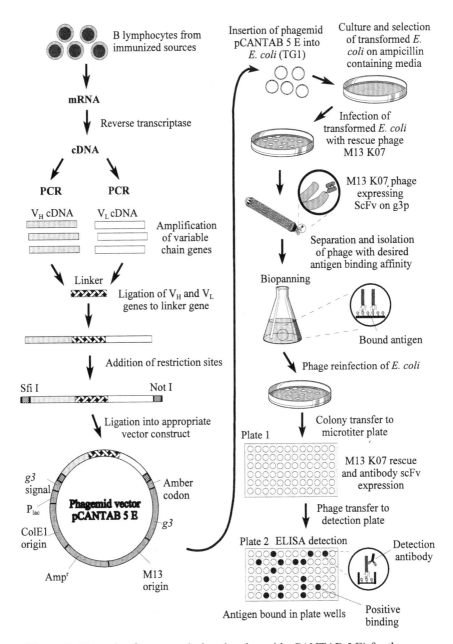

Figure 3. Example of a system (using the phagmid pCANTAB 5 E) for the creation and screening of recombinant phage displayed antibodies (Adaped from refs. 8 and 23)

An inducible *lac* promoter and the *lac* repressor (*lac* I^q) control expression of the ScFv-g3p gene. The presence of the *lac* repressor is essential to avoid the toxic effects of g3p expression prior to infection with the helper phage (M13K07). Since the *lac* repressor allows some expression, further repression can be achieved through catabolic repression in the presence of glucose at 30°C. After infection with the helper phage, glucose is removed and the temperature increased to 37°C, thereby curtailing the catabolic repression and allowing the production of phage-display antibody fragments. Phage displaying the desired antibodies are selected (panned) by binding to antigen coated polystyrene flasks or plates. *E. coli* TG1 are reinfected with the positive phage. In the presence of the helper phage, new phage-displayed ScFvs are produced and panned again. This process is usually repeated several times to enrich for antigen-positive phage-displayed antibodies.

By switching from *E. coli* TG1 to *E. coli* HB2151 in the presence of isopropylthiogalactoside (IPTG) and absence of glucose, soluble ScFv are produced. ScFv are extracted from the cells and medium and purified by affinity chromatography using immobilized antigen.

Selection Of Phage-Display Antibodies

Most phage antibodies exhibit a high degree of non-specific binding. Therefore, different screening methods may be used to isolate specific phage antibodies.

Affinity Chromatography. Phage antibodies can be selected by binding to antigen immobilized on a solid column matrix (17, 18). The use of these affinity columns is particularly suited for the primary and rapid screening of antibodies generated from large combinatorial libraries. These antibodies usually have a wide range of binding affinities. Unbound phage are removed by washing, and bound phage are selectively eluted by a low pH buffer, or by competitive interaction with antigen dissolved in the elution buffer. The desirable high affinity phage are eluted at high antigen concentrations. This process is often repeated to achieve better separation of high affinity antibodies. These phage are then replicated in bacteria to facilitate isolation of the specific antibody gene (4).

Biopanning. Phage antibodies bind to the antigen coated surface of polystyrene micro-titre plates or flasks (Figure 3). Non-specific phage or phage with low binding affinities are washed away. Bound phage-display antibodies are used to infect *E. coli* that are introduced into the flasks or the phage-display antibodies can be eluted with a solution containing high concentrations of salt (4). Coating the panning plates with limiting amounts of antigen has also proven useful for isolating high affinity phage. This procedure is usually repeated several times, and the high affinity phage are used for bacterial infection.

Immortalization of a Specific Genetic Source of Antibodies

The principal objective of both conventional hybridoma (25) and phage-display antibody technology is the production and immortalization of a specific antibody.

However, these techniques use different methods to achieve this goal. The hybridoma technique attempts to immortalize the entire antibody producing cell, whereas, phage-display immortalizes only the antibody genes. The bacterial cultures that support the phage-antibody genes are easier to maintain, and are less expensive than mammalian cell lines (26). This fundamental difference provides the phage-display technique with many practical advantages over conventional monoclonal antibody technology.

Production of Combinatorial Libraries

The principal advantage of phage-display technology is the ability to produce large combinatorial gene libraries from many possible sources (27). Genes can be collected from particular antibody sources such as spleen B cells, plasma cells and peripheral blood lymphocytes from immunized or non-immunized animals. Once the genes which encode the variable heavy and light chain proteins (V_H and V_L) have been amplified and isolated, they can be randomly recombined and expressed on phage particles for specific antigen selection (3).

There are a number of different methods used to recombine the amplified V_H and V_L gene repertoires for the same antigen. PCR can be used *in vitro* to randomly recombine the different V_H genes with V_L genes and then incorporate this combinatorial library into specific vectors (3). In theory, the ability of combinatorial libraries to generate binding fragments to any antigen is dependent on the size and diversity of the library. However in practice, this is limited by the efficiency of bacterial transformation (6). Transformation efficiencies are insufficient to entirely incorporate a large combinatorial library into bacteria for expression.

A more recent approach involves the *in vivo* process known as *combinatorial infection*. With this method, two types of vectors are used. V_H chain repertoires are incorporated into *donor* plasmids, and V_L chain repertoires are cloned into phage vectors which possess an *acceptor* function (6, 28). Bacteria are first transformed with the V_H gene donor plasmids, and subsequently transfected with the V_L gene acceptor phage. Within the bacteria, the two genes are randomly recombined with the aid of a bacteriophage P1 recombinase to form functional fusion protein genes. This method (6, 28) results in the production of a combinatorial antibody library that is approximately one order of magnitude larger than the libraries described above.

Antibody Production Without Immunization

Several researchers have proposed that phage-surface expressed antibody technology will one day eliminate the need for the immunization of animals. While it is much easier to isolate antigen-specific antibody genes from immunized sources, these genes are limited because they have been partially selected for, and are not representative of the potential antibody diversity of the primary immune system. By using non-immunized sources, including humans, it has become possible to create extremely large *naive* or synthetic gene libraries to more effectively harness the vast primary antibody repertoire of the immune system (28). This is exceptionally attractive for generating highly specific and non-immunogenic human antibodies without the obvious ethical dilemmas associated with human immunization (21). It may also be relatively simple and convenient to maintain diverse gene libraries for rapid utilization with the phage-display system.

However, even antibodies from non-immunized sources have predefined binding regions or have been previously selected for an unknown antigen. This poses limitations on the diversity of antibodies that can be generated, especially when using a small gene library, or when bacterial transformation with gene vectors is low. Mature bone marrow lymphocytes may be an excellent gene source as there is no possibility they have been previously subjected to antigen selection (10). These cells would collectively contain all of the individual gene types within the gene families that encode the variable regions. Theoretically these genes could be amplified and randomly recombined *in vitro* to produce functional antibodies to any antigen. This would be the optimal use of phage-display as an *in vitro* mimic of the immune system. Unfortunately, this process may be difficult and complicated. Extensive gene sequence knowledge would be required to produce all of the primers for amplification and to facilitate recombination. (Even with antigen selected antibodies, it can be difficult to isolate the antibody genes from the vast amount of other genetic material.) If the proper genes can be isolated, they may not necessarily result in functional binding proteins. There are genetic processes that are not fully understood at the molecular level, which are required to produce an active antibody.

To date, immunized sources are still the most practical means of generating phage antibodies. With the increase in knowledge of both molecular biology and immunology, production of specific, high-affinity recombinant antibodies from non-immunized sources may become more practical.

Improving Phage Antibody Binding Characteristics.

Unless specific immunization has occurred, current combinatorial libraries usually produce phage antibodies with only low to moderate antigen affinities. After phage antibodies with some antigen binding capacity have been selected and isolated, it is possible to improve their binding properties.

Affinity Maturation In Phage Antibodies. In the natural system, the increase in antibody binding affinity observed after the primary immune response is due to somatic mutation of the antibody variable genes. It is possible to simulate affinity maturation *in vitro* with phage-displayed antibodies. Somatic refinements can be mimicked by selecting desired antibodies that bind to a specific antigen, mutating sections of their genes, and reselecting for improved affinity. Mutations in the antibody genes can be introduce by either error prone PCR (29) or by passing the antibody genes through an error prone strain of *E. coli* (30). Gram *et al.* (10) showed that mouse antibodies specific for progesterone were selected from a naive combinatorial library, and antibody affinities could be improved using error prone PCR.

Mutations can also be precisely incorporated using site-directed mutagenesis. With this method, specific amino acid sequences at or near the antigen binding site can be altered by exchanging them with other amino acid residues (31).

Combining these mutational processes and/or repeating them several times, may greatly improve the generation of specific antibodies from naive combinatorial libraries.

Hierarchical Libraries. The creation of hierarchical libraries involves the exchange and pairing of unique V_H and V_L genes for different phage antibodies (6). These genes (V_H and V_L genes) may come from phage antibodies which have affinity for the same antigen or different antigens. This type of recombination is not possible in the immune system as it occurs subsequent to a primary immune response and, therefore, long after the heavy and light chain genes have been paired within precursor B cells (4).

Environmental Applications for Phage-Display Technology

There are few references in the literature about the use of phage-displayed antibodies or soluble antibodies produced by recombinant technology to identify, isolate, or degrade environmental contaminants. Many researchers have encountered problems with this technology and it has not yet proved to be superior to monoclonal systems. However, once this technology evolves and more researchers begin to utilize it, current problems will likely be solved and the theoretical advantages of recombinant systems will be realized.

Petrenko and Makowski (32) were among the first to report on the potential of this technology for identification of environmental contaminants and bioremediation of contaminated sites. Fusion phage may be constructed that display either enzymes that can degrade environmental contaminants or binding proteins for immobilization and isolation of contaminants. Such phage-display antibodies may be distributed over a contaminated site to contain or enhance degradation of the contaminant. Alternately, these same phage-display antibodies may be used to construct large scale filters or columns through which contaminated water could be processed.

Recombinant antibodies have been used for the detection of pesticides and toxins in food and the environment. Lee *et al.* (33) produced phage-display antibody fragments (ScFv) from the spleen cells of mice immunized with one of two mycotoxins (diacetoxyscirpenal and aflatoxin M1) rather than from a hybridoma cell line. They found several problems in producing phage-displayed antibody fragments with desirable affinities to both mycotoxins when using their combinatorial library. First, they found that the nature of ScFv binding to hapten/protein coated tissue culture flasks was poly-specific. There was considerable binding to both the protein and hapten. Second, regardless of the washing procedure used during the panning process, high non-specific binding occurred. Third, when three different elution solutions (PBS, mycotoxin, or triethylamine mixed with Tris-HCl) were compared during panning, the use of mycotoxin to selectively elute high affinity antibodies did not improve the procedure. Furthermore, when affinity chromatography was used to select high affinity antibodies, low affinity, poly-reactive ScFvs were still obtained. Fourth, they found that ScFvs were unstable when stored. Finally, they had difficulty generating soluble ScFvs when they attempted to overcome some of the problems associated with the phage-displayed ScFvs. Lee *et al.* (33) concluded that the underlying problem appeared to be generating stable antibodies with reasonable affinity when selecting from libraries. We have also experienced similar problems in our laboratory.

Recombinant antibodies from expression systems other than phage have also been produced. For example, Ward *et al.* (1) described the production of a light chain fused to β-galactosidase in *E. coli*. The fusion protein was expressed as a plasmid produced from the genes of an anti-atrazine hybridoma. They could not produce the heavy chain and also had difficulty in producing stable antibody chains in *E. coli*. The baculovirus expression system was examined as a possible expression system to overcome some of the problems. Bell *et al.* (34) discuss the creation of a model for the combining site of the phenylurea herbicide, diuron, using recombinant antibody technology. They successfully created this model by selecting the antibody genes from a diuron-specific hybridoma by cloning them into the M13 phagemid vector pComb3. Two soluble Fabs from separate clones competitively bound free diuron in competition EIAs with the same sensitivity and specificity as Fabs generated from the monoclonal hybridoma.

Comparison of monoclonal and phage antibodies for detection of cyclohexanedione herbicides. The immunoglobulin variable region genes of a murine anti-cyclohexanedione IgG-producing hybridoma were rescued and cloned into a bacterial expression vector in our laboratory. The variable regions of the heavy chain (IgG_1) and the light chain were expressed as a ScFv. The ScFv was constructed by PCR assembly using a $(Gly_4Ser)_3$ linker between the carboxyl end of the variable heavy chain and the amino terminus of the variable light chain. An ELISA was developed to examine the interaction of the ScFv with selected cyclohexanedione analogues (Figure 4) and this cross-reactivity data was compared to the parent monoclonal antibody.

The cross-reactivity profiles of the ScFv and parent monoclonal antibody with selected cyclohexanedione analogues were similar. The structures of the cyclohexanedione analogues tested for cross-reactivity are shown in Figure 4. Analogues 2 and 3 are active inhibitors of acetyl-coenzyme A carboxylase, whereas, analogue 4 is not active. The concentration of analogues 2 and 3 required to inhibit 50% of corn acetyl-coenzyme A carboxylase activity (IC_{50}) were 1 μM and 0.03 μM, respectively. The concentration of analogue 2 required to inhibit 50% of ScFv and monoclonal antibody binding to the coating conjugate (IC_{50}) were 183.9 ± 23.6 nM and 171.0 ± 23.6 nM, respectively (Figure 5). The ScFv and monoclonal antibody IC_{50} values for analogue 3 were, 3.0 ± 0.5 nM and 3.9 ± 0.8 nM, respectively (Figure 5). The dose-response profiles for analogue 4 revealed that this inactive compound did not cross react with either the ScFv or parent monoclonal antibody (IC_{50} were greater than 100 μM). The similarity in the monoclonal antibody and ScFv IC_{50} values suggests the recombinant ScFv possess a similar cross-reactivity pattern and IC_{50} as the parent monoclonal antibody. Analogue 3 is the most potent acetyl-coenzyme A carboxylase inhibitor, and was also the most potent inhibitor of the ScFv and monoclonal antibody. Both ScFv and the monoclonal antibody failed to cross react with the inactive analogue 4.

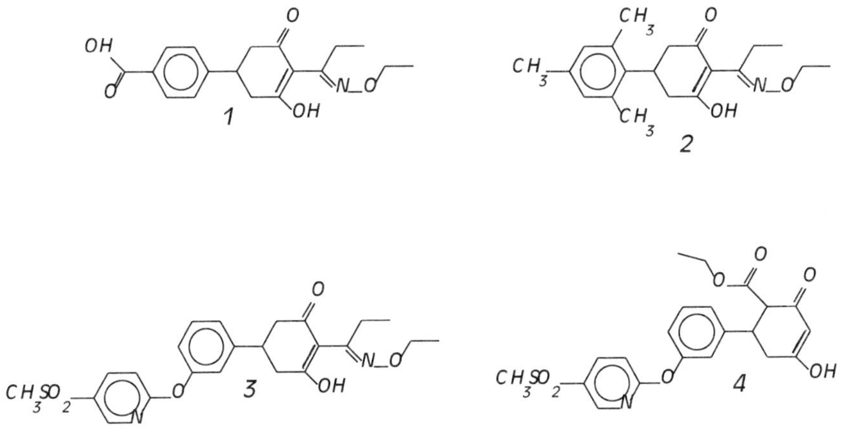

Figure 4. Structures of the various cyclohexanedione analogues. Analogue 1 when conjugated to bovine serum albumin and ovalbumin was the immunogen and ELISA coating conjugate, respectively. Analogues 2 and 3 are active inhibitors of acetyl coenzyme-A carboxylase activity whereas, analogue 4 is a non-active inhibitor.

Other Applications of Phage-Display Technology

Production of novel peptides. Besides expression of antibody fragments on the surface of bacteriophage, other peptides including human growth hormone (35), alkaline phosphatase (11), and bovine pancreatic trypsin inhibitor (13) have been expressed. The ability to rapidly screen phage-expressed peptide variants allowed Lowman *et al.* (36) to develop human growth hormone with improved binding affinities. These results suggest pharmacologically active peptides may be modified to increase their potency or specificity.

Application of phage-expressed peptides may permit the development and selection of peptides with novel catalytic activities. One approach may be to build on existing catalytic peptide structures. Carter and Wells (37) suggest catalytic residues of an enzyme may be retained, but by mutating loop amino acid residues, specificities may be altered. Alternatively, new catalytic peptides may be produced that are similar to catalytic antibodies produced using hybridoma technology. For example, catalytic residues, such as lysine or histidine, may be introduced into antibody binding sites. Modified binding sites possessing catalytic activity may be selected by, binding substrate, products, or transition state analogues. Finally, since Smith (38) and Parmley and Smith (39) first demonstrated small peptides other than antibodies could be displayed on the surface of bacteriophage, there has been an interest in developing a screening system for identifying novel synthetic pharmaceutically active peptides (40, 41, 42). For example, phage-display can be used to express protein fragments from viruses and pathogenic bacteria on their coat

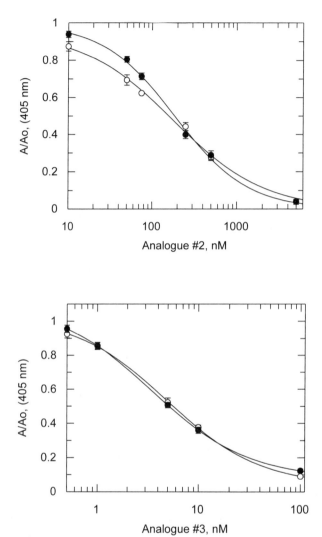

Figure 5. The effect of various concentrations of analogues 2 (top graph) and 3 (bottom graph), shown in Figure 4, on monoclonal antibody (O) and ScFv (●) binding to the coating conjugate in an indirect competitive ELISA. Each figure shown is representative of 3 independent experiments.

to create vaccines. Foreign peptide sequences may be incorporated into *g8* and then large numbers of these peptide/g8p fusion proteins can be expressed on a single phage. This creates a very high epitope density that stimulates a rapid immune response (43). This method is much simpler and less expensive than traditional methods of foreign peptide synthesis and conjugation to a carrier protein (44).

Phage antibodies used as surrogate receptors and shape mimics. An antibody with binding affinity for the pharmacophoric determinants of an active compound may be used as a surrogate receptor or shape mimic to screen large libraries of diverse and unrelated molecules (45). Molecules that bind to the antibody (receptor mimic) may have activity and may be further assayed for activity using the target receptor. This system may increase the screening efficiency because of the speed at which samples can be processed by enzyme immunoassay techniques. This system may be used for screening large chemical inventories for the development of pharmaceuticals and pesticides as well as for other basic research applications (46).

Conclusions

Complicated immunological mechanisms have been utilized to produce antibodies for research. As knowledge of molecular biology has rapidly increased, new and more powerful genetic engineering techniques have been developed. These techniques allow for more efficient use of immunochemistry and have greatly expanded its applications. With the advent of phage-display technology, it has become possible to manipulate the immune processes *in vitro*. As different methods are constantly being introduced, it is now theoretically possible to create a functional immune system *in vitro* without the use of animal immunization. By mimicing specific immune system mechanisms, it may become easier to use large synthetic antibody gene libraries to generate specific antibodies rather than using antibodies produced directly by animals. This technology may have a significant impact on environmental contaminant analysis. However, this technology currently has disadvantages which must be overcome before it will be of practical use for analysis of environmental contaminants.

Acknowledgments

Research grants were provided by AAFC, NSERC (Research Partnership Program), OMAFRA, and URIF. S.R. Webb held an NSERC Postgraduate Award.

Literature Cited

1.   Ward, V.K.; Hammock, B.D.; Maeda, S.; Choudary, P.V. In *New Frontiers in Agrochemical Immunoassay*; AOAC International, Arlington, VA, 1995, pp 197-216.
2.   Lee, H.A.; Alcocer, M.J.C.; Garcia Lacarra, T; Jeenes, D.J.; Morgan M.R.A. In *New Frontiers in Agrochemical Immunoassay;* AOAC International, Arlington, V.A., 1995, pp 187-196.

3. Marks, J.D.; Hoogenboom, H.R.; Bonnert, T.P.; McCafferty, J.; Griffiths, A.D.; Winter, G. *J. Mol. Bio.* **1991**, *222*, 581.
4. Jackson, R.H., J. McCcafferty, K.S. Johnson, A.R. Pope, A.J. Roberts, D.J. Chiswell, T.P. Clackson, A.D. Griffiths, H.R. Hoogenboom and G. Winter. In *Protein Engineering; A Practical Approach;* Rees, A.R.; Sternberg, M.J.; Wetze, R., Eds.; IRL Press: Oxford, U.K., 1992; pp 277-301.
5. Stryer, L. *Biochemistry,* 3rd Ed.;. W.H. Freeman and Company: New York, NY, 1988; pp. 890-900.
6. Winter, G.; Griffiths, A.D.; Hawkins, R.E.; HoogenboomR.H. *Annu. Rev. Immunol.* **1994**, *12*, 433.
7. Alberts, B; Bray, D.; Lewis, J.; Raff,.M.; Roberts, K.; Watson, J.D.; *Molecular Biology of the Cell*; Garland Publishing, Inc.: NewYork, NY; 1994, 1294 pp.
8. Watson, J.D.; Gilman, M.; Witkowski, J.; Zoller, M.; *Recombinant DNA*; W.H. Freeman and Co.: New York, NY, 1992, 626 pp.
9. Bhattacharjee, S.; Glucksman, M.; Makowski, L. Biophys. J. **1992**, *61*: 7625.
10. Gram, H.; Marconi, L.; Barbas III, C.F.; Collet, T.A.; Lerner, R.A.; Kang, A.S. *Proc. Natl. Acad. Sci. USA.* **1992**, *89*, 3576.
11. Marks, J.D.; Hoogenboom, H.R.; Griffiths, A.D.; Winter, G. *J. Biol. Chem.* **1992**, *267*, 16007.
12. Kang, A.S.; Barbas III, C.F.; Janda, K.D.; Benkovic, S.J.; Lerner, R.A. *Proc. Natl. Acad. Sci. USA.* **1991**, *88*, 4363.
13. Markland, W.; Roberts, B.L.; Saxena, M.J.; Guterman, S.K.; Ladner, R.C. *Gene* **1991**, *109*, 13.
14. Chang, C.N.; Landolfi, N.F.; Queen, C. *J. Immunol.* **1991**, *147*, 3610.
15. Huse, W.D.; Sastry, L.; Iverson, S.A.; Kang, A.S.; Alting-Mees, M.; Burton, D.R.; Benkovic, S.J.; Lerner, R.A. *Science* **1989**, *246*, 1275.
16. Crissman, J.W.; Smith, G.P. *Virology* **1984**, *132*, 445.
17. McCafferty, J.; Griffiths, A.D.; Winter, G.; Chiswell, F.J. *Nature* **1990**, *348*, 552.
18. Clackson, T.; Hoogenboom, H.R.; Griffiths, A.D.; Winter, G. *Nature* **1991**, *352*, 624.
19. Garrad, L.J.; Yang, M.; O'Conell, M.P.; Kelley, R.F.; Henner, D.J. *Bio/Technol.* **1991**, *9*, 1373.
20. Barbas III, C.F.; Kang, A.S.; Lerner, R.A.; Benkovic, S.J. *Proc. Natl. Acad. Sci. USA.* **1991** *88*, 7978.
21. Burton, R.D.; C.F. Barbas III. In *Protein Engineering of Antibody Molecules for Prophylactic and Therapeutic Applications in Man;* Clark, M., Ed.; Academic Titles: Nottingham, U.K., 1993; pp. 65-82.
22. Brown, T.A. *Gene Cloning; An Introduction* 2nd Ed., Chapman and Hall, London, 1990; pp. 4-26.
23. Parmacia Biotech; *Recombinant Phage Antibody System Instruction Manual*; Pharmacia Biotech, Molecualr Biology Reagents Division; 1995, 39 pp.
24. Griffiths, A.D.; Hoogenboom, H.R. In *Protein Engineering of Antibody Molecules for Prophylactic and Therapeutic Applications in Man;* Clark, M., Ed.; Academic Titles: Nottingham, U.K., 1993; pp. 45-64.

25. Köhler, G.; Milstein, C. *Nature* **1974**, *265*, 495.
26. Chiswell, D.J.; McCafferty, J. *TIBTECH* **1992**, *10*, 80.
27. Zebedee, S.L.; Barbas III, C.F.; Hom, Y.-L.; Caothien, R.H.; Graff, R.; DeGraw, J.; Pyati, J.; LaPolla, R.; Burton, D.R.; Lerner, R.A.; Thornton, G.B. *Proc. Natl. Acad. Sci. USA.* **1992**, *89*, 3175.
28. Griffiths, A.D; Williams, S.C.; Hartley, O.; Tomlinson; I.M.; Waterhouse, P.W.; Crosby, W.L.; Kontermann, R.E.; Jones, P.T.; Low, N.M.; Allison, T.J.; Prospero, T.D.; Hoogenboom, H.R.; Nissim, A.; Cox, J.P.L.; Harrison, J.L.; Zaccolo, M.; Gherardi, E.; Winter, G. *EMBO.* **1994**, *13*: 3245.
29. Leung, D.W.; Chen, E.; Goeddel, D.V. *Technique* **1989**, *1*, 11.
30. Fowler, R.G.; Schaaper, R.M.; Glickman, B.W. *J. Bacteriol.* **1986**, *167*, 130.
31. Reichmann, L.;Weill, M. *Biochemistry* **1993**, *32*: 8848.
32. Petrenko, V.A.; Makowski, L. In *Bioremediation of Chlorinated and Polycyclic Aromatic Hydrocarbon Compounds;* Lewis Publishers, Boca Raton, 1994, pp 266-270.
33. Lee, H.A.; Wyatt, G.; Garrett, S.D.; Yanguela, M.C.; Morgan M.R.A. In *Immunoanalysis Agrochemicals: Emerging Technologies;* ACS, Washington DC, 1995, pp 22-30.
34. Bell, C.W.; Roberts, V.A.; Scholthof, K.B.G.; Zhang, G.; Karu, A.E. In *Immunoanalysis Agrochemicals: Emerging Technologies;* ACS, Washington DC, 1995, pp 50-71.
35. Bass, S.; Greene, R.; Wells, J.A. *Proteins* **1990**, *8*, 309.
36. Lowman, H.B.; Bass, S.H.; Simpson, N.; Wells, J.A. *Biochem.* **1991**, *30*, 10832.
37. Carter, P.; Wells, J.A. *Science* **1987**, *237*, 394.
38. Smith, G.P. *Science* **1985**, *228*, 1315.
39. Parmley, S.F.; Smith, G.P. *Gene* **1988**, *73*, 305.
40. Scott, J.K.; Smith, G.P. *Science* **1990**, *249*, 386.
41. Cwirla, S.E.; Peters, E.A.; Barett, R.W.; Dower, W.J. *Proc. Natl. Acad. Sci. USA.* **1990**, *89*, 1865.
42. Devlin, J.J.; Panganiban, L.C.; Devlin, P.E. *Science* **1990**, *249*, 404.
43. Minenkova, O.O.; Ilyichev, A.A.; Kishchenko, G.P.; Petrenko, V.A. *Gene.* **1993**, *128,* 85.
44. Willis, A.E.; Perham, R.N.; Wraith, D. *Gene.* **1993**, *137*, 79.
45. Roberts, D.; Guegler, K.; Winter, G. *Gene.* **1993**, *128,* 67.
46. Hogrefe, H.H.; Amberg, J.R.; Hay, B.N.; Sorge, J.A.; Shopes, B. *Gene.* **1993**, *137,* 85.

Chapter 3

Rapid Mercury Assays

Ferenc Szurdoki[1], Horacio Kido[1], and Bruce D. Hammock[1,2]

Departments of [1]Entomology and [2]Environmental Toxicology,
University of California, Davis, CA 95616

We have developed rapid assays with the potential of being applied to mercury analysis in environmental samples. Our methods combine the simple ELISA-format with the selective, high affinity complexation of mercuric ions by sulfur-containing ligands. The first assay format is based on a sandwich chelate formed by a ligand conjugated to a reporter enzyme, a mercuric ion of the analyzed sample, and a protein bound ligand immobilized on the wells of a microtiter plate. The second assay format involves competition between mercuric ions and an organomercury-conjugate to bind to a chelating conjugate. The assays detect mercuric ions in ppb/high ppt concentrations with high selectivity.

Toxicity of Mercury

Mercury is one of the most hazardous toxic metals. The amount of mercury emitted to the environment is continuously increasing mainly due to industrialization and urbanization. This trend has raised concerns about the risk of human exposure to mercury and the resulting adverse health effects (*1*).

Most of the biological effects of mercury can be explained in terms of the very high affinity of mercurials towards endogenous thiols (*1*). Toxicologically, the most important mercury-species are elemental (Hg^0) and mercuric (Hg^{2+}) mercury, and methylmercury (CH_3Hg^+). Different forms of mercury exhibit different bioavailabilities, toxic potencies, tissue and organ distributions, and patterns of biological effects (*1*).

Exposure to Hg^0 vapor is mostly an occupational hazard for dental personnel, workers in the chloralkali industry, plants manufacturing mercury thermometers, and mercury mines (*2-4*). The toxic impact of the Hg^0 vapor released from dental amalgam fillings has recently been discussed (*2, 4, 5*). The initial toxic effects of Hg^0 are observed in the nervous system and later nephrotoxicity develops (*1*). Hg^0

is rapidly oxidized to Hg^{2+} in erythrocytes or in tissues (*1, 4*); thus, the physiological effects of exposure to either of these two forms of mercury are somewhat similar. The kidneys are the primary target organ of deposition and toxicity of Hg^{2+} (*1, 6*). Exposure to Hg^{2+} induces metallothionein synthesis followed by the accumulation of this heavy metal-binding protein and mercury in the kidney (*7*). Hg^{2+} provokes renal injury by damage to proximal tubule cells (*6*). Hg^{2+} also causes strong immunological effects (*5, 8, 9*). Repeated exposure of rats to low doses of mercuric chloride resulted in a number of autoimmune responses (*9*).

Wide public, research, and regulatory interest has been focussed on the toxicity of CH_3Hg^+ since the tragic outbreaks of human poisoning in Japan ("Minamata disease") and Iraq. The highly potent neurotoxicity of CH_3Hg^+ is due, in part, to its lipophilicity, which enables it to pass the blood-brain barrier (*1*). CH_3Hg^+ has been known to produce selective damage to the nervous system in adults and, in the case of prenatal exposure, mental impairment in newborn infants (*6, 10*). CH_3Hg^+ is able to cross the placental barrier and has been found in higher levels in the fetus than in the mother (*10*). CH_3Hg^+ is formed by microbes from inorganic mercury in natural waters and bioconcentrates in fish and other aquatic animals. Prolonged consumption of mercury-contaminated seafood results in the neurological symptoms of CH_3Hg^+-poisoning (*10, 11*). CH_3Hg^+ also has nephrotoxic effects (*1*)

Analysis of Mercury

Mercury analysis and chemical speciation have been extensively reviewed (*12-18*). Mercury speciation often means that only the amounts of Hg^{2+} and CH_3Hg^+ are determined because these two forms of mercury are toxicologically very important and usually the most common in environmental and biological samples. Hg^{2+} and CH_3Hg^+ are frequently identified as inorganic and organic mercury, respectively, for the same reasons. Most conventional methods used for the determination of trace and ultratrace mercury levels require costly and sophisticated instruments and are not readily adaptable for inexpensive laboratory screening of large numbers of samples nor for on-site analyses. In addition, most mercury speciation methods are tedious to apply (*15*).

Several immunoassays have recently been reported which can detect Hg^{2+} (*11, 19, 20*) and other metal ions (*21-23*) either in chelated or free form. Antibodies used in these assays were elicited against chelates of the target metals (*11, 19-23*). These immunoassays open new avenues in the environmental monitoring of hazardous metals due to their high sample throughput and because they may be formatted as sensors for *in situ* analysis. These selective and sensitive assays employ highly specific monoclonal antibodies, possessing high affinity for the target analyte, which are expensive to develop.

Materials and Methods

Safety Note. Carbon disulfide, nitric acid, sodium sulfide, as well as numerous heavy and noble metal salts are hazardous. Care has to be exercised while handling and disposing of these chemicals (*24-26*). Safety precautions, compiled in the literature (*24-26*), are to be followed when dealing with very toxic mercury

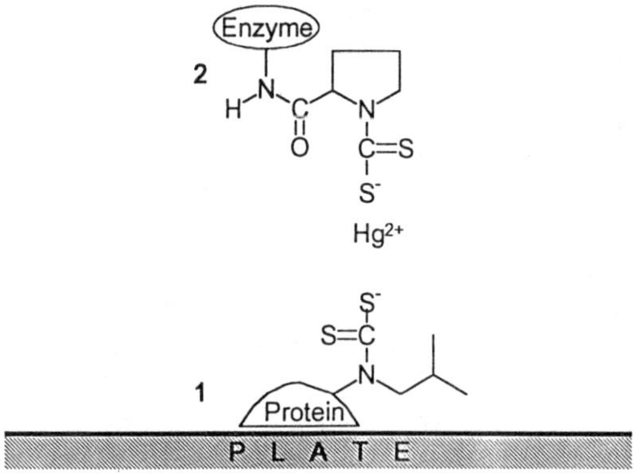

Figure 1.   Chemical structures of the reagents, 1-CONA and 2-AP, involved in assay format 1.

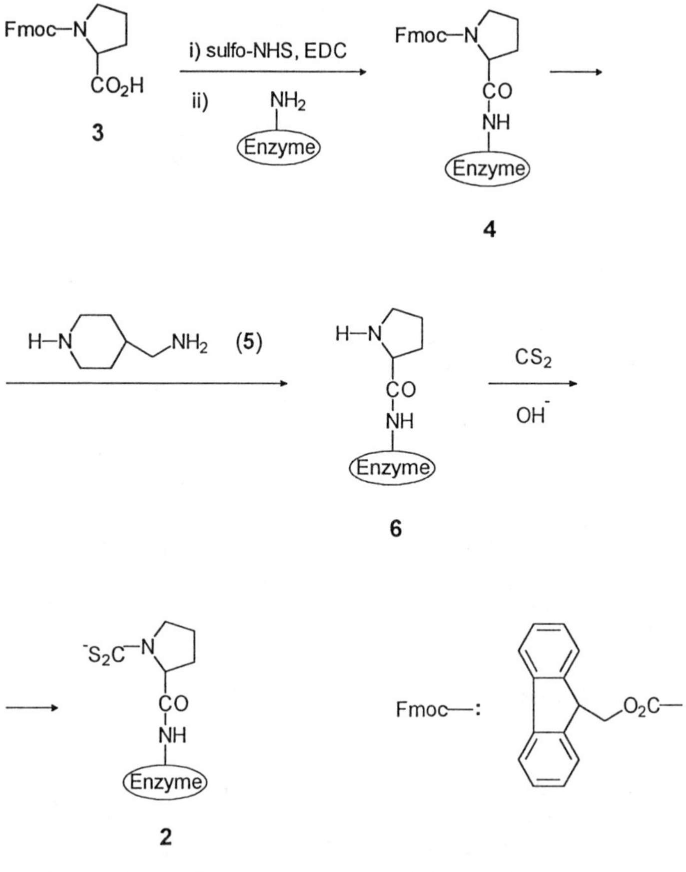

Figure 2.   Synthesis of 2-AP, the chelator linked to the reporter enzyme.

derivatives especially organomercurials. Sodium sulfide is an effective reagent for separation of heavy metal salts from wastewaters by sulfide precipitation (*27, 28*). This compound should not be mixed with acidic solutions because highly poisonous hydrogen sulfide may be formed (*24-26*). [Work with this salt and with carbon disulfide, nitric acid, and volatile organomercurials should be done in a laboratory fume hood (*24-26*).] The pH-value of the mixture of sodium sulfide solution and waste should be greater than 8 to achieve optimal cleanup efficiency (*27*). Other methods (e.g., treatment with ion-exchange resins, thioacetamide, thiourea) to eliminate mercury from wastes before disposal are also recommended in the literature (*29-30*).

New synthesis of tracer 2-AP. (See Figures 1, 2.) This synthetic pathway provided us with a simple tool, UV-spectroscopy, to verify the complete removal of the blocking groups of intermediate 4-AP; thus, the formation of conjugate with secondary amino-groups (6-AP) (*31*). The *N*-(9-fluorenylmethyloxycarbonyl) (Fmoc) protecting group was cleaved by 4-(aminomethyl)piperidine (5) in aqueous solution with only minor loss of enzymatic activity (*31*). Intermediate 4-AP was obtained using the efficient reagent combination of *N*-hydroxysulfosuccinimide (sulfo-NHS) and water soluble carbodiimide (EDC, Figure 2) (*32*).

Assay Procedures. The assays were performed according to the usual solid phase immunoassay methods, the protocols were similar to those routinely used in our laboratory (*33, 34*). Features specific to the mercury assays are indicated below. Spectramax 250 microplate reader (Molecular Devices, Menlo Park, CA) and Softmax Pro (Molecular Devices) software package were used to construct the standard curves, based on Rodbard's four-parameter logistic method (*35*), and for interpolation of unknown sample concentrations. Aqueous solutions were prepared with nanopure water (16.7 megohm/cm) obtained from Sybron/Barnstead Nanopure II system (Barnstead Co., Newton, MA). Trace amounts of heavy metals were removed from buffers used in the assay work by Chelex 100 ion exchange resin (100-200 mesh, sodium form, Bio-Rad Laboratories, Richmond, CA) according to the manufacturer's instructions. Glassware and non-disposable plasticware (e.g., containers for buffers, troughs) were kept in 20% nitric acid at room temperature overnight, followed by thorough rinsing with nanopure water (*36, 37*). Special care was taken to completely eliminate nitric acid because it might interfere with the assay. Microtiter plates (96-well Nunc-Immuno Plate, Nunc InterMed, Denmark) and metal-free micro-pipet tips were used without soaking in acid. Between incubation steps, plates were washed four times with sodium/acetate buffer (0.1 M acetate, pH 5.5).

    Assay Format 1. (See Figures 1, 3.) Conjugate 1-CONA, dissolved in a sodium/carbonate/bicarbonate buffer (50 mM carbonate/bicarbonate, pH 9.6), was immobilized on the wells of a microplate. The wells were washed, incubated with the mixture of the tracer (2-AP) and mercury standard or sample in a sodium/acetate buffer (0.1 M acetate plus 0.05% Tween 20, pH 5.5), and washed again. The wells were treated with the solution of the enzyme substrate (*p*-nitropheny phosphate), and the optical density (OD) values measured following the standard protocol (*34*).

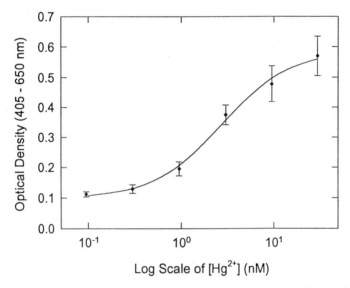

Figure 3. Standard curve of assay format 1 with simultaneous incubation.

Assay Format 2B. (See Figures 4, 5.) The wells of a microtiter plate, coated with conjugate 7-CONA as above, were washed and incubated with the mixture of the tracer (2-AP) and mercury standard or sample in a sodium/acetate buffer (0.25 M acetate plus 0.05% Tween 20, pH 6). After a washing step, the color reaction was performed and the absorbance values were determined as described above.

Results and Discussion

We have utilized the high avidity of sulfur-containing ligands to mercurials and the high sample throughput of the ELISA methodology in the development of sensitive, selective, and rapid assays for Hg^{2+} ion. We have used dithiocarbamate ligands instead of antibodies for the recognition of mercury. Dithiocarbamates show high affinity to a number of heavy metal ions, but Hg^{2+}-bis(dithiocarbamate) complexes have exceptionally high thermodynamic stability (*38*); Hg^{2+} substitutes most other metal ions from their dithiocarbamate chelates (*39*). Only a few noble metal ions (e.g., Au^{3+}) display affinities for dithiocarbamates similar or greater than that of Hg^{2+} (*40*). Organomercurials with the general structure of RHg^+ (R: alkyl or aryl) also form stable mono(dithiocarbamate)-chelates (*41*). Dithiocarbamates obtained from secondary amines are known to be fairly chemically stable; thus, we employed these chelators in our assays.

First Assay Format (Sandwich Complex Assay). In our first assay format (format 1, Figure 1) the target ion forms a sandwich complex or chelate with two complexing agents. Chelator 1-CONA (Figure 1) is a protein conjugate with dithiocarbamate groups. This chelator (1-CONA) is coated on the wells of a microtiter plate. In the structure of reagent 2-AP (Figure 1), the chelator groups are covalently bound to an enzyme. In this work, we used conalbumin (CONA) as carrier protein and alkaline phosphatase (AP) as reporter enzyme (*31*). The activity of AP is only marginally diminished by trace amounts of Hg^{2+} (*42*). To furnish the protein- and enzyme-linked dithiocarbamates, secondary amino groups were formed on the surface of these macromolecules, the intermediates (e.g., 6-AP) were then reacted with carbon disulfide (Figure 2). The preparation of the immobilized chelator (1-CONA) and enzyme-conjugate 2-AP was previously reported (*16*). An improved synthesis of tracer 2-AP (Figure 2) and a modified assay protocol with simultaneous incubation resulted in remarkably better assay performance (*31*). The ionic strength, pH, and detergent concentration of the assay buffer were optimized. Conditions promoting formation of insoluble or weakly solvated material with Hg^{2+}, such as high pH-values, may substantially reduce the signal. Inappropriate Tween 20 concentrations also adversely influence the assay (*16*). The chelator would decompose under strongly acidic conditions. Our results demonstrate that Hg^{2+} ion can be selectively detected at high ppt/low ppb concentrations by this assay (format 1, Figures 1 and 3) (*31*). Figure 3 shows the low-concentration part of the standard curve with an OD_{50} value of 2.7 ± 0.5 nM (0.54 ± 0.1 ppb) Hg^{2+} concentration. (OD_{50} is the analyte concentration at the point of the standard curve where the optical density is the half of the maximum OD minus OD at zero concentration.) The limit of detection (LOD) was 0.375 nM (75 ppt). (LOD is the analyte concentration at the

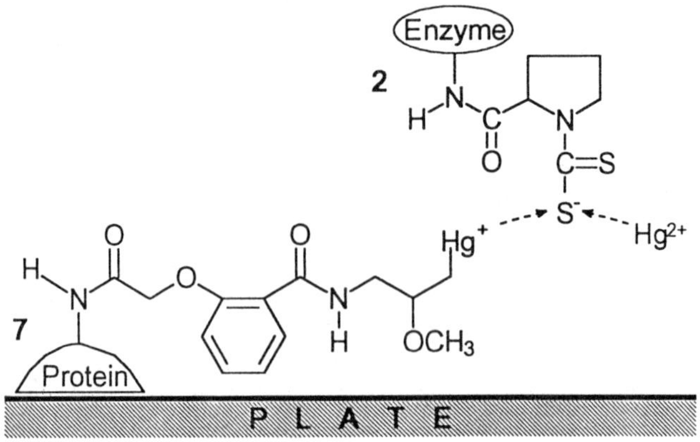

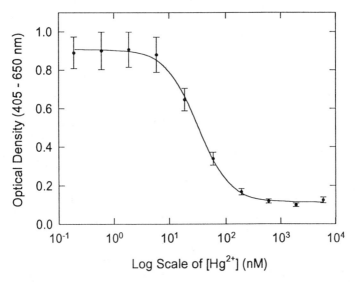

Figure 4.   Chemical structures of the reagents, **2**-AP and **7**-CONA, involved in assay format 2B.

Figure 5.   Standard curve of assay format 2B.

point of the standard curve where the OD is the sum of the signal at zero concentration and three times the standard deviation of the signal at zero concentration.) At very high concentrations of Hg^{2+}, a "hook" effect occurs, the signal decreases with increasing concentration, as it is usual with two site immunoassays (*16*). Standard curves were also constructed with a number of foreign ions up to 3000 nM concentrations. Most cations (e.g., ions of alkaline metals, Al^{3+}, Ca^{2+}, Cd^{2+}, Cr^{3+}, Fe^{2+}, Mg^{2+}, Mn^{2+}, Pb^{2+}) and hexavalent chromium ($Cr_2O_7^{2-}$) demonstrated minimal interference. Ag^+ (7%), Au^{3+} (0.2%), Cu^{2+} (2%), Pd^{2+} (29%), and Zn^{2+} (0.3%) ions displayed some cross-reactivity (CR). The standard curve of CH_3Hg^+ slightly fell below background level at high concentrations which may be due to the competition of CH_3Hg^+ with minute amounts of interfering metal ions contaminating the reagents and buffers used in the assay. An application of this assay, optimized for high-sensitivity detection of Hg^{2+}, could be the determination of the total mercury content of environmental water samples after oxidation of all mercury species to Hg^{2+}. However, CH_3Hg^+ gave essentially background-signal when different assay protocols were employed (*16, 31*), which makes mercury speciation possible by an indirect method (CH_3Hg^+ = total Hg - Hg^{2+}).

Second Assay Format (Competitive Assay). Our second assay format is based upon competition between mercuric ions and an organomercury-conjugate to bind to a chelating conjugate (format 2, Figure 4). One arrangement of this assay format with a mercury-linked enzyme tracer and an immobilized chelator (format 2A) was extensively characterized (*16, 31*) and is currently under development for application to the analysis of environmental samples. The opposite arrangement using the same enzyme labeled chelator (2-AP) as seen in format 1 and an organomercury compound bound to the plate coating protein (7-CONA, format 2B), has also been investigated. The immobilized mercury reagent (7-CONA) was obtained from the corresponding acid by the sulfo-NHS/EDC-reagent (*32*). The Hg^{2+}-standard curve is presented in Figure 5; the IC_{50} value is about 30 ± 2 nM (6 ± 0.4 ppb). There was no significant interference from a panel of metals, similar to one studied with format 1, including Ni^{2+} and Zn^{2+} up to at least 300 nM using this assay; however, Au^{3+} (CR: 11%) and CH_3Hg^+ (CR: 14%) cross-reacted.

Conclusions

We have shown that our new, chelation-based rapid assays can be useful analytical tools for sensitive and selective determination of mercury and mercury speciation. Low-ppt levels of mercury in lakes and oceans and less than 1 ppb of mercury in rivers are commonly found.   The regulatory maximum of the total mercury concentration in environmental water samples is 1 ppb in numerous countries (*16*), but the amount of mercury in river water sometimes exceeds this limit due to industrial and urban waste water and mine runoff. Thus, an anticipated use of our assay systems is the screening of mercury in water. Promising studies with environmental water samples are in progress. Analysis of mercury content in biological samples is another possible application. The action level of the U.S. Food and Drug Administration for mercury in fish is 1 ppm (*43*). The sensitivity of some of our assays seems adequate for the detection of low-ppm amounts of mercury in fish after acidic-oxidative digestion of the sample matrix and appropriate buffering and dilution of the resulting corrosive solution (*11*). Application of our assays for real samples and development of further assay systems and formats (e.g., sensors) are the directions of our present and future work.

Acknowledgements

This work was supported in part by NIEHS Superfund Grant 2 P42 ES04699, U.C. Systemwide Toxic Substances Program, and the U.S. EPA Center for Ecological Health Research at U.C. Davis (R819658). H.K. received a fellowship from NIEHS Center for Environmental Health Sciences ES05707.

Literature Cited

1. Zalups, R. K; Lash, L. H. *J. Toxicol. Environ. Health* **1994**, *42*, 1-44.
2. Skerfving, S. In *Advances in Mercury Toxicology;* Suzuki, T.; Imura, N.; Clarkson, T. W., Eds.; Rochester Series on Environmental Toxicology; Plenum Press: New York, NY, 1991; pp 411-425.
3. Yamamura, Y.; Yoshida, M.; Yamamura, S. In *Advances in Mercury Toxicology;* Suzuki, T.; Imura, N.; Clarkson, T. W., Eds.; Rochester Series on Environmental Toxicology; Plenum Press: New York, NY, 1991; pp 427-437.
4. Falnoga, I.; Mrhar, A.; Karba, R.; Stegnar, P.; Skreblin, M.; Tusek-Znidaric, M. *Arch. Toxicol.* **1994**, *68*, 406-415.
5. Hultman, P.; Johansson, U.; Turley, S. J.; Lindh, U.; Eneström, S.; Pollard, K. M. *FASEB J.* **1994**, *8*, 1183-1190.
6. *Casarett and Doull's Toxicology. The Basic Science of Poisons. 3rd ed.;* Klaassen C. S.; Amdur, M. O.; Doull, J., Eds.; Macmillan Publishing Co.: New York, NY, 1986.
7. Tohyama, C.; Ghaffar, A.; Nakuno, A.; Nishimura, N.; Nishimura, H.; In *Advances in Mercury Toxicology;* Suzuki, T.; Imura, N.; Clarkson, T. W., Eds.; Rochester Series on Environmental Toxicology; Plenum Press: New York, NY, 1991; pp 155-165.
8. Zelikoff, J. T.; Smialowicz, R.; Bigazzi, P. E.; Goyer, R. A.; Lawrence, D. A.; Maibach, H. I.; Gardner, D. *Fundam. Appl. Toxicol.* **1994**, *22*, 1-7.
9. Kosuda, L. L.; Greiner, D. L.; Bigazzi, P. E. *Cell. Immunol.* **1994**, *155*, 77-94.
10. Bellés, M.; Sánchez, D. J.; Gómez, M.; Domingo, J. L.; Jones, M. M.; Singh, P. K. *Toxicology* **1996**, *106*, 93-97.
11. Carlson, L.; Holmquist, B.; Ladd, R.; Riddell, M.; Wagner, F.; Wylie, D. In *Immunoassays for Residue Analysis: Food Safety;* Beier, R. C; Stanker, L. H., eds.; American Chemical Society: Washington, DC, 1996, ACS Symposium Series 621; pp 388-394.
12. Rapsomanikis, S. In *Environmental Analysis Using Chromatography Interfaced with Atomic Spectroscopy;* Harrison, R. M.; Rapsomanikis, S., eds.; Halsted Press: Chicester, West Sussex, UK, 1989; pp 298-317.
13. Lind, B.; Body, R.; Friberg, L. *Fresenius J. Anal. Chem.* **1993**, *345*, 314-317.
14. Jackson, K. W.; Mahmood, T. M. *Anal. Chem.* **1994**, *66*, 252R-279R

15. Ritsema, R.; Donard, O. F. X. *Appl. Organomet. Chem.* **1994**, *8*, 571-575.
16. Szurdoki, F.; Kido, H.; Hammock, B. D. In *Immunoanalysis of Agrochemicals, Emerging Technologies;* Nelson, J. O.; Karu, A. E.; Wong, R. B., eds.; American Chemical Society: Washington, DC, 1995, ACS Symposium Series 586; pp 248-264.
17. MacCarthy, P.; Klusman, R. W.; Cowling, S. W.; Rice, J. A. *Anal. Chem.* **1995**, *67*, 525R-582R.
18. Ombaba, J. M. *Microchem. J.* **1996**, *53*, 195-200.
19. Wylie, D. E.; Carlson, L. D.; Carlson, R.; Wagner, F. W.; Schuster, S. M. *Anal. Biochem.* **1991**, *194*, 381-387.
20. Wylie, D. E.; Lu, D.; Carlson, L. D.; Carlson, R.; Babacan, K. F.; Schuster, S. M.; Wagner, F. W. *Proc. Natl. Acad. Sci. USA* **1992**, *89*, 4104-4108.
21. Reardan, D. T.; Meares, C. F.; Goodwin, D. A.; McTigue, M.; David, G. S.; Stone, M. R.; Leung, J. P.; Bartholomew, R. M.; Frincke, J. M. *Nature* **1985**, *316*, 265-268.
22. Chakrabarti, P.; Hatcher, F. M.; Blake, R. C. 2nd; Ladd, P. A.; Blake D. A. *Anal. Biochem.* **1994**, *217*, 70-75.
23. Boden, V.; Colin, C.; Barbet, J.; Le Doussal, J. M.; Vijayalakshmi, M. *Bioconjugate Chem.* **1995**, *6*, 373-379.
24. Manufacturing Chemists' Association *Guide for Safety in the Chemical Laboratory;* Van Nostrand Reinhold Co.: New York, NY, 1972.
25. *NIOSH Pocket Guide to Chemical Hazards;* DHEW (NIOSH) Publication No. 78-210; U.S. Department of Health and Human Services: Washington, DC, 1985.
26. *Hazards in Chemical Laboratory;* Luxon, S. G., Ed.; Royal Society of Chemistry: Cambridge, Great Britain, 1992.
27. Bhattacharyya, D.; Jumawan, A. B. Jr.; Grieves, R. B. *Separation Sci. Technol.* **1979**, *14*, 441-452.
28. Cowling, S. J.; Gardner, M. J.; Hunt, D. T. E. *Environ. Technol.* **1992**, *13*, 281-291.
29. McGarvey, F. X. In *Mercury and Arsenic Wastes: Removal, Recovery, Treatment, and Disposal;* Pollution Technology Review No. 214; U.S. Environmental Protection Agency; Noyes Data Corporation: Park Ridge, NJ, 1993; pp 36-38.
30. Robins, R. G.; Jayaweera, L. In *Mercury and Arsenic Wastes: Removal, Recovery, Treatment, and Disposal;* Pollution Technology Review No. 214; U.S. Environmental Protection Agency; Noyes Data Corporation: Park Ridge, NJ, 1993; pp 43-45.
31. Szurdoki, F.; Kido, H.; Hammock, B. D. *Bioconjugate Chem.* **1995**, *6*, 145-149.
32. Szurdoki, F.; Bekheit, H. K. M.; Marco, M. P.; Goodrow, M. H.; Hammock, B. D. *New Frontiers in Agrochemical Immunoassay;* Kurtz, D. A.; Skerritt, J. H.; Stanker, L., Eds.; AOAC International: Arlington, VA, 1995; pp 39-63.
33. Schneider, P.; Hammock, B. D. *J. Agric. Food Chem.* **1992**, *40*, 523-530.
34. Bekheit, H. K. M.; Lucas, A. D.; Szurdoki, F.; Gee, S. J.; Hammock, B. D. *J. Agric. Food Chem.* **1993**, *41*, 2220-2227.

35. Rodbard, D. In *Ligand Assay*; Langan, J.; Clapp, J. J., Eds.; Masson Publishing: New York, 1981; pp 45-99.
36. Lansens, P.; Meuleman, C.; Baeyens, W. *Anal. Chim. Acta* **1990**, *229*, 281-285.
37. Lau, O.-W.; Ho, S.-Y. *Anal. Chim. Acta* **1993**, *280*, 269-277.
38. Bond, A. M.; Scholz, F. *J. Phys. Chem.* **1991**, *95*, 7460-7465.
39. Lo, J. M.; Yu, J. C.; Hutchison, F. I.; Wal, C. M. *Anal. Chem.* **1982**, *54*, 2536-2539.
40. Lo, J.-M.; Lee, J.-D. *Anal. Chem.* **1994**, *66*, 1242-1248.
41. Sarzanini, C.; Sacchero, G.; Aceto, M.; Abollino, O.; Mentasti, E. *J. Chromatogr.* **1992**, *626*, 151-157.
42. McComb, R. B.; Bowers, G. N.; Posen, S. *Alkaline Phosphatase;* Plenum Press: New York, NY, 1979.
43. U.S. Food and Drug Administration *Federal Register* **1979**, *44*, 4012.

Chapter 4

Characterization of a Metal-Specific Monoclonal Antibody

D. A. Blake[1], M. Khosraviani[1], A. R. Pavlov[1], and R. C. Blake II[2]

[1]Department of Ophthalmology, Tulane University School of Medicine, 1430 Tulane Avenue, New Orleans, LA 70112–2699
[2]Department of Basic Pharmaceutical Sciences, Xavier University of Louisiana, New Orleans, LA 70112–2699

A new technique was utilized to characterize the binding specificity of a metal-specific monoclonal antibody, 2A81G5. A KinExA automated immunoassay instrument was employed to measure the concentration of unliganded antibody in a mixture of antibody, antigen, and antibody-antigen complex. Unliganded antibody was captured on a resin containing immobilized antigen and quantified by brief exposure to a fluorescein-labeled anti-species secondary antibody. Instrument response was proportional to the concentration of primary and secondary antibody and was also dependent upon the immobilized antigen concentration, reagent flow rate, and duration of flow. Equilibrium dissociation constants for a Cd(II)-ethylenediaminetetraacetic acid complex (Cd-EDTA) and a Cd(II)-EDTA-bovine serum albumin conjugate (Cd-EDTA-BSA) were 10 nM and 72 pM, respectively. The instrument could also be used in a competitive immunoassay format to quantify low levels of antigen in solution; the limit of detection for the Cd-EDTA-BSA conjugate was 40 pM or 5 ppt in ionic cadmium.

The development of any immunoassay is greatly facilitated by the detailed knowledge of the binding characteristics of the antibodies used in the assay. Immunoassay sensitivity is directly related to the antibody affinity for the antigen and assay selectivity is directly related to the binding specificity of the antibody used in the analysis (1). In addition, the performance of the assay is directly related to the kinetic properties (on and off rates) of the antibody binding reaction. Our laboratory has been developing antibodies to metal-chelate complexes and we have recently demonstrated that these antibodies can be used in competitive immunoassays for metal ions (2). Immunoassays offer significant advantages over more traditional methods of metal ion analysis; they are quick, inexpensive, simple to perform, and sufficiently portable to be used at the site of contamination. In contrast, existing technologies to measure heavy metals require complex instrumentation (atomic absorption or inductively coupled plasma emission spectroscopy) in a centralized facility; because of the expense of the instrumentation required for the analysis, samples are often in a queue for long periods of time. As part of our continuing efforts to construct and optimize metal ion immunoassays, we have developed a new technique to rapidly determine the solution equilibrium dissociation constants of antibody-antigen interactions. We have applied these methods in a study of the interactions of the metal-specific monoclonal antibody 2A81G5 with various metal-chelate complexes.

Materials and Methods

Bovine serum albumin (BSA) (fatty acid ultrafree) was a product of Boehringer Mannheim (Indianapolis, IN). Atomic absorption standard metals were purchased from Perkin-Elmer (Norwalk, CT). These standard metals were handled and disposed of according to the directions in the Materials Safety Data Sheets provided by Perkin-Elmer. Fluorescein isothiocyanate conjugate of affinity purified goat-antimouse IgG was a product of Jackson ImmunoResearch Laboratories (West Grove, PA). Polymethylmethacrylate beads (140-170 mesh) were obtained from Bangs Laboratories (Carmel, IN). Reagents for tissue culture and ELISA were obtained as previously described (2). Glassware was mixed-acid washed (3) and plasticware was soaked overnight in 3 N HCl and rinsed liberally with nanopure water before use.

Preparation of protein-chelate conjugates. Protein-chelate complexes were prepared as previously described (2). Extent of substitution of free lysine groups was 55% for the BSA-thioureido-L-benzyl-ethylenediaminetetraacetic acid-cadmium conjugate (Cd-EDTA-BSA) and 30% for the BSA-thioureido-L-benzyl-ethylenediaminetetraacetic acid-mercury conjugate (Hg-EDTA-BSA).

Purification of monoclonal antibody. Isolation and characterization of the 2A81G5 monoclonal antibody will be described elsewhere (D.A. Blake *et al.*, *J. Biol. Chem.*, in press). The concentration of IgG in the purified product (405 µg/ml) and the light chain specificity of the antibody (IgG1) was determined by sandwich ELISA (4).

Determination of equilibrium dissociation constants. Equilibrium measurements of the interactions of chelated metals with the 2A81G5 monoclonal antibody were performed with a KinExA (Sapidyne Instruments, Inc., Idaho City, ID) automated immunoassay system. Polymethylmethacrylate beads (98 ± 8 µm) were coated with antigen by suspending 200 mg (dry weight) of beads in 1.0 ml of phosphate buffered saline (PBS, 137 mM NaCl, 3 mM KCl, 10 mM sodium phosphate buffer, pH 7.4) containing 5-100 µg/ml Cd-EDTA-BSA. For some experiments, the beads were incubated with 100 µg/ml of Hg-EDTA-BSA. After 1.0 h of gentle agitation at 37° C, the beads were centrifuged and the supernatant solution was decanted. Any non-specific protein binding sites remaining on the beads were blocked by the subsequent incubation of the beads with 1.0 ml of 10% (w/v) goat serum in PBS containing 0.03% NaN$_3$ for an additional h at 37° C. The blocked beads were stable at 4° C in the blocking solution for up to 4 weeks.

A bead pack approximately 4 mm high was created in the observation cell of the instrument by drawing 675 μl of a suspension of the blocked beads in PBS (6.7 mg beads/ml) into a capillary cell (i.d. = 1.6 mm) above a microporous screen at a flow rate of 1.5 ml/min and washing the retained beads with sufficient PBS buffer (1.0 ml) to remove the excess goat serum. The beads were then gently disrupted with a brief backflush of PBS (50 μl at 300 μl/min), followed by a 20 s settling period to create a uniform and reproducible pack.

For measurement of equilibrium dissociation constants, the 2A81G5 monoclonal antibody was mixed with varying concentrations of metal ion in PBS supplemented with 5.0 mM EDTA and 1 mg/ml BSA. Once equilibrium was achieved (less than 2 min in all cases), an aliquot of the reaction mixture was drawn past the beads, followed by 166 μl of the PBS-EDTA-BSA buffer to wash out unbound primary antibody and excess metal chelate. One ml of fluorescein-labeled goat anti-mouse antibody (1.0 μg/ml in PBS-BSA) was then drawn past the beads. All solution transfers up to this point were accomplished with a flow rate of 500 μl/min. Unbound labeled secondary antibody was subsequently removed by drawing 3.0 ml of PBS-EDTA-BSA through the bead pack over a period of 2 min. Data acquisition (initiated at the introduction of primary antibody to the beads) and instrument control were accomplished *via* a Toshiba Satellite 486/33 notebook computer interfaced to the KinExA using software provided by Sapidyne. Dissociation constants were obtained from nonlinear regression analyses of the equilibrium binding data using a one-site homogeneous binding model. The contribution of each datum to the analysis was weighted by a factor derived from the application of information theory to protein binding reactions (5).

Results

The KinExA immunoassay instrument is a computer controlled flow spectrofluorimeter designed to achieve the rapid separation and quantification of free, unbound antibody present in reaction mixtures of antibody, antigen, and antibody-antigen complexes. The operating principles of the instrument are illustrated schematically in Fig. 1. Uniform particles were pre-coated with Cd-EDTA-BSA and deposited in a packed bed above a microporous screen fitted in a capillary flow/observation cell. A small volume of the reaction mixture, typically 0.5 ml, is then percolated through the packed bed of microbeads under negative pressure. Antibodies in the mixture with unoccupied binding sites are available to bind to the immobilized antigen coated on the solid phase. Exposure of the antibody-soluble antigen complex to the immobilized antigen is sufficiently brief (typically 240 ms) to insure that negligible dissociation of the soluble complex occurs during the time of exposure to the beads. The soluble reagents are removed from the beads by an immediate buffer wash. Quantification of the primary antibody thus captured is achieved by the brief exposure of the particles to a fluorescein labeled goat anti-mouse secondary antibody, followed by measurement of the resulting fluorescence from the particles after the removal of excess unbound secondary antibody.

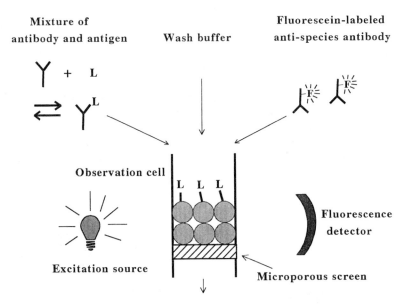

Figure 1. Schematic of KinExA Automated Immunoassay Instrument.

Fig. 2*A* shows representative examples of the time courses for the fluorescence signal when different concentrations of a primary antibody were analyzed in the absence of the soluble antigen. The primary antibody was 2A81G5, a mouse monoclonal raised in response to a Cd-EDTA hapten covalently coupled to keyhole limpet hemocyanin (D.A. Blake *et.al, J. Biol. Chem.,* in press). The immobilized ligand was Cd-EDTA-BSA. Data accumulation was initiated when the beads were first exposed to the unlabeled primary antibody. The fluorescence signal from zero to 90 s thus corresponded to the background signal generated while the primary antibody was exposed to and subsequently washed out of the packed microbead column. The beads were then exposed to a solution of fluorescein labeled goat anti-mouse antibodies (90 to 190 s); excess unbound labeled secondary antibody was then removed from the beads with a buffer wash (190 to 300 s). In the absence of the primary antibody (*curve a*) the fluorescence response approximated a square wave corresponding to the fluorescence of the labeled secondary antibody during its transient passage past the beads in the observation cell. The signal failed to return to that of the background (0.8% higher), indicating a slight nonspecific binding of the fluorescein labeled antibody to the beads. When different concentrations of the primary antibody were analyzed *(curves b* through *d)* the fluorescence response from 90 to 190 s reflected the sum of two contributions: the fluorescence of the unbound labeled antibody in the interstitial regions among the beads; and that of the labeled secondary antibody that had bound to primary antibody captured by the immobilized antigen on the beads. Binding of the secondary antibody was an ongoing process that produced a positive slope in this portion of the curve. When the excess unbound label was removed from the beads, the signal that remained was the sum of both nonspecifically and specifically bound antispecies antibody.

The instrument response as a function of the concentration of primary antibody is shown in Fig. 2*B*. The standard instrument response was taken as the integrated fluorescence signal over the last 90 s of the experimental trace (the 210 to 300 s interval for curves such as those in Fig. 2*A*). The signal in Fig. 2*B* was approximately linear with the concentration of primary antibody up to 0.2 µg/ml. Beyond that concentration deviations from linearity became increasingly apparent. The *inset* of Fig. 2*B* shows the instrument response to concentrations of primary antibody up to 2.5 µg/ml. The data were fit to a rectangular hyperbola with a maximum of 360 volt-seconds and a half-maximal response at 2.1 µg/ml. This dependence, which represented the binding curve of primary antibody to the immobilized antigen on the beads, indicated that only a small fraction of the population of immobilized antigen was involved in antibody binding in the approximately linear portion of the binding curve at low antibody concentrations. A primary antibody concentration of 0.1 µg/ml was chosen for standard assay conditions to insure that the instrument response was well within the linear portion of the curve.

The signal as a function of the concentration of fluorescein-labeled secondary antibody was also determined. The instrument response was approximately linear with the concentration of secondary antibody up to 2.0 µg/ml (data not shown). The slope of a plot of instrument response *versus* the

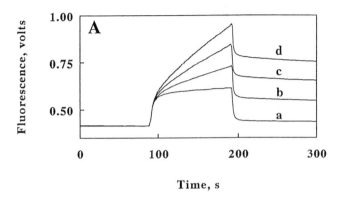

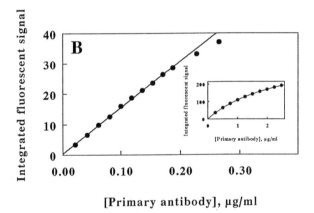

Figure 2. Dependence of the instrument response on antibody concentration. *A*, kinetic traces of fluorescence as a function of time for 0 (*a*), 0.06 (*b*), 0.12 (*c*) and 0.18 (*d*) µg/ml of primary antibody. *B*, linear dependence of the instrument response at low concentrations of primary antibody. *Inset*, hyperbolic dependence of the instrument response on higher concentrations of primary antibody.

concentration of secondary antibody was proportional to the concentration of primary antibody employed in the assay; similarly, changing the concentration of secondary antibody changed the slope of a plot of instrument response *versus* the concentration of primary antibody (data not shown). A secondary antibody concentration of 1.0 µg/ml was chosen as the standard assay condition to insure adequate signal-to-noise characteristics without the excessive consumption of fluorescein labeled protein reagent.

Instrument response was also a function of immobilized ligand concentration, reagent flow rate, and duration of flow. The dependence of the signal on the amount of immobilized antigen is shown in Fig. 3A. Set amounts of the beads were exposed to different concentrations of the Cd-EDTA-BSA coating reagent and subsequently incorporated into the standard assay format. The data in Fig. 3A were fit to a rectangular hyperbola that represented the binding curve for the protein-hapten conjugate to the hydrophobic surface of the beads (half-maximal effect, 3.4 µg/ml). A coating solution of 100 µg of protein-hapten conjugate per ml was chosen as the standard operating condition to insure that an excess of immobilized ligand was present on the beads. The dependence of the instrument response on the flow rate of the primary antibody is shown in Fig. 3B. The higher the flow rate, the lower was the signal. The data in Fig. 3B were fit to a single exponential function of flow rate, with an amplitude of 21.8 volt-seconds and a constant in the exponential term of -1.4 min/ml. The value of the amplitude represented the total instrument response expected if the flow rate were zero, where the amount of primary antibody bound to the immobilized antigen had achieved equilibrium for that set of reagents and assay conditions. The fact that lower instrument responses were obtained at higher flow rates indicated that only a percentage of the primary antibody with unoccupied binding sites was removed from solution by binding to the immobilized antigen. The shape of the exponential curve (*i.e.*, the value of the constant term in the argument of the exponent) was independent of the concentration of primary antibody over a 10-fold range (data not shown). These observations indicated that the kinetics of antibody binding to the beads were driven by the effective concentration of the immobilized antigen, the binding partner in excess. Similar data and conclusions were obtained from the dependence of the instrument response on the flow rate of the secondary antibody (data not shown). Flow rates of 0.5 ml/min for both the primary and secondary antibodies were chosen as the standard assay condition as a compromise between assay sensitivity and expediency.

Finally, experiments were performed to demonstrate that the instrument response was dependent upon the total number of antibody molecules in the solution drawn past the beads. When antibody solution was diluted while concomitantly decreasing the flow rate such that the number of antibody molecules in contact with the beads remained constant, identical experimental traces were obtained (data not shown). For high affinity antibodies, this feature allows the investigator to measure binding constants at very low concentrations of primary antibody.

When soluble antigen was mixed with the primary antibody, the instrument response of the KinExA became a means of quantifying unliganded antibody. The primary data resembled those in Fig. 2A; transformation of these signals into the

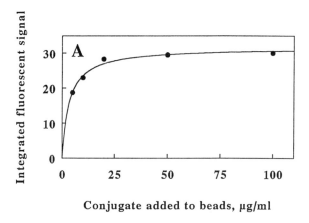

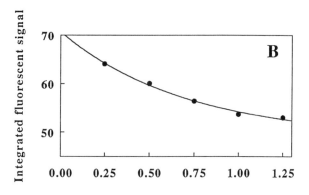

Figure 3. Instrument response as a function of immobilized ligand concentration and reagent flow rate. *A,* individual lots of PMMA beads were exposed Cd-EDTA-BSA at concentrations as shown on the abscissa. *B,* only the flow rate of the primary antibody past the beads was varied; the total volume of antibody remained constant at 0.5 ml.

fraction of occupied binding sites as a function of the concentration of free antigen produced binding curves as shown in Fig. 4. Fig. 4*A* shows a binding curve for Cd-EDTA to antibody 2A81G5. The different symbols displayed in the figure correspond to different immobilized antigens employed in the assays. The 2A81G5 antibody has the ability to recognize both Cd-EDTA-BSA and Hg-EDTA-BSA conjugates (D.A. Blake *et al., J. Biol. Chem.*, in press) and substitution of Hg-EDTA-BSA for Cd-EDTA-BSA on the solid phase had no apparent influence on the outcome of the binding experiments. A single binding curve with an apparent equilibrium dissociation constant of 10 nM was drawn through all of the data points. The concentration of 2A81G5 employed in the binding experiment, 1.3 nM in binding sites, was well below the value of the dissociation constant; thus no corrections in the concentration of free ligand were required.

When the Cd-EDTA moiety was covalently coupled to the ε-amino group of a lysine residue in BSA by the same benzyl-thioureido group as that employed in the immunogen, binding of the antigen to the antibody became much tighter. The equilibrium binding curve for soluble Cd-EDTA coupled to BSA and 2A81G5 is shown in Fig. 4*B*. These binding experiments were conducted by lowering the standard primary antibody concentration 10-fold to 0.01 μg/ml, while increasing the standard volume of analyte 10-fold to 5.0 ml. The *curve* drawn through the data points was that for a homogeneous binding interaction with a dissociation constant of 72 pM, a value some 150-fold lower than that for the Cd-EDTA interaction. It was evident that the binding site of 2A81G5 recognized features of the lysyl-thioureido-benzyl portion of the antigen. Since the concentration of the antibody's binding sites (133 pM) and the value of the dissociation constant were of the same magnitude, the concentration of free antigen on the abscissa of Fig. 4*B* was calculated as the difference between the concentration of the total antigen minus that bound to the antibody according to the following equation:

$$[\text{Cd-EDTA on BSA}]_{\text{free}} = [\text{Cd-EDTA on BSA}] - [\text{Antibody}]_{\text{total}} \text{ x }\text{ fraction of occupied binding sites}$$

The KinExA could also be exploited to devise competitive immunoassays to quantify low levels of the antigen in solution. An apparent binding curve for Cd-EDTA-BSA that resembled a stoichiometric titration could be generated by raising the concentration of 2A81G5 to 0.1 μg/ml. Fig. 5 shows the dependence of the instrument response on the concentration of soluble Cd-EDTA-BSA at an antibody concentration of 0.1 μg/ml. The linear portion of the curve comprised a standard curve for Cd-EDTA-BSA up to about 400 pM. Given that the reproducibility of individual data points in the KinExA was around 5%, a conservative estimate for the limit of detection for Cd-EDTA-BSA using the standard curve in Fig. 5 was approximately 40 pM or 5 parts per trillion in ionic cadmium concentration.

Discussion

Unlike other automated immunoassay instruments where the interaction to be quantified is that between a soluble and an immobilized binding partner (*6-8*) the

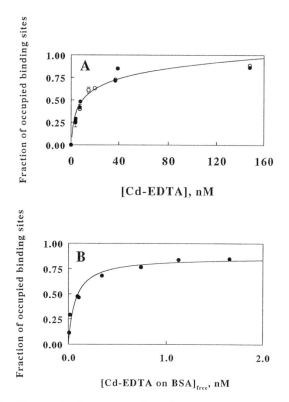

Figure 4. Equilibrium binding curves for Cd-EDTA and Cd-EDTA-BSA. *A*, Concentration of the unliganded antibody in the presence of various concentrations of Cd-EDTA were measured using immobilized Cd-EDTA-BSA (O) or Hg-EDTA-BSA (●) to capture unliganded antibody. *B*, Binding curve for soluble Cd-EDTA-BSA.

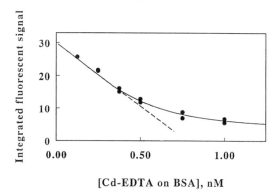

Figure 5. Effect of soluble Cd-EDTA-BSA on the instrument response in the presence of 0.1 µg/ml of 2A81G5. The concentrations on the abscissa are those of the total Cd(II), not those of the multiply-labeled BSA.

immobilized reagent in the KinExA is merely a tool used to separate and quantify the concentration of free, uncomplexed reagent in solution reaction mixtures. When measuring binding interactions in the KinExA, the principal conditions that must be met are: i) binding of the immobilized and the corresponding soluble component must be mutually exclusive and ii) binding to the immobilized component must be sufficiently tight to permit an instrument response with acceptable signal-to-noise characteristics. Once these conditions are met, the quantitative binding information extracted from the data are those of the binding interaction in homogeneous solution between soluble partners. In those alternative cases where one of the principal binding partners is immobilized (6-8), one must always be concerned about the effects of mass transport limitations and surface properties (charge, hydrophobicity, stearic hindrance, etc.) on the apparent strength of the measured interaction (9).

The KinExA can be exploited to quantify any protein-ligand interaction, as long as one of the binding partners can be immobilized in such a way that it still binds to the other, and a means can be devised to convert immobilized complexes quantitatively into fluorescent signals that remain proportional to the concentration of the complex. This laboratory has used the KinExA to determine equilibrium binding constants between 2A81G5 and each of 16 different metal-EDTA complexes, as well as 4 different Cd-chelator combinations (D.A. Blake *et al.*, *J. Biol. Chem.*, in press). The values of individual dissociation constants obtained in this study ranged over 7 orders of magnitude from 10^{-11} to 10^{-4} M. The instrument was also exploited to quantify the values of individual rate constants for the binding of antigens to 2A81G5 (data not shown). Instead of exposing an equilibrium mixture of the binding partners to the immobilized antigen on the beads, a solution of the monoclonal antibody was injected into and mixed with a stream of chelated metal ions that reacted for just 11 s before the mixture encountered the packed beads.

When used as a platform for conventional immunoassay, the KinExA exhibited a limit of detection for Cd-EDTA-BSA equivalent to that for the corresponding competition ELISA (D.A. Blake, unpublished data). It should be noted that no effort was made to adjust the assay parameters in Fig. 5 to optimize the value of the detection limit. Thus, a combination of greater analyte volume, slower flow rates, and a greater concentration of secondary antibody could conceivably lower the observed limit of detection some 50-fold to the sub-ppt level. Of course, since the operating format of the KinExA only permits the analysis of one sample at a time, the KinExA is not likely to replace standard ELISA formats and instrumentation when large numbers of samples must be analyzed on a routine basis. However, because of its unique ability to provide information about antibody-antigen interactions in solution, the KinExA could be a useful adjunct in any analytical immunochemistry laboratory where research and development of new assays is underway.

Acknowledgments

This work was supported by a grant to D.A. Blake from the Environmental Protection Agency (R 824029) and by a grant from the Department of Defense (93-DNA-2) to the Tulane-Xavier Center for Bioenvironmental Research.

Literature Cited

1. Gosling, J.P. *Clinical Chemistry.* **1990**, *36*, 1408-1427
2. Chakrabarti, P.; Hatcher, F.M.; Blake, R.C., II; Ladd, P.A.; Blake, D.A. *Anal. Biochem.* **1994**, *217*, 70-75.
3. Thiers, R.C. *Meth. Biochem. Anal.* **1957**, *5*, 273-335.
4. Harlow, E.; Lane, D. *Antibodies: A Laboratory Manual*; Cold Spring Harbor Laboratory, Cold Spring Harbor, NY, 1988; pp. 553-612.
5. Weber, G. in *Molecular Biophysics*; Pullman, B., and Weissbluth, M., Eds; Academic Press, New York, NY, 1965; pp 369-696.
6. Friguet, B.; Chaffotte, A.F.; Djavadi-Ohaniance, L; Goldberg, M.E. *J. Immunol. Methods.* **1985**, *77*, 305-319.
7. Dill, K.; Lin, M.; Poteras, C.; Fraser, C.; Hafeman, D.G.; Owick, J.C.; Olson, J.D. *Anal. Biochem.* **1994**, *217*, 128-138.
8. Malmqvist, M. *Nature.* **1993**, *361*, 186-187.
9. Kemeny, D.M.; Challacombe, S.J., Eds.; *ELISA and Other Solid Phase Immunoassays: Theoretical and Practical Aspects*, John Wiley and Sons: New York, NY, 1988.

Chapter 5

Development of an Enzyme-Linked Immunosorbent Assay for 1-Nitropyrene

A Possible Marker Compound for Diesel Exhaust Emissions

D. Knopp, V. Väänänen, J. Zühlke, and R. Niessner

Institute of Hydrochemistry, Technical University of Munich, Marchioninistrasse 17, D–81377 Munich, Germany

An enzyme-linked immunosorbent assay (ELISA) was developed for 1-nitropyrene. For hapten syntheses pyrenebutyric acid was nitrated yielding a mixture of 4-(6-nitropyrene-1-yl)- and 4-(8-nitropyrene-1-yl)-butyric acid isomers that could not be separated. High titered antisera were obtained in rabbits after immunization with a KLH immunogen prepared with the mixture of isomers. As coating antigen a corresponding BSA conjugate was used. The calibration curve was linear in the 0.1 to 10 ppb range with a center point at 1.5 ppb and an LOD of 0.02 ppb in 10:90 acetonitrile/water (v/v). The average coefficients of variation were 4.9 and 7.8% over the range of the standard curve for intra- and interassay precision. Interference caused by 1,6- and 1,8-dinitropyrenes was observed. The effect of different concentrations of solvents on the antibody-analyte interaction was studied.

The impact of gasoline- and diesel-powered vehicles in the environment and our lifestyles has been considerable. Although the emissions of engines have been reduced significantly over the last 30 years, the increased number of vehicles in use remain significant contributors of pollutants. Engine emissions have become suspected culprits for some of the health effects observed in urban populations such as increased incidence of respiratory symptoms, decreased lung function, increased hospitalizations and increased cardiopulmonary disease mortality (*1*). Regulated air pollutants include carbon monoxide (CO), nitrogen oxides (NO_x), hydrocarbons (HC), and particulates (*2*). A group of compounds present in diesel and gasoline engine exhausts which are known for its carcinogenic and/or mutagenic potential but unregulated by law are the polycyclic aromatic compounds (PAHs) (*3,4*). Diesel exhaust was classified as probable carcinogen (group 2 A) to humans by the International Agency for Research on Cancer (IARC) in 1989 (*5*). At present it cannot be excluded that the effect of diesel soot results from a combined action of ultrafine carbon particles and carcinogenic PAHs (*6*). Until now, no estimates of general air pollution from diesel exhaust exist. Diesel soot, the most commonly used marker for diesel exhaust is difficult to distinguish from small carbonaceous particles resulting from other sources of combustion. The search for a more specific marker as well as a

Table I. Cross-Reactivity of the 1-Nitropyrene Antiserum at Center Point of theCalibration Curve (IC_{50})

Compound	IC_{50} [%]
1-Nitropyrene*	100
1,8-Dinitropyrene*	1250
1,6-Dinitropyrene*	750
2-Nitropyrene	37
4-Nitropyrene**	20
2-Nitrofluoranthene	20
9-Hydroxy-1-nitropyrene	17
5-Nitroacenaphthene*	16
3-Nitrofluoranthene**	12
1,3-Dinitropyrene*	10
1,3-Dinitronaphthalene*	2
2,7-Dinitrofluorene*	1.8
2-Nitrofluorene**	1
1,5-Dinitronaphthalene*	0.6
1,8-Dinitronaphthalene*	0.5
1-Nitronaphthalene**	<0.5
9-Nitroanthracene**	<0.5
2-Nitronaphthalene**	<0.5
6-Nitrobenzo(a)pyrene*	n.d.[a]
7-Nitrodibenzo(a,h)anthracene**	n.d.[a]
9,10-Dinitroanthracene**	n.d.[a]
6-Nitrochrysene**	n.d.[a]
2-Nitrobiphenyl**	n.d.[a]
3-Nitrobiphenyl*	n.d.[a]
4-Nitrobiphenyl**	n.d.[a]
Phenanthrene*	<1.0
Anthracene*	<1.0
Fluorene*	<1.0
Fluoranthene*	<1.0
Benzo(a)anthracene*	<1.0
Chrysene*	<1.0
Pyrene*	<1.0
Naphthalene*	<0.1
Benzo(a)pyrene*	<0.1
Dibenzo(a,h)anthracene*	<0.1
Indeno(1,2,3-c,d)pyrene*	<0.1
6 PAH Standard (GDWA)**	<0.1
16 EPA PAH Standard**	<0.1

[a] No inhibition detectable. * Compound was supplied by Aldrich (Steinheim, FRG) or ** Dr. W. Schmidt (PAH Research Institute, Greifenberg, FRG).

unique biomarker has been unsuccessful so far. (*7*). Compared with gasoline engines with catalytic converters, diesel engines produce as much as 30-100 times more particulate emissions, 20-30 times more nitro-PAHs, and 10 times as much total PAHs, depending on the driving conditions (*8*).

NO$_2$

1-nitropyrene

The most abundant nitroarene in diesel particulate organic matter is 1-nitropyrene. Pitts et al. have found this compound to be responsible for 20-27% of the mutagenicity of a diesel exhaust particulate extract (*9*). Therefore, the question arised whether 1-nitropyrene (1-NP) could be exploited for source identification. However, exhaust particulate extract represents a very complex matrix which contains 1-NP only in trace amounts. Thus traditional analysis involves time-consuming purification and enriching steps followed by highly sophisticated chromatographic separation and detection systems (for HPLC, common detection systems are UV, chemical reduction/fluorescence, photoreduction/fluorescence, chemiluminescence, reductive electrochemical and differential pulse detection; GC often is coupled with MS, ECD, NPD, flame ionization, thermal energy analyzer, tandem MS and chemiluminescence detection) (*10-14*).

Immunochemical methods offer the advantage of being rapid, sensitive, and cost-effective analytical tools that need only microliter amounts of sample volume. During the last years, environmental immunoassays were developed mainly for pesticides. Immunochemical determination of PAHs have been reported for benzo(a)pyrene (*15-19*), pyrene (*20*), and nitrofluoranthenes (*21*). Immunochemical test kits for screening of PAH contamination in soil and/or water are also available commercially (PAH-RIS Test/EnSys; D Tech PAH Test Kit/Merck; EnviroGard PAH Test Kit/Millipore; PAH Test/Quantix). The objective of this investigation was to prepare polyclonal antibodies for 1-NP and to apply these reagents for the development of an enzyme-linked immunosorbent assay (ELISA).

Methods

Reagents and Equipment. All reagents were of analytical-reagent grade unless specified otherwise. Compounds cited in Table I by an * were purchased from Aldrich (Steinheim, FRG) and by an ** from Dr. W. Schmidt (Institute of PAH Research, Greifenberg, FRG). 2-Nitropyrene and 2-nitrofluoranthene were supplied by Campro Scientific (Veenendaal, Netherlands). Goat anti-rabbit IgG-horseradish peroxidase conjugate (goat anti-rabbit IgG-POD) was from Pierce (Rockford, USA). Bovine serum albumin (BSA) crystalline, research grade, 3,3',5,5'-tetramethylbenzidine (TMB) and Visking dialysis tubing were purchased from Serva Feinbiochemica (Heidelberg, FRG). Acetonitrile (MeCN), acetone (ACET), N,N-dimethylformamide (DMF), ethylene glycol (EG), methanol (MeOH) and ethanol (EtOH) were obtained from Merck (Darmstadt, FRG). Freund's complete and incomplete adjuvants were purchased from Difco Labs (Detroit, USA).

The ELISA analysis was carried out in polystyrene plates with 96 F bottom wells with high binding capacity (Greiner Labortechnik, Frickenhausen, FRG). Plate washing and absorbance readings were made by using Easy Washer 812 SW1 and Easy Reader 340 ATC, both from SLT Labinstruments (Crailsheim, FRG).

The following buffers and solutions were used: (1) Coating buffer, pH 9.6, prepared by adding 1.59 g of Na_2CO_3, 2.93 g of $NaHCO_3$ and 0.2 g of NaN_3 to 1 L of distilled water; (2) phosphate-buffered saline, pH 7.6 (PBS), prepared by adding 8.5 g of NaCl, 1.56 g of $NaHPO_4$ x 2 H_2O and 12.46 g of Na_2HPO_4 x 2 H_2O to 1 L of distilled water; (3) washing buffer, pH 7.6 (PBS-Tween), prepared by adding 51 g of NaCl, 5.89 g of KH_2PO_4, 65.32 g of K_2HPO_4 and 30 mL of Tween 20 to 1 L of distilled water. 30 mL of that stock solution were added to 1.8 L of distilled water; (4) blocking buffer, pH 7.6, PBS containing 2% casein; (5) citrate buffer, pH 4.4, prepared by adding 46.04 g of potassium dihydrogen citrate and 0.1 g of sorbic acid to 1 L of distilled water; (6) substrate solution, 500 µL of TMB (375 mg in a mixture of 5 mL of DMSO and 20 mL of MeOH) and 100 µL of 1% H_2O_2 were added to 25 mL of citrate buffer.

The HPLC system for the analysis of 1-NP in air particulate matter consisted of two pumps, models S1000 and S1300, S3400 fluorescence detector, Promis II autosampler, and S4110 column oven (Sykam, Gilching, FRG). 1-NP was catalytically reduced to 1-AP using packed alumina that was coated with 5% platinum (Merck, Darmstadt, FRG).

Nitration of 1-Pyrenebutyric Acid (PBA). A mixture of nitric acid (d 1.39, 0.85 mL, 12.2 mmol) was dropped to an acetic acid solution of PBA (2.86 g, 9.9 mmol) stirred at 50 °C. After 30 min, reaction mixture was cooled to room temperature and water (30 mL) was added. Precipitated crystals were separated, washed and frite successively by cold water (3 x 25 mL), acetone (5 mL) and benzene (3 x 5 mL) and dried in vacuum exsiccator for 15 h. Raw compound (3.13 g, 94.7%, calculated for mono-nitroderivative) melting mainly in interval 178-186 °C, was recrystallized twice melting in interval 172-205 °C. HPLC analysis of raw as well as recrystallized material exhibited a mixture of two compounds. The elemental analysis gave the following results: $C_{20}H_{15}NO_4$ (333.1), percent calculated vs percent found, C, 72.06 vs 72.13; H, 4.54 vs 4.65; N, 4.20 vs 4.09. Part of this product (0.93 g) was esterified with diazomethane in diethylether by usual way. Then the obtained methyl ester was purified by chromatography on silica gel column by elution with 100:1 benzene/ethanol (v/v). Chromatographically pure compound (660 mg, mp 80-83 °C) was recrystallized from acetone (5 mL) to give 616 mg of methyl ester, mp 84-88 °C, which was also a mixture of two compounds as revealed by HPLC analysis. The elemental analysis gave the following results: $C_{21}H_{17}NO_4$ (347.4), percent calculated vs percent found, C, 72.61 vs 72.25; H, 4.93 vs 5.1; N, 4.03 vs 4.05.

^1H NMR ($CDCl_3$, 400 MHz, Bruker AM-400, tetramethylsilane as internal standard, δ-scale) of methyl ester isomeric mixture; signals of minor isomer: δ 8.85 (d, J=9.8, 1H, A_1^*), 8.58 (d, J=8.5, 1H, E_1^*), 8.47 (d, J=9.8, 1H, A_2^*), 8.03 (d, J=9.0, 1H, E_2^*), 8.11 (d, J=9.1, 1H, F_1^*), 8.2 (d, J=8.0, 1H, H_1^*), 7.94 (d, J=9.8, 1H, F_2^*), 7.92 (d, J=7.9, 1H, H_2^*), 3.7 (s, 3H, OCH_3), 3.38 (t, 2H, α-CH_2), 2.48 (t, 2H, γ-CH_2), 2.17 (m, 2H, ß-CH_2); signals of major isomer: δ 8.74 (d, J=9.5, 1H, B_1^*), 8.42 (d, J=9.2, 1H, D_1^*), 8.16 (d, J=9.2, 1H, B_2^*), 8.03 (d, J=9.0, 1H, D_2^*), 8.57 (d, J=8.5, 1H, C_1^*), 8.16 (d, J=8.3, 1H, G_1^*), 8.04 (d, J=8.5, 1H, C_2^*), 7.92 (d, J=7.9, 1H, G_2^*), 3.71 (s, 3H, OCH_3), 3.38 (t, 2H, α-CH_2), 2.49 (t, 2H, γ-CH_2), 2.17 (m, 2H, ß-CH_2) [*The non-assigned proton signals of the aromatic skeleton are marked by symbols in respect to their coupling relations. The atoms marked by the same alphabetical character belong to couple of nuclei which is coupled by the same coupling constant. So, J(A_1) should have the same value as J(A_2) and so on. Some differences from this could be explained by inaccuracy in the recordings of spectra due to overlap of signals of two compounds. All values of J of this couples (7-10 Hz) correspond to the hydrogen atoms in ortho positions].*

^{13}C NMR (CDCl$_3$, 100 MHz, Bruker AM-400, tetramethylsilane as internal standard, δ-scale, attached proton test) of methyl ester isomeric mixture; signals of minor isomer: δ 120-143 aromatic part (atoms without bond hydrogens: 142.5, 136.5, 135.3, 129.6, 128.5, 127.8, 125.0, 124.6; atoms with one bond hydrogen: 130.9, 128.4, 127.6, 127.4, 126.1, 123.9, 122.5, 121.6); 168.2 (C=O), 51.9 (OCH$_3$), 33.5 (C$_\alpha$H$_2$), 32.9 (C$_\gamma$H$_2$), 26.0 (C$_\beta$H$_2$); signals of major isomer: δ 120-143 aromatic part (atoms without bond hydrogens: 142.7, 138.9, 134.6, 128.8, 125.1, 125.1, 125.0, 124.1; atoms with one bond hydrogen: 131.6, 128.5, 127.0, 126.8, 126.6, 123.8, 122.5, 120.9); 173.7 (C=O), 51.9 (OCH$_3$), 33.5 (C$_\alpha$H$_2$), 32.9 (C$_\gamma$H$_2$), 27.0 (C$_\beta$H$_2$).

HPLC-Analysis of the Isomeric 4-(Nitropyrene-1-yl)-Butyric Acid and Their Methyl Esters. The analysis were performed on a CGC Separon SGX C$_{18}$ column (150 x 3 mm ID, particle size 5 μm, Tessak, Czech Republic). Beckman 100A pump and UV detector Pye Unicam PU 4020 were used. The mixture of free isomeric acids exhibited two peaks of capacity factors 8.44 and 9.0, when eluted with 1:5 acetonitrile/water (v/v) as mobile phase at flow rate of 12 mL/h. The mixture of corresponding methyl esters was eluted with 7:3 acetonitrile/water (v/v) at flow rate of 18 mL/h and showed two peaks with capacity factors 10.1 and 10.7.

Preparation of Immunogen and Coating Antigen. The isomeric mono-nitropyrenebutyric acids (NPBA) as obtained after nitration of PBA were covalently attached through the carboxylic acid moiety to the lysine groups of KLH (immunogen) and BSA (coating antigen) as described for pyrene (*20*). Briefly NPBA (0.2 mmol) was dissolved in 3 mL of anhydrous dioxane. Then equimolar amounts of tri-n-butylamine and isobutyl chloroformate were added. After mixing, the solution was placed in a 4 °C refrigerator for about 20 min, after which it was added dropwise to 60 mg of aqueous protein solution. The pH value was set to 8.5 with 0.1 M NaOH. After standing for 4 h at room temperature, the solution was exhaustively dialyzed for 48 h against four 2 L changes of 0.1 M glycine buffer (pH 9.0), followed by 24-h dialysis against two changes of distilled water. After dialysis, immunogen and coating antigen were freeze-dried.

Antisera Production and Characterization. Two adult random-bred rabbits weighing 3-4 kg were immunized with NPBA-KLH conjugate. Primary immunization was performed subcutaneously and intradermally at 10 multiple sites by injecting 0.1 mL of the emulsified immunogen. The latter was prepared by dissolving 3 mg of NPBA-KLH in 1.5 mL saline solution emulsified with 1.5 mL of Freund's complete adjuvant. Booster injections were administered at 3, 16, 26, 32 and 36 weeks after first immunization using Freund's incomplete adjuvant. Rabbits were bled through the ear vein within 7-14 days after each booster injection. The blood was incubated at room temperature for 1 h, stored at 4 °C overnight, and after equilibration at room temperature was centrifuged at 2000 g for 20 min. The assay was performed using serum samples from bleedings after the second booster injection. Antisera were dispensed in 1-mL aliquots and stored undiluted at -20 °C until use. The sera were used without further purification.

Analysis of Titer. The titer of the serum from each animal was determined by checkerboard titration (two-dimensional titration method) measuring the binding of serial dilutions (1:500 to 1:100,000) to microtiter plates coated with several concentrations of NPBA-BSA (10-0.1 μg/mL). Optimal concentrations for coating antigen and antisera dilution were determined.

Effect of Solvents on the Antibody-Coating Antigen Interaction. To study the effect of different solvents such as MeOH, EtOH, ACET, MeCN, and DMSO on the antibody-coating antigen interaction, optical densities (OD) of the blank value (without analyte) in solvent (v/v) at 0, 0.17, 0.33, 1.7, 3.3, 6.7, 10, and 13% final concentration were measured. Additionally, 1-NP calibration curves were obtained using standards prepared at different composition of MeCN (10, 20, 30, and 40%). These were compared to the calibration curve prepared in aqueous buffer.

Cross-Reactivity Determinations. The relative sensitivity of the immunoassay towards the compounds listed in Table I was determined by assaying a dilution series of each compound in water containing 10% MeCN. The IC_{50} values (concentration of inhibitor that produces a 50% decrease in the maximum normalized response) were compared and expressed as a percent IC_{50} based on 100% response of 1-NP.

ELISA Procedure. An indirect competitive ELISA format using polystyrene microtiter plates was optimized for 1-NP. The assay was performed as follows: Microtiter plates were coated overnight at 4 °C with NPBA-BSA at 0.2 µg/mL (40 ng/well) in coating buffer. The plates were covered with adhesive plate sealers to prevent evaporation. After 15-18 h the plates were washed with PBS-Tween using an automatic plate washer set for four 300-µL washes. Sites not occupied by the NPBA-BSA conjugate were blocked with 300 µL of blocking buffer at room temperature with agitation on a horizontal shaker for 60 min. Plates were then washed as before. Sample or standard (50 µL/well) and diluted rabbit antiserum (dilution 1:30,000 in PBS; 100 µL/well) were added and incubated at room temperature with agitation for 30 min. After washing goat anti-rabbit IgG-POD conjugate at a 1:50,000 dilution in PBS was added (200 µL/well). The plates were incubated with agitation at room temperature for 60 min and washed as before. Substrate solution (200 µL/well) was added and gently agitated for about 15 min. Finally, the enzyme reaction was stopped with 100 µL of 0.5 M sulphuric acid and the absorbance of each well was read at 450 nm. All determinations were performed at least in triplicate. The sigmoidal standard curve was set up by plotting the means of the absorbances against the logarithms of standard concentrations and interpolating by Rodbard's four-parameter logistic model *(22)*.

1-Nitropyrene HPLC-Analysis. Briefly the 1-NP separation and reduction is accomplished in a reversed phase HPLC-system with column-switching using 70:30 methanol/water (v/v) as mobile phase under isocratic elution. The optimal operating temperature for the catalytic column was found to be 80 °C. In the first step, the effluent fraction which contains 1-NP and compounds of similar retention times is transferred from a short analytical column (25 x 4 mm ID stainless steel column packed with LiChrospher 5 µm; Merck, Darmstadt, FRG) to the reduction column. The resulting 1-AP is then separated on a longer analytical column (250 x 4 mm ID LiChrospher 5 µm; Merck, Darmstadt, FRG) and finally measured by fluorescence detection (excitation at 360 nm, emission at 430 nm).

Results and Discussion

Synthesis of Haptens and Conjugates. An essential prerequisite for covalent coupling of haptens to proteins is the presence of an appropriate chemical function for conjugation that is distal from the necessary functional groups for antibody binding. In the intention of his study, PBA was nitrated to get a 4-(6-nitropyrene-1-yl)-butyric acid containing a carboxylic function. The use of a C_4-spacer arm makes the nitro group of 1-NP far from the protein surface. This strategy was designed to make the whole 1-NP molecule available for epitope recognition by the immune cells. As a result, nitration of PBA afforded a mixture of two isomers of nitropyrenebutyric acid, whose elemental analysis confirms the presence of only one nitro group in the molecule. The isomers differ in the position of nitro group on the aromatic skeleton which was determined from ^1H and ^{13}C NMR of the corresponding methyl esters.

Mainly, the signals of nuclei in aromatic parts of the molecule were decisive for determination of the number of isomers in the mixture, their abundance ratio and position of the nitro group. On the other hand, the used NMR experiments did not permit to assign the certain ^1H or ^{13}C signals to definite atoms of the aromatic skeleton and also to assign the certain set of signals to a definite isomer in the mixture. The total number of the ^1H or ^{13}C signals in „aromatic parts" of spectra agreed with the presence of two isomeric nitropyrene compounds in the mixture. This conclusion was confirmed also by HPLC analysis. The ^1H signals clearly exhibited the double magnitude in their integrals, which showed approximately 1:1.8 abundance ratio of both isomers. All signals of aromatic protons are doublets and their coupling constants (7-10 Hz) correspond to the hydrogen atoms with one other hydrogen on vicinal carbon. Coupling of hydrogen was confirmed by help of ^1H-^1H COSY experiment. By this was found four couples of the vicinal hydrogens in the set of the higher signals as well as in the set of the lower ones. From these facts results impossible binding of the nitro group in the positions 2, 3, 4, 5, 7, 9, and 10 of pyrene. Hence, the measured sample is a mixture of methyl esters of 4-(6-nitropyrene-1-yl)-butyric acid and 4-(8-nitropyrene-1-yl)-butyric acid. Because separation of the two isomers in free acids mixture as well as in methyl ester mixture by help of crystallization or partial extraction or column chromatography was unsuccessful, the mixture of 4-(6-nitropyrene-1-yl)-butyric acid and 4-(8-nitropyrene-1-yl)-butyric acid was used for immunogen and coating antigen preparation.

4-(6-nitropyrene-1-yl)-butyric acid   4-(8-nitropyrene-1-yl)-butyric acid

Response of Rabbits to the NPBA-KLH Conjugate. Ten days after the first booster injection, anti-1-NP antibodies were detectable in the serum of both animals pointing to strong immunogenicity of the hapten-KLH conjugate. The antibody titer increased so rapidly that ELISA development was started using serum samples taken after the second booster immunization. The results presented in this paper were achieved with antisera from bleeding after 122 days after primary immunization.

Hapten-specific antibody titer and optimal concentration of coating antigen were determined by checkerboard titration on ELISA plates coated with homologous NPBA-BSA conjugates and blocked with casein. Casein was superior to bovine serum albumin, ovalbumin or gelatin as a blocking agent to prevent nonspecific binding. The use of casein results in nonspecific binding lower than ten percent of the blank value (zero analyte). BSA served as control. The dilutions providing highest sensitivity and sufficient color development (absorbance of 0.5 in about 10 min) were determined. The antisera from both rabbits required an optimal coating antigen concentration of 0.2 µg/mL (40 ng/well) and antibody titers of 1:20,000 (rabbit no. 15/90) and 1:30,000 (rabbit no. 20/90).

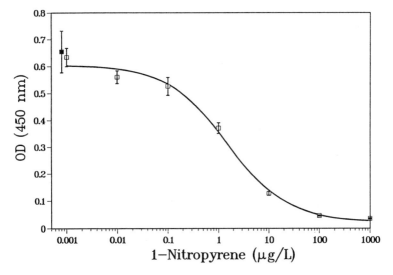

Figure 1. Standard curve for 1-nitropyrene in 10:90 acetonitrile/water (v/v)
(Error bars: 1 SD, blank 3 SD; n = 12).

Assay Optimization. Recognition of free 1-NP was tested using optimal concentrations for the NPBA-BSA conjugate and antiserum 20/90, determined from the checkerboard titration described above. The immunoassay uses an indirect competitive assay format. Since the haptenated coating antigen competes with the sample analyte for the antibody binding sites, the antibody binding to the microtiter plate is inversely proportional to the concentration of analyte in the sample. After addition of the horseradish peroxidase-labeled goat anti-rabbit antibody and color development, analyte concentration was measured indirectly by reading the absorbance of the colored product. Figure 1 shows the calibration graph for 1-NP which was established with twelve replicates of each standard. The assays' limit of detection (LOD) was 81 pM (0.02 µg/L) and was calculated as the concentration corresponding to the absorbance of the zero dose minus three standard deviations. The linear measuring range was from 0.1 to about 10 µg/L, with a center point at 1.5 µg/L. Three different 1-NP concentrations (0.1, 1, and 10 µg/L) were used to assess the intra- and interassay precision. Twelve aliquots of each sample were measured in one assay (intraassay) and the same number of aliquots were run on three separate days (interassay). The average percent coefficients of variance (CVs) were 4.9 and 7.8%, respectively. As assay optimization revealed, a blocking time of 30 min was sufficient to prevent antibodies from nonspecific binding to the solid phase. Increasing the blocking time did not improve the results. The same was true for the incubation step with the goat anti-rabbit IgG-POD. There was almost no difference after an incubation time of thirty and sixty minutes. Another result was obtained if the incubation time of the competition step (coating antigen/sample analyte/1-NP antibody) was changed. Shortening this time resulted in better LOD from 0.02 to 0.2 using 30 min and 60 min respectively. We continued using a 30 min incubation period.

Specificity of Antibodies. Antibody specificity was tested by performing competition assays with several parent PAHs (11 compounds), defined mixtures (16 EPA PAH Standard, 6 PAH Standard according to the German Drinking Water Act), nitrobiphenyls (3 compounds), and a high number of NPAHs (21 compounds) as inhibitors (Table I). All of the chemicals were tested in the concentration range 0.001 to 1000 µg/L in 10% MeCN/water (v/v). For unsubstituted PAHs, corresponding IC_{50} values on a molar basis were well below 1%. This is true for the single compounds as well as for the two defined PAH mixtures. Therefore, the nitro group seems to play an important role in analyte recognition by the antibody. Only marginal antibody binding, if there is any, was observed with nitrobiphenyls and mono- or dinitro-PAHs that contain at least four aromatic rings. With the exception of 5-nitroacenaphthene and 2- or 3-nitrofluoranthene, IC_{50} values higher than five percent were recorded only with some nitropyrenes (Figure 2). A 10-fold increase in cross-reactivity relative to 1-NP was observed with the 1,6- and 1,8-dinitropyrenes. This may be found explained by the structure of the immunogen which was composed of a mixture of two NPBA-compounds containing alkyl and nitro functions either in 1,6- or 1,8-positions. Therefore, the presence of parent PAHs in airborne particulate matter should not constitute a serious problem with the immunochemical determination of 1-NP. In contrast, 1,6-and 1,8-dinitropyrenes will interfere even at very low levels. Whether other nitroarenes which are formed in the atmosphere, such as 2-nitropyrene and 2-nitrofluoranthene, can constitute a problem in the 1-NP ELISA is a subject of future studies.

Matrix Effects. Ultimately this assay may be used to measure 1-NP in airborne particulate extracts. Thus, one of the aims of this study was to determine the effect of different compositions of solvents on the antibody-analyte(coating antigen) interaction. The effect of some solvents was tested on the indirect competitive ELISA. The results are outlined in Figure 3. The addition of any solvent tested led to an increase in the absorbance of the blank. The solvent effect was minimal for DMSO and MeOH; for ACET and EtOH, 10% of organic solvent caused a 2-fold increase in

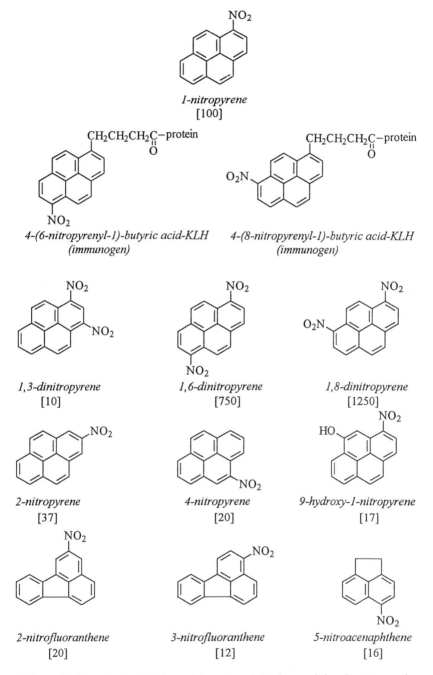

Figure 2. Chemical structures of target analyte, immunizing haptens and compounds with cross-reactivities higher than 5% in the 1-NP ELISA. Number in brackets corresponds to the percent IC_{50} value based on 100% response of 1-NP.

OD of the blank. Addition of MeCN caused up to a 7-fold increase in OD at 13.3.% organic solvent. Similar effects were found by other authors, such as the indirect molinate ELISA where color intensity was increased with 5% MeCN in the incubation mixture (*23*). In addition, a competitive inhibition ELISA for simazine, resulted in an increase in the final color intensity of the assay mixture with 3.4-5.1% of MeOH (*24*). A moderate amount of MeOH was reported to have a very strong positive effect on antibody-antigen binding, decrease matrix effect, and reduce binding of analyte to interfering surfaces (*25*).

In contrast to the effect of organic solvents on the 1-NP ELISA, an extremely different result was observed in our pyrene ELISA (Figure 4) (*20*). In general, a decrease in OD was observed for all organic solvents at concentrations higher than 3%.

Nevertheless, the addition of many organic solvents at higher amounts leads to a shift of the calibration curve to higher analyte concentrations. This is demonstrated in Figure 5 with MeCN in the 1-NP ELISA. Despite the strong positive effect of MeCN on the ODs the assay detection limit was significantly decreased. Since the first report on the behavior of antibodies in anhydrous organic solvents (*26*), related investigations are of growing interest (*27,28*). It is known that antibody-antigen binding is dependent on noncovalent interactions, such as hydrophobic interactions, van der Waals forces, and hydrogen bonds. Therefore, the composition of the mixture, where the immunochemical reaction takes place should affect the antibody conformation and specificity. The same effect should be observed with antibody-coating antigen interaction when performing an indirect competitive ELISA with organic solvents. The activity, substrate specificity, enantioselectivity, prochiral selectivity, regioselectivity, and chemoselectivity of enzymes have been found to dramatically depend on the nature of the solvent (*29-32*).

Despite the increasing research activities in this field, the issue of solvent dependence of antibody-antigen interactions is still unexplored to a great extent. Obviously, there is no single physicochemical property of a nonaqueous solvent, such as dielectric constant, polarity, partition coefficient between n-octanol and water, and dipole moment, that can be used solely to predict antibody function in this solvent. Moreover, the influencing mechanism will be of a complex nature. Beside looking through the existing literature of still manageable size in that field to compare data, decisive progress can be expected after more extended investigation of solvent dependence for several types of antibodies. As a final result, the control of antibody selectivity simply through addition of distinct solvents offers exciting new applications of antibody-based reactions.

Air Particulate Samples. Immunochemical and HPLC analysis of 1-NP in air particulate samples from a busy intersection in central Munich were compared. Airborne particulates were collected over a 24-h period on glass fiber filters (type Whatman GF/C, Bender & Hobein, Munich, FRG) using a low-volume sampler (type GS 050/3-C, Derenda, Berlin, FRG) equipped with an open face sampler head (type K4, Derenda, Berlin, FRG). Air was sampled at a flow rate of 2.3 m^3/h. Loaded filters were extracted in 15 mL of dichloromethane by sonication for 30 min. The organic solvent was removed under a gentle stream of N_2 and the remaining residue was dissolved in 1 mL of methanol. ELISA determination in crude methanol extracts exhibited an about 600% overestimation of 1-NP compared to the HPLC method. To verify, whether this overestimation was attributable mainly to the presence of 1,6- and 1,8-dinitropyrenes, a short clean-up procedure was developed to remove these compounds. For that, after evaporation of dichloromethane, the residue was dissolved in 200 µL of cyclohexane and loaded onto a glass 50 mm x 8 mm I.D. column, filled with 0.5 g of silica (LiChroprep Si 60, 40-63 µm, Merck, Darmstadt, FRG) and conditioned with 7 mL of cyclohexane. Next, the sample was eluted with 5 mL of cyclohexane followed by seven 1-mL portions of 5:1 cyclohexane/dichloromethane (v/v). Under these conditions, the dinitropyrenes were retarded on the column as was

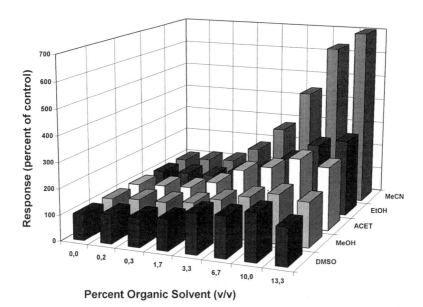

Figure 3. Effect of organic solvents on the optical density (OD) of the blank value at different solvent concentrations in the 1-nitropyrene ELISA: ACET (acetone), DMSO (dimethyl sulfoxide), EtOH (ethanol), MeCN (acetonitrile), MeOH (methanol).

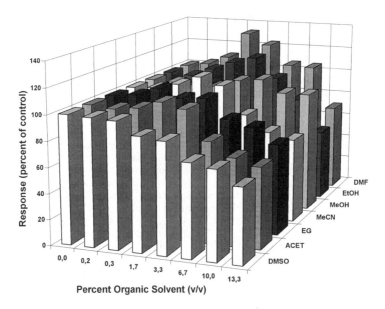

Figure 4. Effect of organic solvents on the optical density (OD) of the blank value at different solvent concentrations in the pyrene ELISA: ACET (acetone), DMF (dimethylformamide), DMSO (dimethyl sulfoxide), EG (ethylene glycol), EtOH (ethanol), MeCN (acetonitrile), MeOH (methanol).

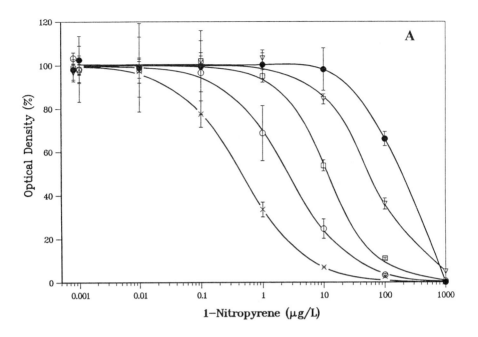

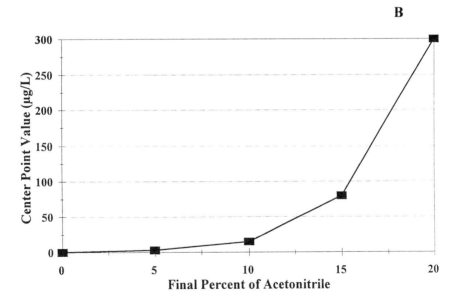

Figure 5. Effect of acetonitrile on the 1-nitropyrene standard curve. A: (x) no acetonitrile, (o) 5 %, ( ) 10 %, (∇) 15 %, (•) 20 % final acetonitrile concentration. Values are the mean ± SD of six replicates. B: Effect of acetonitrile on the center point value.

found with calibration standards. The last three 1-mL fractions, which contain 1-NP, were combined, evaporated to dryness, dissolved in 1-mL of methanol, and analyzed with HPLC and ELISA. By this clean-up, ELISA overestimation was reduced to 50-60% leading to the conclusion that dinitropyrenes can strongly interfere with the immunochemical determination of 1-NP in airborne particulate samples. To prove this finding, high-volume air samplers must be applied to get higher amounts of air particulate matter, together with high-resolution GC-MS determination of dinitropyrenes.

Conclusions

A sensitive indirect competitive ELISA using polyclonal antibodies was developed for 1-NP. In comparison to the monoclonal antibody-based nitrofluoranthene ELISA found in the literature (21), the present assay is almost three orders of magnitude more sensitive to 1-NP and shows far less cross-reactivity with parent PAHs and NPAHs. Preliminary investigations with crude airborne particulate extracts exhibited about 600% overestimation of 1-NP by ELISA compared to HPLC. This overestimation was reduced to about 50% with a short clean-up of the filter extract to remove the dinitropyrene fraction. However, this finding has to be proved with high-volume air samples and high-resolution GC-MS determination of dinitropyrenes. Further work is in progress to study the applicability of the 1-NP ELISA for airborne and diesel particulate samples. Whether the high cross-reactivity of the 1-NP antibodies for 1,6- and 1,8-dinitropyrenes can be exploited for the determination of the latter in femtomole concentrations in these samples will warrant further investigations. According to Hayakawa et al. (33), diesel-engined vehicles contribute up to 99% of dinitropyrenes in the air as compared to about 80% contribution to 1-NP.

The anti-1-NP antibodies are being used in the preparation of sol-gel immunoadsorber material for quick and selective sample clean-up (34). Another interesting application of the 1-NP ELISA is the analyses of nitroarene protein and/or DNA adducts for human biomonitoring (35-37). The high potential of immunochemical methods for individual exposure measurements has been outlined recently (38).

Effects of organic solvents on the antibody-analyte interaction in the 1-NP ELISA and another pyrene ELISA further support the studies on solvent-controlled antibody based reactions.

Acknowledgments

This work was supported in part by Deutsche Bundesstiftung Umwelt. We are very grateful to Dr. Karel Kefurt and Dr. Jitka Moravcová from the Department of Chemistry of Natural Compounds, Institute of Chemical Technology in Prague, Czech Republic, who prepared and characterized nitropyrenebutyric acid and to Prof. Akio Koizumi from the Department of Hygiene, Akita University School of Medicine, Japan, who kindly provided 9-hydroxy-1-nitropyrene.

Literature Cited

1. Pope, C.A. III; Bates, D.V.; Raizenne, M.E. *Environ. Health Perspect.* **1995,** *103*, 472-480.
2. Westerholm, R.; Egebäck, K.E. *Environ. Health Perspect.* **1994,** *102*, 13-23.
3. Ball, J.C.; Young, W.C. *Environ. Sci. Technol.* **1992,** *26*, 2181-2186.
4. Mücke, W.; Fiedler, H. *Z. Umweltchem. Ökotox.* **1990,** *2*, 170-176.

5. Diesel *and Gasoline Engine Exhaust and Some Nitroarenes;* IARC Monographs the Evaluation of the Carcinogenic Risks to Humans; International Agency for Research on Cancer: Lyon, France, **1989;** Vol. 46.

6. Heinrich, U. Abstracts. 15th International Symposium on Polycyclic Aromatic Compounds, Belgirate, Italy, **1995,** p150.

7. Wegman, D. Proceedings. International Workshop: Setting Priorities in Environmental Epidemiology, World Health Organization: Rome, Italy, **1992,** 1-22.

8. Nielsen, P.S.; Autrup, H. *Clin. Chem.* **1994,** *40,* 1456-1458.

9. Pitts, J.N., Jr.; Lokensgard, D.M.; Harger, W.; Fisher, T.S.; Mejia, V.; Schuler, J.J.; Scorziell, G.M.; Katzenstein, Y. A. *Mutat. Res.* **1982,** *103,* 241-249.

10. Scheepers, P.T.J; Velders, D.D.; Martens, M.H.J.; Noordhoek, J.; Bos, R.P. *J. Chromatogr. A* **1994,** *677,* 107-121.

11. Veigl, E.; Posch, W.; Lindner, W.; Tritthart, P. *Chromatographia* **1994,** *38,* 199-206.

12. Librando, V.; Fazzino, S.D. *Chemosphere* **1993,** *27,* 1649-1656.

13. Hayakawa, K.; Murahashi, T.; Butoh, M.; Miyazaki, M. *Environ. Sci.* Technol. **1995,** *29,* 928-932.

14. Murayama, M.; Dasgupta, P.K. *Anal. Chem.* **1996,** *68,* 1226-1232.

15. Kado, N.Y.; Wei, E.T. *J. Natl. Cancer Inst.* **1978,** *61,* 221-225.

16. Hagen, I.; Herikstad, B.V. *Hereditas* **1988,** *108,* 119.

17. Herikstad, B.V.; Overebo, S.; Haugen, A.; Hagen, I. *Carcinogenesis* **1993,** *14,* 307-309.

18. Gomes, M.; Santella, R.M. *Chem. Res. Toxicol.* **1990,** *3,* 307-312.

19. Roda, A.; Bacigalupo, M.A.; Ius, A.; Minutello, A. *Environ. Technol.* **1991,** *12,* 1027-1035.

20. Meisenecker, K.; Knopp, D.; Niessner, R. *Anal. Methods Instrumentation* **1993,** *1,*114-118.

21. Haas, R.A.; Hanson, C.V.; Monteclara, F. Proceedings. 79th Annual Meeting of the Air Pollution Control Association, Minneapolis, USA, **1986,** 1-13.

22. Rodbard, D. In *Ligand Assay;* Langan, J., Clapp, J.; Eds.; Masson Publishing: New York, USA, 1981, Chapt. 3; pp 45-101.

23. Gee, S.J.; Miyamoto, T.; Goodrow, M.H.; Buster, D.; Hammock, B.D. *J. Agric. Food Chem.* **1988,** *36,* 863-870.

24. Goh, K.S.; Spurlock, F.; Lucas, A.D.; Kollman, W.; Schoenig, S.; Braun, A.L.; Stoddard, P.; Biggar, J.W.; Karu, A.E.; Hammock, B.D. *Bull. Environ. Contam. Toxicol.* **1992,** *49,* 348-353.

25. Li, Q.X.; Hammock, B.D.; Seiber, J.N. *J. Agric. Food Chem.* **1991,** *39,* 1537-1544.

26. Russell, A.J.; Trudel, L.J.; Skipper, P.L.; Groopman, J.D.; Tannenbaum, S.R.; Klibanov, A.M. *Biochem. Biophys. Res. Commun.* **1989,** *158,* 80-85.

27. Francis, J.M.; Craston, D.H. *Analyst,* **1994,** *119,* 1801-1805.

28. Stöcklein, W.; Gebbert, A.; Schmid, R.D. *Anal. Letters* **1990,** *23,* 1456-1476.

29. Wescott, C.R.; Klibanov, A.M. *Biochim. Biophys. Acta* **1994,** *1206,* 1-9.

30. Wescott, C.R.; Klibanov, A.M. *J. Am. Chem. Soc.* **1993,** *115,* 1629-1631.

31. Almarsson, Ö.; Klibanov, A.M. *Biotechnol. Bioeng.* **1996,** *49,* 87-92.

32. Louwrier, A.; Drtina, G.J.; Klibanov, A.M. *Biotechnol. Bioeng.* **1996,** *50,* 1-5.

33. Hayakawa, K.; Murahashi, T.; Miyazaki, M. *Jpn. J. Toxicol. Environ. Health,* **1995,** *41,* 13.

34. Zühlke, J.; Knopp, D.; Niessner, R. *Fresenius J. Anal. Chem.* **1995,** *352,* 654-659.

35. El-Bayoumy, K.; Johnson, B.; Roy, A.K.; Upadhyaya, P.; Partian, S.; Hecht, S.S. *Environ. Health Perspect.* **1994,** *102,* 31-37.

36. El-Bayoumy, K.; Johnson, B.; Partian, S.; Upadhyaya P.; Hecht, S.S. Carcinogenesis, **1994,** *15,* 119-123.
37. Fu, P.P.; Herreno-Saenz, D.; Von Tungeln, L.S.; Lay, J.O.; Wu, Y.S.; Evans, F.E. *Environ.* Health *Perspect.* **1994,** *102,* 177-183.
38. Knopp, D. Anal. *Chim. Acta* **1995,** *311,* 383-392.

Chapter 6

Development of Immunoassays for Detection of Chemical Warfare Agents

David E. Lenz, Alan A. Brimfield, and Lara A. Cook

Biochemical Pharmacology Branch, U.S. Army Medical Research Institute of Chemical Defense, 3100 Ricketts Point Road, Aberdeen Proving Ground, MD 21010–5425

With the advent of enzyme linked immunoabsorbent assays (ELISA) and monoclonal antibodies, there has been considerable effort devoted to the development of antibodies to detect and quantify low molecular weight toxic substances in environmental or biological fluids. Monoclonal antibodies developed against a structural analogue of the chemical warfare agent soman when used in a competitive inhibition enzyme immunoassay (CIEIA) were capable of detecting soman in buffer solutions at a level of 10^{-6} M (~180 ng/mL). These antibodies were found to be highly specific for soman even in the presence of its major hydrolysis product. Subsequent studies with antisoman monoclonal antibodies extended the level of sensitivity to ~80 ng/mL These antibodies did not cross react with other chemical warfare nerve agents such as sarin or tabun. In all cases, the time for a confirmatory test was two hours or less. These reagents offer a sensitive, rapid and low cost approach to the diagnosis or detection of the presence of toxic chemical substances. More recent efforts have focussed on developing antibodies specific for sulfur mustard a highly reactive vesicating agent.

The standard methods of analysis for detection of organophosphorus chemical warfare agents either require time-consuming isolation and cleanup procedures and expensive analytical equipment such as gas chromatography (GC) or gas chromatography-mass spectrometry (GC/MS), or they rely on rather nonspecific color reactions that result from changes in the activity of the enzyme acetylcholinesterase. The former procedures are quantitative but slow and expensive; while the latter rapid approach is rapid, qualitative or semi-quantitative but can give ambiguous results. What would be most desirable is a method that could detect a specific chemical warfare agent in a rapid, quantitative manner. In addition, the results should be subject to minimal interference from hydrolysis products or structurally related compounds. Such a method would have value not only for detecting exposure to a toxic material, but it could also be used to diagnose the type of toxicant so appropriate medical treatment could be administered.

Antibodies as Analytical Reagents

One approach that provides many of the features of such an analytical method is based on the use of antibodies specific for the analyte of interest. This has shown considerable applicability in the analysis of insecticides in environmental samples (*1-6*) and also has application in clinical and forensic settings. For example, Hunter and Lenz reported that paraoxon, the active metabolite of parathion, could be detected at a level of 1 nM in biological fluids (*3*).

Antibodies as Reagents for Detecting Chemical Warfare Agents. More recent efforts have expanded this approach and applied it to developing antibodies against various chemical warfare agents. Initially, polyclonal and subsequently monoclonal antibodies were produced against the highly toxic organophosphorus chemical warfare agent soman (*7-9*). These proved to be useful in the establishment of several immunoassays capable of quantitatively determining the amount of soman present in aqueous solution (*7*) as well as mammalian serum and milk (*10, 11*). In all cases a competitive immunoassay was the method of choice for quantification. More recently antibodies specific for a variety of chemical warfare agents to include polyclonal (*12*) and monoclonal (*13*) antibodies against VX, polyclonal antibodies against sarin (*14-16*) and polyclonal (*17*) and monoclonal antibodies (*18*) against sulfur mustard and sulfur mustard DNA adducts (*19,20*) have been developed.

Considerations for Development of an Immunoassay

An immunoassay, to be most effective, should be designed to detect the analyte of interest (the hapten) and, ideally, nothing else. While this may seem obvious, it must be remembered that other analytical approaches such as GC, HPLC or GC/MS often detect a host of species from which the analyte of interest must then be uniquely identified. As with most analytical techniques, it is useful to estimate the expected concentration of the analyte in the milieu chosen for analysis. In the case of immunoassays, if reasonable estimates can be made regarding the concentration and binding constant characteristics of the antibody being developed, then theoretical calculations can be carried out to determine if an assay will have the necessary sensitivity to detect the analyte of interest.

Immunogen Development. Since most small molecules (<1000 Daltons) are not capable of eliciting an immune response that would lead to antibody production, the hapten must be attached to a carrier molecule such as a protein to make it immunogenic. In carrying out the reaction to covalently join the hapten to the carrier protein, it must be ensured that the most important structural features of the hapten that unique, are retained after the immunogen is synthesized. Unfortunately, the strongest immune response will be to the carrier protein molecule, rather than to the hapten attached to it. To identify those antibodies that are specific for the hapten, the hapten must be covalently attached to a second carrier protein molecule that is not related to that used for the immunogen. This second protein-hapten molecule is often called an antigen and is used as the test compound in an enzyme-linked immunoassay (EIA) to ensure that the antibodies identified are specific for the hapten of interest and not for the carrier protein.

Type of Antibody to be Developed. It is also useful to consider if polyclonal or monoclonal antibodies will be required. If the intent is to qualitatively determine the presence of a class of compounds,   polyclonal antibodies may not only be acceptable, but they may be preferable. What polyclonal antibodies lack in specificity they can often make up for in sensitivity since these antibodies have a range of binding constants. In contrast, if a more quantitative assay is required, then the higher specificity found in monoclonal antibodies would be a requisite. In this case while some sensitivity may be sacrificed, selectivity will be increased.

Screening for Selection of Monoclonal Antibodies of Interest. Lastly, a decision must be made with respect to the test hapten that will be used in screening the antibodies produced by hybridoma cells when developing monoclonal antibodies. If high specificity and affinity are desired, then the test hapten for screening the antibody producing cells should be the specific analyte itself. If that compound is unavailable or unstable under the condition of the assay, then a compound that has maximum structural similarity with the analyte should be chosen. Since the antibodies being produced at this point represent a wide repertoire of binding affinity and specificity, it is incumbent upon the scientist to limit the possibilities in order to enhance specificity; otherwise, the resulting antibody may not be suited to the analytical task at hand. To obtain a rapid quantitative estimate of the amount of analyte present, a simple competitive inhibition enzyme immunoassay (CIEIA or CIA) is often the method of choice during assay development.

Antibodies against Chemical Warfare Agents

Antibodies against Soman. In the early 1980's Hunter et al. (*7*) developed monoclonal antibodies against the chemical warfare agent soman. They immunized animals with a soman analogue where the fluorine leaving group was replaced with a p-aminophenol moiety that was covalently attached to keyhole limpet hemocyanin (KLH) (Figure 1). The resultant monoclonal antibodies bound to a soman analogue attached in the same manner to a bovine serum albumin (BSA) test antigen (Figure 1; P-linked). This binding was inhibited in a competitive manner by free soman in solution (Figure 2). Further studies with these antibodies allowed for the elucidation of their structural and stereochemical specificity (*9*). The antisoman antibodies did not cross react either with sarin, another toxic chemical warfare agent, or with hydrolyzed soman wherein the fluorine had been replaced by a hydroxyl group (C; Table I). Based on their studies Hunter et al. (*7*) concluded that these antibodies could be used to quantitate levels of soman at 1 μM (200 ppb). Unpublished estimates of the dose of soman that would cause lethality in one-half of a human population are approximately 6 μg/kg. If all of the soman was distributed in a blood volume of five liters, the circulating concentration of soman would be 0.5 μM, (Table II) or below the limits of detection with the antibody developed by Hunter et al. (*7*) This represents 'best case' estimate. In all probability the concentration of soman in circulation would decrease with time due to pharmacodynamics and pharmacokinetics. Using a similar approach for unpublished date for sarin and VX, estimates with respect to the expected concentrations of those nerve agents also are included in Table II. Attempts to improve the affinity of the anti-soman monoclonal antibodies by changing the immunogen to one wherein the fluorine was replaced by a methoxy group and the hapten was attached to the carrier protein through a terminal carbon of the pinacolyl group (Figure 3; C-linked) were carried out. Antibodies against this immunogen had an equal affinity for soman as those linked through the phosphorus atom (P-linked) and were equally selective for soman but had a reduced affinity for the p-aminophenol derivative (F; Table I) as well as for dipinacolyl methylphosphonate (E; Table I). The differences in relative affinities for soman analogues served to demonstrate the need for careful immunogen design.

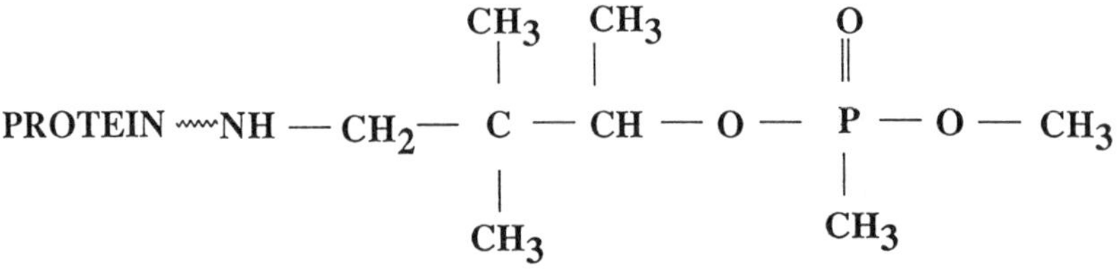

$$\text{PROTEIN} \sim\!\!\!\sim\!\!\text{NH} - \text{CH}_2 - \underset{\underset{\text{CH}_3}{|}}{\overset{\overset{\text{CH}_3}{|}}{\text{C}}} - \underset{}{\overset{\overset{\text{CH}_3}{|}}{\text{CH}}} - \text{O} - \underset{\underset{\text{CH}_3}{|}}{\overset{\overset{\text{O}}{\|}}{\text{P}}} - \text{O} - \text{CH}_3$$

Figure 1. Structure of soman-protein conjugate employed for immunization and/or immunoassay, coupled to protein through the phosphonate phosphorus (P-link).

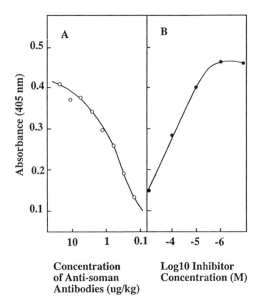

Figure 2. (A) Titration of the binding of anti-soman antibodies to soman-KLH. (B) Inhibition of the binding of anti-soman antibody to soman-KLH by various concentrations of free soman. Reproduced with permission from reference 7.
Copyright 1982, with kind permission of Elsevier Science - NL,
Sara Burgerhartstraat 25, 1055 KV Amsterdam, The Netherlands.

Table I. Differences in IC50 Depending on Atom Used to Link Hapten to Carrier Protein

Inhibitor	P LINK (uM) IC50	C LINK IC50
A. $(CH_3)_3 - C - CH(CH_3) - O - \overset{\overset{\displaystyle O}{\|}}{\underset{\underset{\displaystyle CH_3}{\|}}{P}} - F$	6.0	8.0
B. $(CH_3)_2 - CH - O - \overset{\overset{\displaystyle O}{\|}}{\underset{\underset{\displaystyle CH_3}{\|}}{P}} - F$	NI[a]	NI
C. $(CH_3)_3 - C - CH(CH_3) - O - \overset{\overset{\displaystyle O}{\|}}{\underset{\underset{\displaystyle CH_3}{\|}}{P}} - OH$	NI	5000
D. $(CH_3)_3 - C - CH(CH_3) - OH$	NI	NI
E. $(CH_3)_3 - C - CH(CH_3) - O - \overset{\overset{\displaystyle O}{\|}}{\underset{\underset{\displaystyle CH_3}{\|}}{P}} - O - CH(CH_3) - C - (CH_3)_3$	1.8	NI
F. $(CH_3)_3 - C - CH(CH_3)_3 - O - \overset{\overset{\displaystyle O}{\|}}{\underset{\underset{\displaystyle CH_3}{\|}}{P}} - O - \bigcirc - NH_2$	4.5	100

[a]NI = No Inhibition

Table II.   Estimate of Nerve Agent in Circulation at a 1LD50 Exposure

Structure	Est. Human LD_{50} Subcutaneous
$(CH_3)_2CH-O-\overset{\displaystyle O}{\underset{\displaystyle CH_3}{\overset{\|}{P}}}-F$ **Sarin**	2×10^{-6} M
$(CH_3)C-CH(CH_3)-O-\overset{\displaystyle O}{\underset{\displaystyle CH_3}{\overset{\|}{P}}}-F$ **Soman**	5×10^{-7} M
$CH_3CH_2-\overset{\displaystyle O}{\underset{\displaystyle CH_3}{\overset{\|}{P}}}-S-CH_2CH_2N-CH\overset{\displaystyle CH_3}{\underset{\displaystyle CH_3}{}}$ **VX**	7×10^{-7} M

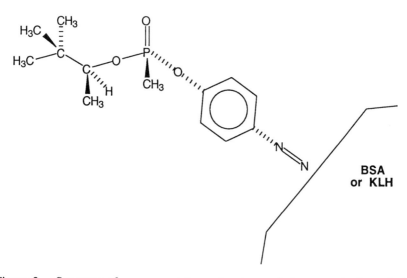

Figure 3.   Structure of soman-protein conjugate coupled to protein through the terminal pinacolyl group (C-link). To enhance the stability of the immunogen after administration to an animal, the fluorine was replaced with the much less labile methoxy group.

Subsequent work by Erhard et al. (*10, 11*) led to the development of additional anti-soman monoclonal antibodies that were capable of detecting pure soman in solution at a level of 0.5 μM, which is within the range estimated as necessary for detecting a potentially toxic concentration of soman in humans (*vide supra*). The same antibodies were found to be capable of detecting soman in mammalian serum at a level of 1.3 μM (11) which was in good agreement with the levels of detection reported by Hunter et al. (*7*). Clearly, if monoclonal antibodies are to be used in an immunochemical detection system for serum samples, an increase in the binding constant of at least one order of magnitude, and preferably two, will be required.

Antibodies against VX. In addition to soman, antibodies specific for VX (Table II), another highly toxic organophosphorus chemical warfare agent, have been developed. Initially, polyclonal anti-VX antibodies were developed by Rong and Zhang (*12*) using an undefined VX analogue. These antibodies, raised in rabbits, afforded the immunized animals full protection against the lethal effects of an LD_{95} dose of VX at both 7 days post immunization and again at 31 days post immunization. When the serum from immunized rabbits was administered to mice in a passive immunization, the mice were afforded protection against an LD_{95} dose of VX for up to 10 days after immunization (*12*). Unfortunately, the authors did not carry out any *in vitro* experiments to define the specificity of these antibodies. Neither did they attempt to quantitate the relative affinity of these polyclonal antibodies for VX versus other chemical warfare nerve agents.

More recently, Grognet et al. (*13*) reported the development of monoclonal anti-VX antibodies against a series of structurally similar VX analogues. Most of the monoclonal antibodies produced were specific for VX, with only low cross reactivity with soman, sarin or tabun. Two of the haptens, however, elicited antibodies with affinity for soman or sarin in the micromolar range. While these antibodies could neutralize the inhibition of acetylcholinesterase by VX *in vitro*, they afforded no *in vivo* passive protection, even though they had affinity constants of 9 nM (*13*). Based on the estimated concentration of nerve agents in circulation in a 70 kg human, these antibodies should have the requisite sensitivity for detecting VX at a sub lethal concentration (Table II). The authors reported that they were working on the development of an immunoassay for VX using these antibodies.

Antibodies against Sarin. Sarin (Table II) is the smallest of the chemical warfare agents on a molecular weight basis (MW = 140). Initial attempts to make antibodies against sarin proved to be unsuccessful (*14, 15*). More recent attempts by Zhou and co-workers (*16*) have led to the production of polyclonal antibodies against sarin. They reported that by covalently linking two moles of diisopropyl phosphate to L-lysine which in turn is covalently bound to a carrier protein as an immunogen, polyclonal antibodies were elicited in rabbits. When used in an ELISA, these antibodies were inhibited by sarin at concentration as low as 1 μM and the standard curve was linear over three orders of magnitude. The ability of soman, VX, isopropyl alcohol and isopropyl methyl phosphonic acid to compete with sarin also was determined. In each case when the concentration of these compounds was less than 2 mM, none of them interfered with the antibodies' abilities to bind to sarin. The authors noted that there was a need to detect sarin at concentrations below 1 μM (see also Table II) and they suggested that a more sensitive assay, such as a time-resolved fluorescence immunoassay might satisfy that requirement (*16*).

Antibodies against Sulfur Mustard. Polyclonal antibodies have been developed against sulfur mustard analogues (*17*) using a 4-(2-chloroethyl)benzoic acid (K; Table III) conjugate with keyhole limpet hemocyanin. The polyclonal antibodies against sulfur mustard had a lower limit of detection of 5 μM (*17*). Subsequently, Cook et al. (*18*) using the same immunogen were able to produce monoclonal antibodies that could bind to sulfur mustard and several related molecules (Table III). Of particular note is the low cross reactivity of these antibodies with the hydrolysis product of sulfur mustard, thio-2,2'-diethanol (thiodiglycol) (L; Table III) and the structurally similar molecule 2-hydroxyethyl disulfide (M; Table III). Given that unpublished estimates of the concentration of sulfur mustard that would cause a vesicant injury to human skin range between 0.1 and 0.2 mM, the authors feel that the monoclonal antibody probably lacks the required selectivity to serve as a useful reagent in an immunoassay. These results demonstrate once again that antibodies to a highly reactive toxic molecule can be produced and that the antibodies are capable of distinguishing between chemical substances that are structurally related to the antigen and those chemical structures associated with hydrolysis products.

Antibodies against Sulfur Mustard Adducts. Using an alternate approach, Benschop and coworkers (*19, 20*) have developed antibodies against the major reaction product of sulfur mustard and DNA. They confirmed prior findings that the major adduct on DNA of sulfur mustard was at the N7 position of guanine to form N7-[2-[(hydroxyethyl)thio]ethyl]guanine (N7-HETE-Gua) (*19*). Antiserum with antibodies to this adduct was obtained when rabbits were immunized with calf thymus DNA that had been reacted with sulfur mustard (*20*). Using this serum a CIEIA was developed to detect sulfur mustard adducts to DNA at a minimum level of 1-5 fmol per well. Adducts in white blood cells could be detected after exposure of human blood to sulfur mustard concentrations of ≥2 μM. In both cases, the level of sensitivity was in excess of that needed to identify the presence of sulfur mustard at concentrations below those that are estimated to cause a vesicant injury on human skin. Immunization of mice with N7-HETE-GMP coupled to keyhole limpet hemocyanin resulted in several hybridomas that produced monoclonal antibodies that recognized the N7-HETE-Gua. Using these antibodies in a CIEIA gave sensitivity comparable with that using the rabbit serum (*20*).

Summary

Current efforts have demonstrated that monoclonal and polyclonal antibodies that are specific for three of the chemical warfare organophosphorus nerve agents, soman, sarin and VX can be developed. All of these antibodies have been investigated to a sufficient degree to determine if they could be used in an immunoassay to detect the presence of soman, sarin or VX respectively in an aqueous sample. In the case of soman and sarin, the detection limits were in the micromolar range which, while fairly sensitive, would only be able to detect about twice the estimated concentration that would be expected following exposure to an LD_{50} dose of soman in humans and just equal to the estimated LD_{50} dose of sarin in humans (Table II). The antibodies for VX have not yet been utilized in an immunoassay, but to be useful they should also be capable of detecting VX in the sub micromolar range. Based on the reported binding of VX in the nM range (*13*), these antibodies should be capable of being developed into an assay with sufficient sensitivity to detect VX at concentrations less than those expected after exposure to a toxic dose. In most of the reports to date, the total assay time was two hours or less with an incubation time of the antibody with the nerve agent of 10 to 30 minutes. The anti-soman, anti-sarin and anti-VX antibodies all had the required specificity and exhibited a little or no cross reaction with either the metabolites or hydrolysis products of the respective nerve agents.

Table III. IC50 Values for Inhibitors of Sulfur Mustard antibodies

Inhibitor	Structure	IC50
G. Sulfur Mustard (HD)	$Cl - CH_2 - CH_2 - S - CH_2 - CH_2 - Cl$	1.00×10^{-4} M
H. 2-Chloroethyl Ethyl Sulfide (CEES)	$Cl - CH_2 - CH_2 - S - C_2H_5$	7.85×10^{-5} M
I. 2-Chloroethyl Methyl Sulfide (CEMS)	$Cl - CH_2 - CH_2 - S - CH_3$	3.92×10^{-4} M
J. 8-Chloro-Caprylic Acid (8CCA)	$HO - \overset{O}{\overset{\|}{C}} - CH_2 - (CH_2)_6 - Cl$	7.65×10^{-3} M
K. 4-(2-Chloroethyl) Benzoic Acid (4-CBA)	$HO - \overset{O}{\overset{\|}{C}} -$ ⬡ $- CH_2 - CH_2 - Cl$	4.57×10^{-3} M
L. 2-2 Thiodiethanol (TDG)	$HO - CH_2 - CH_2 - S - CH_2 - CH_2 - OH$	$> 1.00 \times 10^{-2}$ M
M. 2-Hydroxyethyl Disulfide (HED)	$HO - CH_2 - CH_2 - S - S - CH_2 - CH_2 - OH$	$> 1.00 \times 10^{-2}$ M

In the case of antibodies developed directly against sulfur mustard, neither the polyclonal nor the monoclonal antibodies (*17,18*) have the specificity needed for a useful immunoassay (Table III). The antibodies against the N7-HETE-Gua developed by Benschop and co-workers (*19-20*) do have very high level of sensitivity for the N7-guanine adduct that occurs following the reaction of sulfur mustard with DNA. In this case the authors indicate that they are in the process of developing a single cell assay using immunofluorescence microscopy to quantify adduct formation in sulfur mustard exposed skin (18).

The results to date clearly show that antibodies can be developed against various chemical warfare agents or their specific adducts. These antibodies have been used to demonstrate that immunoassays capable of detecting chemical nerve agents in the micromolar (ppb) range and adducts in the nanomolar are quite feasible. Practically, immunoassays for chemical warfare nerve agents or vesicating agents have potential application in the emerging area of chemical weapons remediation or destruction. Workers required to verify the destruction of chemical weapons will need rapid semi-quantitative tests to determine that storage sites and surrounding areas have not been contaminated. There will also be a need for the same types of tests to screen workers for potential exposures to chemical weapons that would require medical treatment. The current antibodies do not have the requisite affinity for detecting nerve agents at a nanomolar concentration, which would be needed to determine levels of contamination, for determining potential exposure or to diagnose the type of exposure pursuant to making decisions on medical treatment. There are, however, no technical barriers to developing antibodies of higher affinity. Given that several reports of immunoassays for nerve agents have already been published, there is every reason to believe that future efforts will result in the development of immunoassays for nerve agents or their metabolites with the required sensitivity, selectivity and speed of response. In addition, detection of adducts of sulfur mustard at levels sensitive enough to confirm exposure to sub-vesicating dose are already available.

LITERATURE CITED

1. Hammock, B. D., Gee, S. J., Cheung, P. Y. K., Miyamoto, T., Goodrow, M. H., Van Emon, J., and Seiber, J. N. in *Pesticide Science and Biotechnology;* Greenhalgh, R. and Roberts, T. R., Eds.; Blackwell Scientific Publications, Oxford, 1987; pp 309-316.
2. Brimfield, A. A., Lenz, D. E., Graham, C. and Hunter, Jr., K. W. *J. Agr. Food Chem.* **1985,** *33*, 1237-1242.
3. Hunter, Jr, K. W. and Lenz, D. E. *Life Sci.* **1982,** *30*, 355-361.
4. Al-Rubae, A. Y. The enzyme-linked immunosorbent assay, a new method for the analysis of pesticide residues. *PhD Dissertations*, The Pennsylvania State University, University Park, PA, 1978.
5. Van Emon, I., Seiber, I., and Hammock, B. *Bull. Environ. Contam. Toxicol.* **1987,** *39*, 490-497.
6. Heldman, E., Balan, A., Horowitz, O., Ben-Zion, S., and Torten, M. *FEBS Lett.* **1985,** *180*, 243-248.
7. Hunter, Jr., K. W., Lenz, D. E., Brimfield, A. A., and Naylor J. A. *FEBS Lett.* **1982,** *149*, 147-151.
8. Lenz, D. E., Brimfield, A. A., Hunter, Jr., K. W., Benschop, H. P., de Jong, L. P. A., van Dijk, C., and Clow, T. R. *Fund. Appl. Toxicol.* **1984,** *4*, S156-S164.
9. Brimfield, A. A., Hunter, K. W., Lenz, D. E., Benschop, H. P., van Dijk, C., and de Jong, L. P. A. *Mol. Pharmacol.* **1985,** *28*, 32-39.
10. Erhard, M. H., Schmidt, P., Kuhlmann, R., and Losch, U. *Arch. Toxicol.* **1989,** *63*, 462-468.
11. Erhard, M. H., Kuhlmann, R., Szinicz, L., and Losch, U. *Arch. Toxicol.* **1990,** *64*, 580-585.
12. Rong, K.-T. and Zhang, L.-J. *Pharmacology and Toxicology* **1990,** *67*, 255-259.
13. Grognet, J.-M., Ardouin, T., Istin, M., Vandais, A., Noel, J.-P., Rima, G., Satge, J., Pradel, C., Sentenac-Roumanou, H., and Lion, C., (1993) *Arch. Toxicol.*, *67*, 66-71.
14. Dean, R. G. Defense Advanced Research Projects Agency, R & D Status Report, USA, 1982: AD-A122300.
15. Dean, R. G. Defense Advanced Research Projects Agency, R & D Status Report, USA, 1987: AD-A188149.
16. Zhou, Y.- X., Yan, Q.- J., Ci, Y.- X., Guo, Z.-Q., Rong, K.-T., Charg, W.-B. and Zhao, Y.-F. *Arch. Toxicol.* **1995,** *69*, 644-648.
17. Lieske, C. N., Klopcic, R. S., Gross, C. L., Clark, J. H., Dolzine, T. W., Logan, T. P. and Meyer, H. G. *Immuno. Let.* **1992,** *31*, 117-122.
18. Cook, L., Lieske, C. N. and Brimfield, A. A. *Proceedings of the 1996 Medical Defense Bioscience Review*; Baltimore, MD, 1996, pp 136-137.
19. Fidder, A., Moes, W. H., Scheffer, A. G., van der Schans, G. P., Baan, R. A., de Jong, L. P. A., Benschop, H. P. *Chem. Res. Toxicol.* **1994,** *7*, 199-204.
20. van der Schans, G. P., Scheffer, A. G. , Mars-Groenendijk, R. H., Fidder, A., Benschop, H. P., Baan, R. A. *Chem. Res. Toxicol* **1994,** *7*, 408-413.

Chapter 7

Development of Various Enzyme Immunotechniques for Pesticide Detection

B. B. Dzantiev, A. V. Zherdev, O. G. Romanenko, and J. N. Trubaceva

A. N. Bach Institute of Biochemistry, Russian Academy of Sciences, Leninskiy prospect 33, 117071 Moscow, Russia

Different enzyme immunotechniques were developed for the following pesticides: 2,4-dichlorophenoxyacetic acid (2,4-D), 2,4,5-trichlorophenoxyacetic acid (2,4,5-T), simazine, atrazine, permethrin, phenothrin and their derivatives. The sensitivities obtained using solid-phase techniques such as ELISA varied from 0.02 to 8 ng/mL, and the assay duration ranged from 1 to 2 hrs. A new homogeneous assay technique based on the inhibition of the catalytic activity of amylase-pesticide conjugate by anti-pesticide antibodies was examined. Two types of immunosensors that could measure 2,4-D and 2,4,5-T for 12 min with sensitivities close to ELISA were also developed. Furthermore, a new visual membrane immunoassay based on polycation-polyanion interaction is proposed. The latter allows detection of 2,4-D up to 10 ng/mL and is useful for on-site analysis.

Determination of pesticides in soil, water, and food is an actual problem nowadays. Enzyme immunoassay methods are very effective solutions to analytical problems because they combine the specificity of antigen-antibody interactions and the high sensitivity characteristic of enzyme detection. To date, there are enzyme immunoassay systems developed for more than 70 pesticides, as mentioned in a number of reviews (1-6).

Applications of a new analytical technique depend substantially on its quickness and simplicity. If a technique can reduce analysis time without significant loss of assay sensitivity, it becomes appropriate for a variety of tasks. Therefore, simple and fast assays, including the ones with instrument-free detection, are of great interest for pesticide residue monitoring and environmental quality studies.

This paper will describe different enzyme immunotechniques developed in our laboratories for some of the widely used pesticides.

Solid-Phase Immunoassays

Enzyme-linked immunosorbent assay (ELISA) is the most traditional and extensively developed type of enzyme immunotechniques. We have developed ELISAs for four herbicides, namely 2,4-dichlorophenoxyacetic acid (2,4-D), 2,4,5-trichlorophenoxyacetic acid (2,4,5-T), simazine and atrazine (Figure 1).

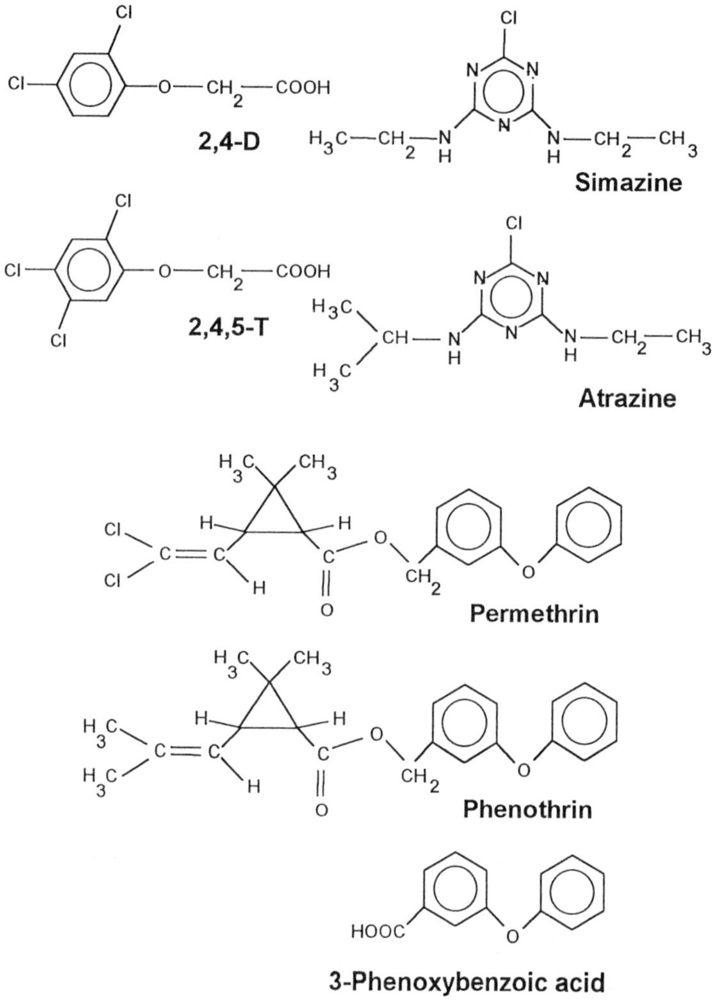

Figure 1. Chemical structures of selected pesticides.

Direct vs. Indirect ELISA. Monovalent antigens can be detected only by competitive ELISA formats, where a pesticide-protein conjugate and a free pesticide in the sample compete for the binding sites of the antibodies. Competitive ELISAs can either have labeled antibodies and immobilized pesticide-protein conjugate (indirect ELISA) or labeled antigens and immobilized antibodies (direct ELISA). Both formats were examined (7). For this purpose we have produced specific anti-pesticide polyclonal antibodies and have synthesized pesticide-protein conjugates. Then the ELISAs have been optimized, i.e. reagent concentrations and stages duration have been chosen in order to reach maximal assay sensitivities.

In the cases of 2,4-D and 2,4,5-T, higher sensitivities were achieved using the scheme with labeled antigen (i.e. with pesticide-peroxidase conjugates), whereas for simazine and atrazine the scheme with labeled antibodies had 12-30 times better sensitivities (Table I).

Table I. Detection Limits of Two Standard ELISA Formats

Pesticide	ELISA Formats	Detection Limit
2,4-D	The scheme with labeled antigen	5 ng/mL
	The scheme with labeled antibodies	20 ng/mL
2,4,5-T	The scheme with labeled antigen	8 ng/mL
	The scheme with labeled antibodies	50 ng/mL
Simazine	The scheme with labeled antigen	10 ng/mL
	The scheme with labeled antibodies	0.8 ng/mL
Atrazine	The scheme with labeled antigen	30 ng/mL
	The scheme with labeled antibodies	1 ng/mL

In addition to the traditional assay optimization methods, some additional approaches for improving ELISA sensitivity have been developed (8-9).

Effect of Pre-incubation. First, the anti-pesticide antibodies were incubated with pesticide-containing sample before addition of the competitor (pesticide-protein conjugate). This approach proved to be useful for simazine and atrazine. The pre-incubation of simazine with the antibodies increased ELISA sensitivity 16 fold (from 0.8 to 0.05 ng/mL), and in the case of atrazine the gain was 10 fold (from 1 to 0.1 ng/mL). On the other hand, the sensitivities of 2,4-D and 2,4,5-T assays did not change after the pre-incubation.

Use of Protein A. Assay sensitivities using labeled antigen may be increased (2-5 fold for different pesticides) if antibodies are immobilized on polystyrene by means of staphylococcal protein A. The same effect was observed in earlier studies on testosterone ELISA (10). This effect may be attributed to the fact that protein A binds to the Fc region of the antibody molecule, resulting in a favorable orientation of the antibody for antigen binding in solution.

Effect of Conjugation Ratios. Composition of the pesticide-protein conjugates significantly influences the assay sensitivity. Conjugates with minimal loading of antigenic groups have low rate of antibody-binding which prolongs the assay duration. On the other hand, increase in pesticide:protein molar ratio leads to an increased conjugate-antibody equilibrium binding constant due to bivalent interactions. In contrast, free pesticide is not capable of bivalent binding, therefore, the sensitivity of the competitive assay falls.

We have synthesized 2,4-D-peroxidase and 2,4,5-T-peroxidase conjugates via hydroxysuccinimide and carbodiimide technique with different pesticide:peroxidase ratios. The compositions of the resulting conjugates were determined by comparing the ultraviolet spectra and the number of surface amino groups of the free protein and the hapten-protein conjugates. Amino groups were detected using 2,4,6-trinitro-benzensulfonic acid.

Figure 2 shows that the conjugates with pesticide:peroxidase molar ratios of 2:1 and 5:1 are optimal for competitive ELISAs. To obtain such conjugates the initial molar ratio of the pesticide and peroxidase should be within the range of 4:1 to 16:1.

Use of Monovalent Antibodies. The bivalent complex formation may be eliminated by using antibody monovalent derivatives, which can be prepared by reducing the IgG (antibody) S-S-bonds using hydrosulfite in the presence of cystein. The reduced IgG (half-antibody) may be purified by gel-filtration on *Sephadex G-100*. When the half-antibodies are immobilized on the solid support using protein A, the sensitivity of the 2,4-D ELISA with labeled antigen becomes 3 times greater than the sensitivity of the ELISA using native IgG (it changes from 5 to 1.5-2 ng/mL).

Analyte Activation. The affinity of an antibody to a pesticide-conjugate may be significantly higher than its affinity towards free pesticide. For example, the affinities of the anti-2,4-D antibodies towards 2,4-D-BSA, 2,4-D-ovalbumin or 2,4-D-soybean trypsin inhibitor conjugates were about 10 times higher than the affinity towards free 2,4-D. However, this negative effect may be minimized if the free 2,4-D in the sample is conjugated with an amino acid or protein (8). Preliminary incubation of water samples containing 2,4-D with activation agents (N-hydroxysuccinimide and 1-cyclohexyl-3(2-morpholinoethyl)-carbodiimide) and carrier (amino acid or protein: glycine, lysine or BSA) results in 10 to 30 times increase in sensitivity of the direct ELISA technique (using 2,4-D-peroxidase conjugate as competitor). The same gain in the ELISA sensitivity has been demonstrated for 2,4,5-T. Conjugation reaction was allowed to proceed for 15 min at 37°C to reach optimum sensitivity. The proposed technique does not require additional steps of separation because excess reagents do not interfere in the assay. The ratio of the sample:activation reagent:carrier was 4:1:1.

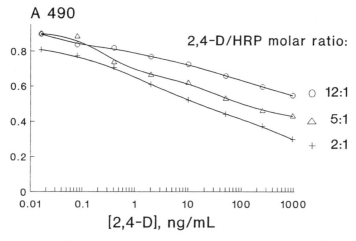

Figure 2. Competitive curves for 2,4-D ELISA with 2,4-D-peroxidase conjugates of different composition.

Addition of the activation step slightly increases the total duration of the assay from 60 to 80 min.

The resulting limits of detection for the modified immunoassay techniques are presented in Table II. The coefficients of variation and recoveries were determined using 10 to 25 samples for each format. The pesticide concentrations varied from 0.5 to 5 ng/mL and each sample was analyzed in four repetitions. Measured C.V. in seria are 7-12%, and between series - 12-20%. Addition of pesticides in known concentrations (0.5-5 ng/mL) to water or milk, results in recoveries ranging from 80% to 115%.

The modified enzyme immunoassays described above have considerable advantages. Whereas traditional ELISA optimization has permitted to reach only a few ng/mL as a limit of detection, the above modifications improved the detection to levels that correspond with the requirements for environmental monitoring and agricultural production control. The principles described here are universal and may be used in immunoassay development for other pesticides and other low-molecular weight compounds.

Table II. Detection Limits of Optimum ELISA Formats

Pesticide	ELISA Format	Detection Limit
2,4-D	The scheme with labeled antigen using immobilization of IgG monovalent fragments through protein A and the sample activation	0.1 ng/mL
2,4,5-T	The scheme with labeled antigen using immobilization of IgG monovalent fragments through protein A and the sample activation	0.02 ng/mL
Simazine	The scheme with labeled antibodies using the sample pre-incubation	0.05 ng/mL
Atrazine	The scheme with labeled antibodies using the sample pre-incubation	0.1 ng/mL

Homogeneous Immunoassays

Homogeneous immunoassays are based on modulation of the catalytic activity of enzyme-antigen conjugates by the formation of complexes with antibodies. Although homogeneous immunoassays are known to be fast and simple, its use has been hindered by the limited availability of suitable enzyme labels and conjugation chemistry that will cause significant changes in the catalytic activity of the enzyme. Therefore, it is important to find new labels that are appropriate for homogeneous assays.

We have shown the suitability of alpha-amylase from *Bacillus subtilis* on the examples of homogeneous immunoassays for 3-phenoxybenzoic acid (derivative of some pyrethroid pesticides, see Figure 1) and 2,4-D (7). The necessary amylase-pesticide conjugates were prepared via succinimide reaction, and purified from unreacted compounds of low molecular weight by gel-filtration using *Sephadex G-25* and dialysis. They keep 60-70% of the initial enzymatic activity and contain about 5 to 10 pesticide groups (for different preparations) on one amylase molecule.

The main advantage of a homogeneous assay is its methodical simplicity. In contrast to ELISA, the homogeneous enzyme immunoassay does not demand preliminary immobilization or long incubations. Thus, the homogeneous assay developed in our laboratory includes three simple steps: 1) incubation of the pesticide-containing sample and pesticide-amylase conjugate with the anti-pesticide antibodies, 2) addition of the enzyme substrate (starch hydrolysis), and 3) detection of the extent of hydrolysis.

Specific anti-pesticide antibodies at high concentrations inhibit the catalytic activity of the pesticide-enzyme conjugate. On the other hand, if the pesticide is present in the sample, the pesticide reacts with the antibodies and displaces the pesticide-enzyme conjugate, restoring the initial activity of the enzyme conjugate (Figure 3). Therefore, the higher the pesticide concentration in the sample, the greater the starch hydrolysis. The amount of starch hydrolysis may be quantified using either the starch-iodine test, or the test for the presence of free aldhydes. In the former test, iodine is added to the test solution to indicate the presence of unhydrolyzed starch. If there is pesticide present in the sample, the enzyme amylase will be free and active, resulting in starch hydrolysis. This will be indicated by the discoloration of the iodine solution or low absorbance at 630 nm. On the other hand, a high absorbance at 630 nm means high amount of unhydrolyzed starch, and thus, low pesticide concentration in the sample.

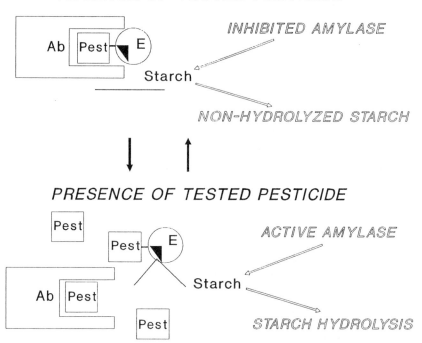

Figure 3. Principle of the proposed homogeneous immunoassay technique. Ab - anti-pesticide antibodies, Pest - pesticide in the tested sample, Pest-E - conjugate of bacillary alpha-amylase with the pesticide.

The occurrence of starch hydrolysis may also be monitored by detecting the presence of free aldehydes released during the breakage of glucose links in starch. The aldehyde test uses 3,5-dinitrosalicylic acid for color development. The color intensity is measured at 570 nm; a high absorbance corresponds to a high concentration of free aldehydes (greater starch hydrolysis), thus higher pesticide concentration. In both tests, the colored products are formed quickly after stopping of the starch hydrolysis; within 1 min for iodine probe and 5 min for interaction with 3,5-dinitrosalicylic acid. Concentrations of the colored products obtained were measured using standard photometers for ELISA.

The amylolytic activity modulations prove to be sufficient for reliable detection of 3-phenoxybenzoic acid and 2,4-D. The assay time is 40 min, which is half the time required for the improved ELISA described above. However, the sensitivity of the homogeneous assay for 2,4-D (about 50-100 ng/mL) is poorer compared to ELISA, thus, its use is limited only to samples with high pesticide concentration. Nonetheless, such system is appropriate for testing solid and heterogeneous samples particularly when the pesticide had been extracted before assay.

Determination of 3-phenoxybenzoic acid by the homogeneous assay results in better sensitivity as compared with the 2,4-D homogeneous technique. 3-phenoxyben-zoic acid can be detected at concentrations as low as 5 ng/mL (Figure 4). As indicated in the figure, native pesticides permethrin and phenothrin, both containing 3-phenoxy-benzoic group, also react with the antibodies and thus influence the conjugate catalytic activity. So the proposed technique may be used both for the native pyrethroid pesticides and products of their degradation.

Electrochemical Immunosensors

Immunosensors have become popular as immunoanalytical techniques. Their principal advantage is possibility to provide many assays in a short time with minimal handling procedures. We have developed two different types of immunosensors for 2,4-D and 2,4,5-T detection.

The first type uses graphite electrodes with antibodies adsorbed on their surface (*11*). This assay is based on the competitive binding of pesticide and pesticide-peroxidase conjugate with the antibodies, followed by the detection of activity of the bound peroxidase. An automatic potentiometric device measures the change in redox potential during the peroxidase reaction using 5-aminosalicylic acid and hydrogen peroxide as enzyme substrates. The optimum conditions for this assay have been determined, such as conditions for antibody immobilization and concentrations of reagents. The developed immunosensor allows detection of 2,4-D at concentrations as low as 40 ng/mL (both in water and biological fluids). The total time of the assay including electrode regeneration is 12 min. One electrode can be used for 60 sequential analyses. The measured signal retains no less than 85% of its amplitude after 20 analyses or after storage under the operating conditions at room temperature. The same parameters have been reached by the sensor for 2,4,5-T.

The principle of the second immunosensor is based on potentiometric detection of the enzyme label by field-effect transistor (FET). The FET-sensor preparation includes the following stages (*12*). Aldehyde groups are introduced onto membranes (80-100 nm pore diameter) from regenerated cellulose on nylon net by photo-activation. Then, specific antibodies react with active groups at the membrane surface. After washing, the membrane is attached to the electrode.

In this assay the quantity of formed complexes between the immobilized antibodies and 2,4-D-peroxidase conjugate reflects the amount of 2,4-D in the sample. Enzymatic activity of label on the solid phase is detected potentiometrically from pH changes during catalytic reaction. Peroxidase substrate solution used for this purpose contains 5-aminosalicylic acid, ascorbic acid, and hydrogen peroxide.

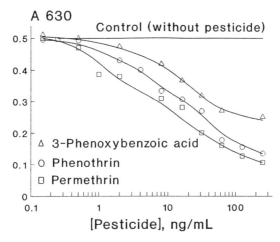

Figure 4. Competitive curves for the homogeneous enzyme immunoassay of pyrethroid pesticides. Enzymatic activity of the 3-phenoxybenzoic acid-amylase conjugate is detected by starch-iodine reaction.

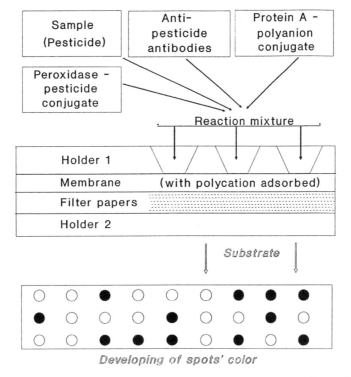

Figure 5. Principle of the proposed visual membrane immunoassay for pesticides detection.

At optimum conditions, this system permits detection of 2,4-D at concentrations as low as 1 ng/mL. The total time of testing is 45 min. The use of changeable membranes reduces the FET preparation and eliminates the electrode regeneration. The potentiometric measurement demands no more than 5 min.

Visual Membrane Enzyme Immunoassays

The instrument-free assay with visual detection is a very promising direction in pesticide immunoanalysis. We have proposed a new visual method based on "dot"-assay principle employing water-soluble linear polyelectrolytes (*13-14*).

The polyelectrolyte interaction is used for quick reagent separation after immunochemical reaction. A pair of such polyelectrolytes was selected: salts of poly-N-ethyl-4-vinylpyridine (polycation) and polymethacrylate (polyanion). These ions interact with each other with extremely high affinity due to cooperative binding between their chains' links. Insoluble interpolymeric complexes are formed instantly. This reaction and its applications for enzyme immunoassays have been described (*15*).

The scheme of the visual immunoassay for pesticides is presented in Figure 5. First, four solutions are mixed: 1) sample containing the pesticide to be determined, 2) pesticide-peroxidase conjugate, 3) anti-pesticide antibodies, and 4) staphylococcal protein A covalently bound to the polyanion. During the incubation, complexes are formed, such as polyanion-protein A : antibodies : pesticide and polyanion-protein A : antibodies : pesticide-peroxidase. Then, the reaction mixture is applied onto the nitrocellulose porous membrane where the polycation is already adsorbed (the membrane has been placed in special holder with microwells. The polyanion (and the immune complex bound to it) reacts with the immobilized polycation almost instantaneously. The other components are removed through the membrane into layers of filter paper by washing. Finally the membrane is placed in peroxidase substrate solution, giving insoluble colored product. In the absence of pesticide in the test sample, only polyanion-protein A : antibodies : pesticide-peroxidase complexes bind with the membrane; the quantity of the peroxidase immobilized is high, and therefore, the resulting spots are darker. In contrast, a decrease in the spot intensity (up to pale) corresponds to the presence of pesticide in the sample.

This principle was realized for 2,4-D detection in water. The optimum assay conditions are presented at Table III. The total duration of the assay is no more than 20 min. The proposed qualitative technique allows detection of 2,4-D in water samples at concentrations as low as 10 ng/mL.

**Table III. Optimum Conditions of 2,4-D Detection
by the Proposed Visual Membrane Enzyme Immunoassay**

Concentration of protein A-polyanion conjugate	5 µg/mL
Concentration of specific antibodies against 2,4-D	1 µg/mL
Concentration of 2,4-D-peroxidase conjugate	0.5 µg/mL
Duration of the reaction mixture incubation	10 min
Membrane pores diameter	450 nm
Duration of the filtration and washing	3 min
Peroxidase substrate solution	N,N'diaminobenzidene $+ CoCl_2 + H_2O_2$
Duration of the spot color developing	3 min

Conclusion

The developed enzyme immunotechniques allowed the determination of 2,4-D, 2,4,5-T, simazine, atrazine, permethrin, phenothrin and 3-phenoxybenzoic acid. Solid-phase techniques (ELISA) proved to be the best format for quantitative measurements; their proposed modifications improved the assay sensitivities to 0.02-0.1 ng/mL. Homogeneous techniques and immunosensors reduce the assay duration significantly. Measurements for the homogeneous technique using alpha-amylase can be realized by means of standard photometers for ELISA. The main advantages of immunosensors include reduction of handling procedures and possible automation of the analytical procedure. The proposed visual membrane immunoassay is useful for on-site qualitative determination of pesticide. Nowadays, we apply the developed rapid enzyme immunoassays (immunosensors, visual membrane assay) for the detection of s-triazines and pyrethroids in environmental samples.

Literature Cited

1. Mumma, R.O.; Brady, J.F. In: *Pesticide Science and Biotechnology*; Greenhalg, R., Roberts, T.R., Eds.; Blackwell Scientific, Oxford, **1987**, 341-348.
2. Jung, F.; Gee, S.J.; Harrison, R.O.; Goodrow, M.N.; Karu, A.E.; Braun, A.E.; Li, Q.X.; Hammock, B.D. *Pestic Sci.* **1989**, *26*, 303-317.
3. Van Emon, J.M.; Mumma, R.O. *ACS Symp. Ser.* **1990**, *442*, 112-139.
4. Gazzaz, S.S.; Rasco, B.A.; Dong, F.M. *Crit. Rev. Food Sci. Nutr.* **1992**, *32*, 197-229.
5. Hock, B. *Acta Hydrochim. Hydrobiol.* **1993**, *21*, 71-83.
6. Knopp, D. *Anal. Chim. Acta* **1995**, *311*, 383-392.
7. Dzantiev, B.B.; Zherdev, A.V.; Moreva, I.Yu.; Eremin, S.A.; Romanenko, O.G.; Sapegova, L.A. In: *Modern Enzymology: Problems and Trends*; Kurganov, B.I., Kochetkov, S.N., Tishkov, V.I., Eds.; Nova Sci. Publ., Commack, **1995**, 803-807.
8. Dzantiev, B.B.; Zherdev, A.V.; Moreva, I.Yu.; Romanenko, O.G.; Sapegova, L.A.; Eremin, S.A. *Appl. Biochem. Microbiol. (Moscow)* **1994**, *30*, 752-759.
9. Dzantiev, B.B.; Zherdev, A.V.; Romanenko, O.G.; Titova, N.A.; Trubacheva, J.N.; Cherednikova, T.V.; Eremin, S.A. *Appl. Biochem. Microbiol. (Moscow)* **1995**, *31*, 120-125.
10. Rukavishnikova, G.E.; Dzantiev, B.B.; Liozner, A.L.; Eremin, S.A.; Sigal, E.R. In: *Advances in Steroid Analysis `90*; Gorog, S., Ed.; Akademiai Kiado, Budapest, **1991**, 103-108.
11. Dzantiev, B.B.; Zherdev, A.V.; Yulaev, M.F.; Sitdikov, R.A.; Dmitrieva, N.M.; Moreva, I.Yu. *Biosensors & Bioelectronics* **1996**, *11*, 179-185.
12. Khomutov, S.M.; Zherdev, A.V.; Dzantiev, B.B.; Reshetilov, A.N. *Anal. Lett.* **1994**, *27*, 2983-2995.
13. Dzantiev, B.B.; Zherdev, A.V.; Romanenko, O.G.; Izumrudov, V.A.; Zezin, A.B. In: *Advances in Steroid Analysis `93*; Gorog, S., Ed.; Akademiai Kiado, Budapest, **1994**, 119-125.
14. Dzantiev, B.B.; Choi, M.J.; Park, J.; Choi, J.; Romanenko, O.G.; Zherdev, A.V.; Eremin, S.A.; Izumrudov, V.A. *Immunol. Lett.* **1994**, *41*, 205-211.
15. Dzantiev, B.B.; Blintsov, A.N.; Bobkova, A.F.; Izumrudov, V.A.; Zezin, A.B. *Doklady Biochemistry (Moscow)* **1995**, *342*, 77-80.

Chapter 8

Detection of 2,4-Dichlorophenoxyacetic Acid by Non-Instrumental Latex Immunoassay

Yu. V. Lukin[1], A. N. Generalova[1], T. V. Tyrtysh[1], and S. A. Eremin[2]

[1]Shemyakin & Ovchinnikov Institute of Bioorganic Chemistry, Russian Academy of Sciences, Miklukho-Maklaya St 16/10, 117871 Moscow, Russia
[2]Department of Chemistry, Division of Chemical Enzymology, M. V. Lomonosov State University, 000958 Moscow, Russia

A simple and rapid immunoassay technique called microtiter particle agglutination inhibition (MPAI) for monitoring the level of 2,4-dichlorophenoxyacetic acid (2,4-D) in cereals has been developed. MPAI is a non-instrumental assay based on the agglutination between colored polyacrolein particles sensitized with anti-2,4-D monoclonal antibodies (MAb) and novel synthetic agglutinator 2,4-D-polyacrylamide. The agglutination picture could be detected after 1 to 1.5 h by the naked eye. The detection limit of 2,4-D in water was 0.25 ng/mL for MPAI in 25 mL sample. The cross-reactivity with structurally related herbicides was not more than 10%. A good correlation between MPAI and polarization fluoroimmunoassay was shown when analysing 2,4-D in cereals.

The analytical control of drinking water, food and soil for herbicide contamination is of great importance for a healthy environment. Conventional analytical methods used for herbicide detection, such as gas or liquid chromatography, are time-consuming and require sophisticated equipment to perform the analysis (1). Alternative immunoanalytical approach is of special interest for environmental monitoring due to its simplicity, rapidity and high sensitivity (2). Interest in enzyme immunoassay has grown dramatically because it allows the analysis of a large number of samples with high sensitivity and low cost. This method, however, requires separation of immunoreagents, involves multiple washing steps and enzymatic detection. One important requirement for a successful screening method for pesticides is simplification of the assay. Therefore a rapid "homogeneous" immunoassay technique which do not require any separation or washing step is of special interest.

This paper is devoted to the development of a method, namely microtiter particle agglutination inhibition (MPAI), for detection of 2,4-dichlorophenoxyacetic acid (2,4-D) as a model analyte.

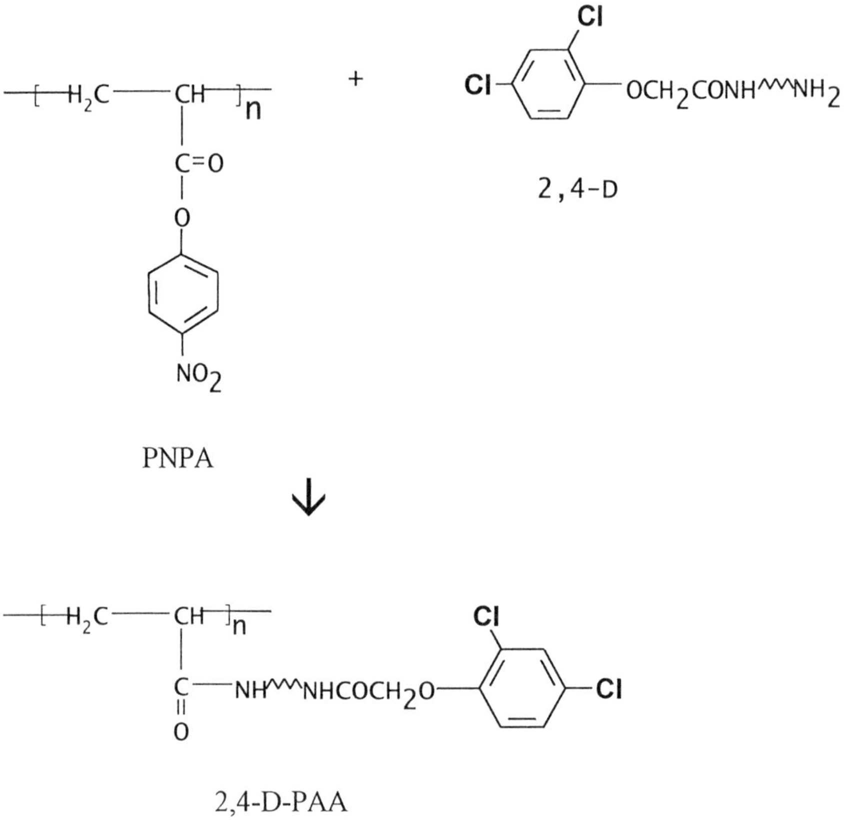

PNPA

2,4-D

2,4-D-PAA

Figure 1. The synthesis of 2,4-D-PAA agglutinators.

Materials and Methods

Reagents. Chemicals, 2,4-D and its structurally related chlorophenoxyacids were supplied by Sigma (St.Louis, MO). Monoclonal antibodies (MAb) against 2,4-D (clone E2/G2) were obtained from The Veterinary Research Institute, Brno, Czech Republic (*3*). The extracts from cereals were obtained from the station for agricultural production control (Krasnodar, Russia) and prepared according to certified method for GC detection of 2,4-D.

Preparation of latex-antibody conjugates. Colored polyacrolein latex (PAL) particles with the diameter of 2 μm were obtained by polymerization of acrolein in water under alkaline conditions in the presence of various dyes (*4*). Particle-antibody conjugates were obtained by incubation of 1% (w/v) PAL with MAb in phosphate buffered saline (PBS), pH 7.4 at room temperature (RT) for 2 h (10-30 μg MAb per 10 mg PAL). The conjugates were washed 3 times with PBS containing 0.2% BSA by centrifugation (3000 rpm x 5 min) and were lyophilized.

Preparation of 2,4-D agglutinators. The agglutinator based on the ovalbumin (2,4-D-Ov) was described earlier (*5*). The synthesis of 2,4-D-polyacrylamide (2,4-D-PAA) agglutinators is shown on Figure 1. The poly(4-nitrophenyl)acrylate was prepared by polymerization of 4-nitrophenylacrylate in dry benzene in the presence of a common oil-soluble radical initiator, azo-isobutyronitrile (AIBN) (*6*). The aminoderivative of 2,4-D was synthesized by adding dicyclohexylcarbodiimide (100 mg, 0.48 mmol) to the solution containing 2,4-D (50 mg, 0.23 mmol) and N-hydroxysuccinimide (35 mg, 0.3 mmol) in dry dichloromethane (2 mL). The mixture was stirred at RT for 20 h and filtered. N,N-diisopropylethylamine (62 μL, 0.6 mmol) and 1-(tert-butoxycarbonylamino)-6- aminohexane hydrochloride (76 mg, 0.3 mmol) were added to the filtrate and the reaction mixture was kept at RT for 1 h and filtered. After evaporation of the solvent the mixture was chromatographed on silica gel using toluene/ethylacetate (3:1). The yield of pure 1-(2,4- dichlorophenoxy-acetylamino)-6-(tert-butoxycarbonylamino)-hexane was 72 mg (75%). TLC (toluene-ethylacetate 1:1) showed Rf 0.50. N-(tert-butoxycarbonyl) group was removed with 1 mL of 95% trifluoroacetic acid in 1 mL chloroform. The reaction mixture was kept at RT for 30 min, evaporated and obtained as crystalline substance. The yield of 1-(2,4-dichlorophenoxyacetylamino)-6-aminohexane-trifluoroacetate was 70 mg (95%). The purity and structure of this substance was characterized by TLC and confirmed by ^{1}H-NMR. The poly(4-nitrophenyl)acrylate and solution of aminoderivative of 2,4-D were condensed in different ratio as described previously (*6*).

Agglutination assay. The assays were carried out in 96-well U-bottom microtiter plates (Nunc) using PBS-BSA (0.2%) buffer for the dilution and titration of the samples.

Microtiter particle agglutination (MPA) assay was conducted by serial two-fold dilutions of the agglutinator in the wells of each row (50 μL per well). Then 25 μL of 0.15% PAL-MAb conjugate were added to each well. The plate was shaken by hand for 30 sec and left at RT without vibration. After 1.0 - 1.5 h the wells were checked visually for the agglutination picture. A positive result is revealed by a diffuse colored film of agglutinated particles evenly covering at least 1/3 of the well bottom (1+) or more (2+, 3+). On the other hand, a distinct spot in the well center means a negative result.

For microtiter particle agglutination inhibition (MPAI) assay, serial two-fold dilutions (25 μL per well) of 2,4-D water solution or other samples were prepared in the wells of each row. Then 25 μL of the agglutinator and 25 μL of PAL-MAb conjugate were added successively to each well. The following steps were identical to those described for MPA assay. In these tests the concentration of agglutinator was at least 5 times higher than the least detectable concentration in MPA. A sample with a known concentration of 2,4-D was used as a reference probe. Interpretation of the results in the inhibition test was opposite to that for MPA assay, that is, a positive sample is shown by the absence of the diffuse film of agglutinated particles, and vice versa. All the experiments were performed in triplicate.

Results and Discussion

The latex agglutination immunoassay has been used in clinical chemistry for many years. The most familiar application of this technique is the qualitative pregnancy slide test. More recently, semi-quantitative or quantitative latex agglutination methods which are precise and accurate, have been developed for the measurement of both macromolecules and haptens (7,8). Latex particles can be used as a "tag" or label in any immunoassay. This is analogous to the radioactive, enzyme or fluorescent tags in an appropriate immunoassay. The reaction occurs when the latex particles sensitized with antibody, agglutinate by cross-linking in some manner. This method requires no separation or washing steps. A reading of agglutination reaction can be performed either visually using slide, tube or microtiter tray formats, or with the help of an instrument such as spectrophotometer (7).

Microtiter particle agglutination is one of the most simple and sensitive latex immunoassay method that does not require instrumentation (4,8). The reaction is performed in a microtiter plate with U-bottom wells where the particles form the appropriate agglutination patterns similar to those known for hemagglutination reaction (9). The use of colored polymer particles significantly simplifies visual reading of the patterns. The agglutination results in the formation of a diffuse film of agglutinated particles, while non-agglutinated particles form a distinct button after settling to the bottom of the well.

A simple microtiter particle agglutination inhibition (MPAI) assay for analyzing haptens such as 2,4-D has been recently developed (5). In this study, we describe the optimization of the assay using novel polymeric agglutinator consisting of 2,4-D coupled to polyacrylamide. The MPAI technique involves a polyvalent antigen (agglutinator) prepared by conjugation of a hapten with a large soluble carrier molecule. In principle, various proteins, polysaccharides or synthetic polymers can be used as a carrier. The agglutinator is capable of promoting cross-linking of the latex particles coated with the appropriate antibodies which are either covalently attached or physically adsorbed. The covalent attachment of antibodies is preferred because of the increased stability of the antibody-latex conjugate. When the agglutinator and the antibody-coated latex particles are mixed, agglutination takes place. Conversely, if free hapten is present, it will bind to the antibody on the latex, preventing the agglutination reaction from taking place. Thus, the hapten and agglutinator compete with each other in binding with the antibodies on the latex particles.

MPAI assay has been developed using colored polyacrolein latex microspheres with the diameter of 2 μm and specific gravity of 1.25 g/cm³ (5). The microspheres settle in the well of a microtiter plate within 1.0 to 1.5 h making the assay relatively rapid. For comparison, sedimentation of conventional polystyrene latex particles takes about 15 to18 h (8). The polyacrolein latex was synthesized by heterophase polymerization of acrolein in water under alkaline conditions in the presence of different water-soluble dyes that produce latex suspensions of a desired color (4). The surface aldehyde groups of the particles will react directly with primary amines, thus making the covalent attachment of proteins relatively easy. The coupling reaction does not need any activation step and gives a high yield of protein immobilization, reaching 70% after 2-3 hours of incubation period (10). In practice, the process of preparing antibody-coated latex simply involves incubating polyacrolein latex with antibody under the proper conditions, blocking any unreacted aldehyde groups with inert amines, and removing any uncoupled antibody by centrifugation. The conjugates are then lyophilized and stored for about one year.

The preparation of the agglutinator involves the attachment of a hapten to a protein molecule or synthetic polymer carrier. Previously, 2,4-D agglutinator was prepared by coupling 2,4-D with ovalbumin carrier (2,4-D-Ov) through carbodiimide/N-hydroxysuccinimide activation (11). The resulting conjugate was characterized and found to have hapten:protein molar ratio of 25:1. As little as 0.6 ng/mL of 2,4-D may be determined by MPAI using ovalbumin-based agglutinator and polyacrolein particles sensitized with anti-2,4-D monoclonal antibody (clone E2/G2) (5).

The nature of the agglutinator plays a principal role in promoting extensive cross-linking of the particles in the presence of the trace quantity of the reagent. The smaller the amount of agglutinator, the lower the concentration of hapten detected. To fulfil this condition an agglutinator with the desired molecular weight and optimal amount of hapten residues is required.

We have prepared an agglutinator based on polyacrylamide (PAA) by condensation of amino-spacered 2,4-D with poly-n-nitrophenylacrylate (PNPA) e.g. fully activated polyacrylic acid (6) (Figure 1).This reaction has several advantages. Firstly, the reaction is very simple and is carried out in one step under mild conditions. Secondly, the amount of hapten residues coupled to the polymer may be controlled to obtain the desired hapten:polymer ratio. Up to 25% substitution proceeds quantitatively at room temperature without a catalyst. PAA-based conjugates differ from ovalbumin-based ones in terms of chemical and immunological inertness. PAA-based agglutinators have lower non-specific reactions with proteins. Furthermore, the flexibility of the polymer, which is a random coil, allows the hapten residues to rearrange themselves and to interact with the antibodies in an optimal way (12).

In order to optimize the structure of synthetic agglutinator, a number of 2,4-D-PAA conjugates with substitution ranging from 2.5 to 25% and MW ranging from 40 up to 800 kDa were synthesized and tested in parallel with ovalbumin-based agglutinator by MPA assay. Table I shows the upper and lower detection limits for the tested agglutinators. The lower detection limit was defined as the lowest concentration of agglutinator giving a positive reaction (1+ agglutination). As little as 0.02 ng/mL of 2,4-D-PAA conjugate with the same parameters as for 2,4-D-Ov can be detected by MPA. This is about 2 orders of magnitude lower than the minimum detectable dose attained for the ovalbumin-based agglutinator. Besides, assays using 2,4-D-PAA agglutinators exhibit the prozone effect (e.g. false negative reaction when the amount of the antigen in the solution is much higher than the concentration of antibody immobilized on the particles) at concentration greater than 50 ng/mL or 200 ng/mL, that is defined as the upper detection limit. In assays using ovalbumin-based agglutinator this effect was not observed at concentrations up to 1 μg/mL. The sensitivity and speed of MPA reaction were not affected significantly by varying % substitution and molecular weight of the 2,4-D-PAA agglutinators. However, the differences becomes more pronounced when the agglutination reaction is measured by spectrophotometer (data not shown).

Table I. Studies of 2,4-D agglutinators by microtiter particle agglutination (MPA) assay using latex-MAb (clone E2/G2)conjugate

Agglutinator			Detection limit,	
	substitution	MW	upper	lower
	%	kDa	ng/ml	
2,4-D-PAA				
	2.5	40	50.0	0.01
	5	40	50.0	0.02
	10	40	50.0	0.02
	25	40	100.0	0.02
	10	80	200.0	0.04
	10	400	200.0	0.04
	10	800	100.0	0.08
2,4-D-Ovalbumin				
	25	40	no	3.00

The effect of the concentration of 2,4-D-PAA agglutinator on the sensitivity of MPAI for 2,4-D was studied. Table II shows a typical MPAI assay picture where double titrations of 2,4-D and 2,4-D-PAA (substitution 10% and MW 40 kDa) were prepared in the horizontal and vertical rows of the plate, respectively. After addition of the latex-anti-2,4-D conjugate to the wells and settling of the particles a classical "stair" agglutination inhibition picture was observed. The picture shows a decrease in the lower detectable concentration of 2,4-D from 2.0 to 0.06 ng/mL with decreasing agglutinator concentration (from 1.0 to 0.03 ng/mL). As the working concentration of the agglutinator in these experiments was 5 times higher than the least detectable concentration in MPA (0.015 ng/mL), the detection limit of 2,4-D was estimated to be 0.25 ng/mL. Thus, the use of PAA-based agglutinator results in an increase in sensitivity of MPAI assay, as compared to ovalbumin-based agglutinators. Another advantage is the low concentration of PAA-based agglutinator required for the analysis. This makes the assay more economical.

Table II. Detection of 2,4-Dichlorophenoxyacetic acid (2,4-D) by microtiter particle agglutination inhibition (MPAI) assay

Agglutinator (2,4-D-PAA) concentration ng/ml	*2,4-D concentration (ng/ml)*						
	2.0	1.0	0.5	0.25	0.12	0.06	0
1.00	⊕*	2+	3+	3+	3+	3+	3+
0.50	±	⊕	2+	3+	3+	3+	3+
0.25	-	±	⊕	2+	3+	3+	3+
0.12	-	-	±	⊕	2+	3+	3+
0.06	-	-	-	±	⊕	2+	3+
0.03	-	-	-	-	±	⊕	2+
0.015	-	-	-	-	-	±	⊕
Buffer	-	-	-	-	-	-	-

* Latex reaction graded as degree of particle agglutination inhibition from positive (- or ±) to negative (+, 2+, 3+).
⊕ the end point of the reaction.

The study of cross-reactivity with various substances structurally related to 2,4-D, revealed rather high specificity of MPAI.. The highest cross-reacting compounds, 2,4,5-trichlorophenoxyacetic acid and 4-chloro-o-tolyloxyacetic acid, showed cross-reactivity at concentrations at least one order of magnitude greater than that for 2,4-D. The specificity of developed MPAI is similar to ELISA of 2,4-D using the same Mab (*3*).

The MPAI was applied for detection of 2,4-D in cereals. The extracts from nine cereal samples were dissolved in water and analysed by the developed MPAI and recently described polarization fluoroimmunoassay (PFIA) (*13*). Four of these samples had 2,4-D concentration between 30 to 80 ng/mL, which is below the detection limit of PFIA (100 ng/mL). The other 5 samples with detectable level of 2,4-D by PFIA were also positive by MPAI. The recalculated results of 2,4-D in cereals by both methods are presented in Figure 2, which shows a good correlation between MPAI and PFIA. These results clearly indicate that MPAI could be used for semi-quantitative detection of 2,4-D in actual food samples.

Conclusion

The potential usefulness of MPAI technique for pesticides detection in food samples has been demonstrated. Previously, we described MPAI for determination of free D-biotin in liquid samples (*14*). In this paper, the possibility of 2,4-D monitoring in

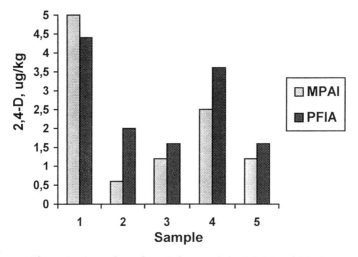

Figure 2. Detection of 2,4-D in cereals by MPAI and PFIA.

water and cereals is shown. MPAI is a sensitive and simple one-step method which require neither any laboratory equipment nor specially trained personnel and is suitable for non-laboratory settings.

Acknowledgements

This work has been partially supported by grants from John D. And Catherine T. MacArthur Foundation as well as the Logovaz Foundation. We wish to thank Dr. N.V.Bovin for providing polyacrylamide-based reagents and fruitful discussions. We also gratefully acknowledge the generous gift of 2,4-D antibodies by Dr. Milan Franek.

Literature Cited

1. Smith, A.E.; Hayden, B.J. *J. Chromat.* **1979**, *171*, 482-485.
2. Van Emon, J.M.; Seiber, J.N.; Hammock, B.D. In *Analytical Methods for Pesticides and Plant Growth Regulators: Advanced Analytical Techniques;* Sherma, J.,Ed.; Academic Press: New York, 1989, Vol. XVII; pp.217-263.
3. Franek, M.; Kolar, V.; Granatova, M.; Nevorankova, Z. *J. Agric. Food Chem.*, **1994**, *42*, 1369-1374.
4. Pavlova, I.S.; Lukin, Yu.V.; Avdeev, D.N.; Kulshin, V.A. *J.Immunoassay*, **1995**, *16*, 199-212.
5. Lukin, Yu.V.; Dokuchaev, I.M.; Polyak, I.M., Eremin, S.A. *Anal. Let.*, **1994**, *27*, 2973-2982.
6. Bovin, N.V.; Korchagina, E.Yu.; Zemlyanukhina, T.V.; Byramova, N.E.; Galanina, O.E.; Zemlyakov, A.E.; Ivanov, A.E.; Zubov, V.P.; Mochalova, L.V. *Glycocon. J.*, **1993**, *10*, 142-151.

7. Kusnetz, J.; Mansberg, H.P. In *Automated Immunoanalysis, Part1*; Ritchie, R.F., Ed.; Marcel Dekker: New York, 1978, pp.1-43.
8. Quash, G.; Roch, A.-M.; Niveleau, A.; Grange, J.; Keolouangkhot, T.; Huppert, J. *J. Immunol. Meth.*, **1978**, *22*, 165-174.
9. Schuurs, A.H.W.M.; Gribnau, T.C.J.; Leuvering, J.H.W. In *Affinity Chromatography and Related Techniques*; Gribnau, T.C.J.; Visser, J.; Nivard R.J.F., Eds.; Elsevier Press: Amsterdam, 1982, pp.343-356.
10. Zhorov, O.V.; Preigerzon, V.A.; Lukin, Yu.V.; Zubov, V.P.; Martsev, S.P. *Russ. J. Bioorgan. Chem.* , **1995**, *21*, 224- 229.
11. Hall, J.C.; Deschamps, R.J.A; Krieg, K.K. *J. Agric. Food Chem.*, **1989**, *37*, 981-985.
12. Bovin, N.V.; Gabius, H.-J. *Chem. Soc. Rev.*, **1995**, *24*, 413- 421.
13. Eremin, S.A. In *Immunoanalysis of Agrochemicals: Emerging Technologies;* Nelson, J.O.; Karu, A.E.; Wong, R.B. Eds.; ACS Symposium Series No.586; American Chemical Society, Washington, DC, 1995, pp. 223-234.
14. Pavlova, I.S.; Lyubavina, I.A.; Lukin, Yu.V. *Russ. J. Bioorgan. Chem.*, **1995**, *22*, 196-200.

Chapter 9

Rapid Screening of Immunoreagents for Carbaryl Immunosensor Development

S. Morais[1], M. A. González-Martinez[1], R. Puchades[1],
Angel Montoya[2], A. Abad[2], and A. Maquieira[1]

[1]Departamento de Quimica and [2]Laboratorio Integrado de Bioingenieria,
Universidad Politécnica de Valencia, Apdo. de Correos 22012,
Camino de Vera s/n, 46071 Valencia, Spain

A fast, accurate selection of antibodies, solid supports, assay format and dissociating agents for immunosensor systems may be achieved by means of a screening tool based on the Enzyme Linked Immuno Filtration Assay principles. This approach is demonstrated in this paper using the immunological reaction between the pesticide carbaryl as a model analyte and a set of monoclonal antibodies raised against carbaryl. Future development of immunosensors may be simplified using this method.

Environmental contamination is partly caused by the pesticides applied to crops to protect them against pests. Consequently, the presence of pesticide residues needs to be monitored to protect both human health and environment. To make constant monitoring feasible, a greater effort should be made to develop sensitive, rapid, cost-effective and automated methods for routine analyses (1).

Suitable immunoassays are being tested as alternative techniques to traditional methods (2) when high sample throughput or on–site screening analyses are required. For example, there has been an increasing interest in the development of immunosensors because of its great promise for on–site analysis. An immunosensor is an analytical device composed by an immobilized immunoreagent in intimate contact with a physical transducer that converts the product of the immunoreaction into a quantificable signal. Flow Injection Immuno Analysis provide attractive approaches to immunosensor development (3). In this type of immunosensor the sample is incorporated in a carrier stream which enters a reaction chamber containing the immobilized immunoreagent, where the immunological reaction takes place. In immunosensor development, the following aspects have to be optimized: appropriate immobilization support material, suitable immunoreagents, and feasible dissociating agents. In this report, the rapid optimization of these parameters using the principles of the Enzyme Linked ImmunoFiltration Assay (ELIFA) (4) is described.

Experimental section

Reagents. Carbaryl analytical standard was purchased from Dr. Ehrenstorfer (Augsburg, Germany). Haptens 6-[[(1-naphthyloxy)carbonyl] amino] hexanoic acid (CNH), 3-[[(1-naphthyloxy)carbonyl] amino] propanoic acid (CNA) (Figure 1), and monoclonal antibodies (MAbs) were obtained, purified, and characterized as previously described (*5*). Horseradish peroxidase (HRP) was from Boehringer (Mannheim, Germany) and rabbit anti-mouse peroxidase-conjugated immunoglobulin from Dako (Glostrup, Denmark). All other reagents were of analytical grade.

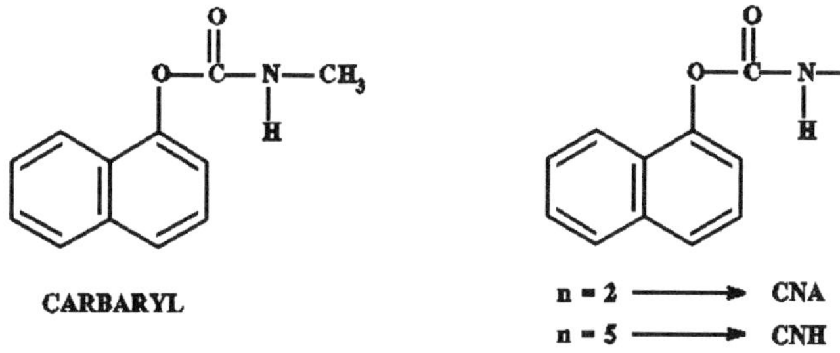

CARBARYL

n = 2 ⟶ CNA
n = 5 ⟶ CNH

Figure 1. Structures of Carbaryl and the haptens CNA and CNH.

Solid supports. Support selection criteria were: good dynamic properties, ease of reagent immobilization (antibodies or haptens), high stability to regeneration conditions and the variety of geometries available, i.e. spectroscopic cells, fiber optics, membranes, and electrodes. Beaded supports included controlled pore glass (CPG) and derivatized agarose gels. CPG (PG240-200) was purchased from Sigma–Aldrich Química (Madrid, Spain). Affi–gel hydrazide (Affi–gel Hz) and Affi–gel 15 were from Bio–Rad laboratories (Richmond, CA). Several membranes of different chemical compositions were tested: Immobilon–P transfer membrane was from Millipore Corporation, (Bedford, MA), glass microfibre filters from Whatman International Ltd. (Maidstone, Kent, U.K), Loprodyne (Pall, Madrid, Spain), and homemade polyetherimide membranes (*6*).

Apparatus. The Easy Titer ELIFA device was acquired from Pierce Chemical Company (Rockford, IL). A Gilson Minipuls-3 peristaltic pump (Villiers, Le Bel, France) was attached to the ELIFA device to control the flow rate. Polystyrene 96-well microtiter plates (Nunc, Roskilde, Denmark) were used, and absorbance was read and recorded with a Dynatech MR-700 microplate reader (Sussex, U.K.).

Procedures

Preparation of hapten–protein conjugates. To visualize the antigen–MAb reaction in direct immunoassays, a CNH–HRP conjugate was prepared by a variation of the anhydride mixed method (7). Briefly, to 4 mg of CNH (13.3 μmol) in 197 μL of N–N'dimethylformamide (DMF) was added 2.9 μL of tri–n–butylamine (12.2 μmol) followed by 1.6 μL of isobutyl chloroformate (12.2 μmol). The mixture was stirred for 1 hour at room temperature, and 200 μL of the resulting activated hapten was diluted in 1.8 mL DMF. A volume of 136 μL of this mixture was added to 3 mg of HRP diluted in 1.4 mL of 50 mM carbonate buffer, pH 9.6. The coupling reaction was incubated at room temperature for 5 h with stirring, and the resulting conjugate was purified by gel filtration on Sephadex G–25 using 100 mM sodium phosphate buffer, pH 7.4, as eluant.

For indirect immunoassay, CNH was covalently attached to bovine serum albumin (BSA) using the modified active ester method (8). A mixture of 30.2 mg of CNH (0.1 mmol), 12.5 mg of N–hydroxysuccinimide (0.1 mmol), and 20.6 mg of N,N'dicyclohexylcarbodiimide (0.1 mmol) in 1 mL of DMF was stirred at room temperature for 2 h to convert the carboxyl group into the succinimide ester function. After centrifuging, (12,000 g; 15 min), 600 μL of the clear supernatant containing the active ester was added dropwise to 3 mL of a 15 mg/mL BSA solution in 50 mM carbonate buffer, pH 9.6. The mixture was allowed to react at room temperature for 4 hours with stirring. After centrifuging as above, futher purification of the CNH–BSA conjugate was carried out using the gel filtration technique described for the CNH–HRP conjugate.

Although both CNH and proteins show absorbance peaks at almost the same wavelength (280 nm), the CNH spectrum displays little characteristic shoulders in this spectral zone, so when conjugation occurred, the modification of the protein spectra was very evident (5). Then, the extent of coupling of CNH to proteins was determined by UV absorbance spectrophotometry at 280 nm by assuming additive absorbance values of CNH and proteins in the conjugates. The estimated hapten to protein molar ratios were 17 and 2 for the BSA–CNH and HRP–CNH conjugates, respectively.

MAbs immobilization for direct immunoassays. MAbs to carbaryl raised from CNH (5) and CNA (9) were immobilized on both alkylamined CPG and Affi–gel Hz beads. Periodate oxidation of vicinal hydroxyl groups of the carbohydrate results in the formation of aldehydes for specific coupling to both supports, resulting in an oriented and covalent immobilization. Finally, the immunosorbent suspension was prepared by the addition of storage buffer, 20 mM sodium phosphate, pH 7.4, 0.02% NaN$_3$, and stored at 4°C till use.

Conjugate immobilization for indirect immunoassays. Affi–gel 15 was used to immobilize CNH–BSA conjugate through free amino groups of BSA, following the manufacturer's protocol. In addition, CPG was prepared (10) by glutaraldehyde activation prior to immobilization of the CNH–BSA conjugate. After coupling, the remaining active binding sites were blocked with 1 M ethanolamine, pH 8.0/HCl and left for 1 hour at room temperature for the blocking reaction to be completed. Both coupled blocked gel and CPG were washed with deionized water.

Finally, the antigen–coated immunosorbent suspensions were prepared and stored as described above for the Mab–coated immunosorbents.

Screening of Immunoreagents. An inert membrane sheet wetted with 2.5% BSA solution was sealed over two gaskets covering the 96 wells, and the sample application plate clamped on top of the membrane (Figure 2). First, 200 μL/well of 2.5% BSA solution was pipetted and left to flow through the membrane for 5 min. Then, 50 μL of immunosorbent suspension which contained aproximately 1 mg of immunosorbent were added and the following solutions sequentially pipetted: For the direct competitive ELIFA assays, 200 μL of a mixture solution in phosphate–buffered saline (PBST; 10 mM phosphate, 137 mM NaCl, 2.7 mM KCl, pH 7.4 containing 0.05% Tween 20) of Carbaryl:CNH–HRP (1:1 v/v), and wash solution (5 times with 200 μL PBST). For the indirect competitive ELIFA assays, 200 μL of a mixture solution of Carbaryl:MAbs (1:1 v/v), wash solution (5 times with 200 μL PBST), peroxidase conjugated rabbit anti–mouse immunoglobulins solution (200 μL) and, wash solution (6 times with 200 μL PBST).

A peristaltic pump controlled the flow rate as well as the vacuum required to ensure that all wells were emptied simultaneously (in 5 min for immunoreagent solutions and 2 min for the wash solution). Then, a microtiter plate was hold in the lower chamber of the device in order to collect the product of the enzymatic reaction. The enzyme substrate solution consisting in 75 μL, 2 mg/mL OPD and 0.012% H_2O_2 in 25 mM citrate, 62 mM phosphate buffer, pH 5.4 was added. After 5 min the reaction was stopped by adding 75 μL of 2.5 M sulfuric acid. Finally, the plate was removed from the device and the absorbance at 490 nm was read in the microplate reader.

Both direct and indirect competitive formats were studied. The concentration of carbaryl which inhibited 50% of the maximum signal (I_{50}) was used to estimate the strength of the apparent antibody affinity in each format. The lower the I_{50} value the greater is the binding between the MAb and antigen. Negative controls (zero analyte) were included to estimate the maximum signal for I_{50} determination. Competition curves were obtained by plotting absorbances vs the logarithm of competitor concentration and were fitted to four–parameter logistic equations (*11*).

Results and Discussion

Optimization of the system parameters. The nature of the filtration membrane was the first parameter optimized. An inert membrane without nonspecific physical adsorption of the immunoreagents was required to ensure that the immunological reaction ocurred exclusively on the immobilization support. Several membranes such as Immobilon–P, Loprodyne, glass fiber membrane and homemade polyetherimide membranes were tested by ELIFA. First, homemade membranes with available pore size were eliminated because of sporadic flow impediment when the wells are filled. For the rest of the membranes, the nonspecific adsorption was measured by means of adding enzymatic conjugates to the membrane which had been previously treated with BSA solutions. After washing the membrane with PBST, the presence of

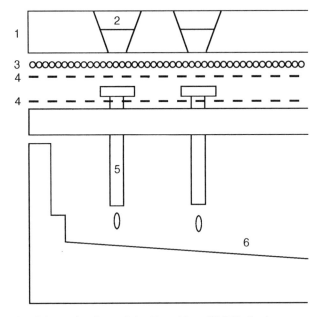

Figure 2. Schematic view of the Easy Titer ELIFA device.
(1: Sample application plate, 2: Sample well, 3: Membrane, 4: Gaskets,
5: Transfer canula, 6: Collection chamber).

nonspecifically adsorbed enzyme conjugates was determined enzymatically as described above.

Immobilon–P, a nitrocellulose membrane designed for high protein binding was tested. BSA solutions of 1–5% were used as blocking agents. In spite of using high BSA solution (5%) to block active sites, high levels of the immunoreagents (CNH–HRP conjugate as well as rabbit anti–mouse peroxidase–conjugated immunoglobulin) were adsorbed in both direct and indirect formats, respectively. The Loprodyne membrane, designed for low protein adsorption, precoated with the above blocking agent also showed high nonspecific adsorption capacity for both labeled immunoreagents. These high nonspecific signals prevented the use of both Immobilon–P and Loprodyne membranes. The glass fiber membrane blocked with 2.5% BSA gave the lowest nonspecific signals, i.e. 0.03 and 0.11 optical density (O.D) units under the best conditions of the method in direct and indirect formats, respectively, and was therefore used throughout this work. Flow rate parameters ranging from 1–10 mL/min were investigated. A flow rate of 4 mL/min was chosen as a trade–off between speed and sensitivity of the assay. In these conditions, wells were emptied simultaneously in approximately 5 min.

Direct format. The I_{50} of eight anti–carbaryl MAbs, immobilized on both CPG and Affi–gel Hz supports, were measured. Different solvents were tested to determine the regeneration of the antibodies after disrupting the MAb–antigen complex. Optimal concentrations of immobilized MAbs (0.03–1 mg/g bead) and HRP–conjugates ranging from 0.1–0.5 µg/mL were selected by non–competitive ELIFA to give absorbances in the range of 0.8–1.2 O.D.

Table I shows the carbaryl I_{50} obtained by direct competitive ELIFA for the MAbs immobilized on both Affi–gel Hz and CPG supports. The activity of immobilized MAbs was dependent on the type of support used. The enzyme conjugate concentration required to achieve optimal absorbance was lower on Affi–gel Hz than on CPG. I_{50} values were 3–6 times higher on CPG, probably due to the fact that physical adsorption dominates the chemical orientation of immobilized MAbs (*12*). The MAbs LIB–CNH 3.2, LIB–CNH 3.6, LIB–CNH 4.5, LIB–CNH 10.3, and LIB–CNA 3.6 displayed the lowest I_{50} to carbaryl, corresponding to the highest apparent affinity in this assay format.

Table I.   Carbaryl I_{50} values (nM) determined by direct competitive ELIFA format for MAbs immobilized on Affi–gel Hz and CPG

	MAb (LIB–)							
Support	CNH 3.2	CNH 3.6	CNH 3.7	CNH 4.5	CNH 8.9	CNH 10.1	CNH 10.3	CNA 3.6
Affi–gel Hz	10	25	29	25	74	34	25	23
CPG	60	85	150	115	200	150	125	130

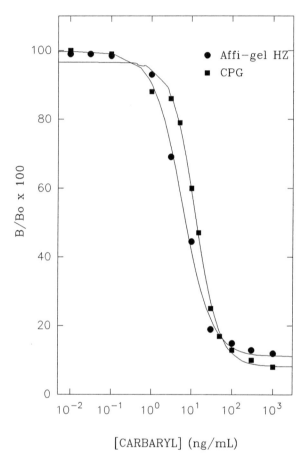

Figure 3. Inhibition curves of monoclonal antibody LIB–CNA 3.6 which was covalently immobilized on Affi–gel Hz (●) and CPG (■) (direct format). Each point represents the average of four wells in the same application plate.

For immunosensor to become a practical quantitative analytical tool, regeneration of the immunoreagents must be accomplished after disrupting the antigen–MAb complex. One drawback of using dissociating agents is the eventual decrease in antibody activity. Different reagents for antibody regeneration, such as highly concentrated NaSCN (3M), NaCl (0.5M–3M), methanol in water (50:50), and buffers with low pH (glycine, acetate, phosphate) were studied with regard to the fast and complete regeneration of immobilized MAbs. 0.1 M glycine/HCl, pH 2.0 turned out to be the best dissociating reagent, since it allowed the effective immunocomplex disruption after the application of few (1–3) short–time (5 min) regeneration cycles.

Most of the MAbs tested lost their native abilility (>50%) to recognize the enzyme conjugate after 4 or 5 runs on both supports (Table II). Only LIB–CNA 3.6 was able to maintain its native activity (100%), estimated as the maximum signal in non–competitive conditions (zero analyte), after at least 15 regenerations without signal decline. Figure 3 shows the inhibition curves displayed by LIB–CNA 3.6 on Affi–gel Hz and CPG. The I_{50} values for both immunosorbents in the competitive direct format were 5 and 13 ng/mL, respectively.

As indicated by the manufacturer, the binding of antibodies to Affi–gel Hz is highly stable under low pH conditions. Therefore, at least for this immunosorbent, the MAb activity decrease should not be attributed to dissociation of the immunoreagent (MAb) from the solid support, but to MAb denaturation.

Table II. Remaining activity (%) of MAbs immobilized on both Affi–gel Hz[a] and CPG[b] supports, after dissociating steps in the direct format

	MAb (LIB–)							
Run	CNH 3.2	CNH 3.6	CNH 3.7	CNH 4.5	CNH 8.9	CNH 10.1	CNH 10.3	CNA 3.6
1	80[a]–60[b]	75–70	72–65	45–70	70–80	30–30	45–55	100–100
2	80–55	70–65	45–60	35–70	60–60	30–20	40–50	100–100
3	40–50	60–50	45–50	25–55	50–50	30–20	40–50	100–100
4	35–50	48–50	35–25	25–50	50–50	25–15	25–50	100–100
5	25–45	45–50	20–20	20–45	50–45	25–15	20–30	100–100
15	<25–45	<45–50	<20–20	<20–45	<50–45	<25–15	<20–20	100–100

Indirect format. Indirect assays were based on the immobilization of CNH–BSA conjugate on both CPG and Affi–gel 15 as described in the procedures section. This format allowed the characterization and selection of MAbs with regard to both their affinity and ease of dissociation from the immobilized conjugate. A complete and rapid dissociation process is required in order to reuse the immunosensor. The concentration of MAb ranging from 0.5–2 μg/mL and the dilution of a secondary

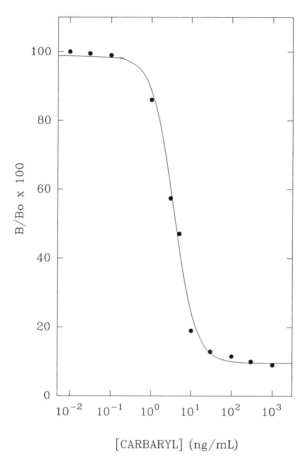

Figure 4. Inhibition curve of monoclonal antibody LIB–CNH 4.5 using the indirect format. Points represent the mean of four determination in the same assay.

antibody (1/8000 dilution of peroxidase–conjugated rabbit anti–mouse immunoglobulins in PBST) to be used in the competitive assays were previously optimized by noncompetitive ELIFA to give absorbances in the range 0.8–1.2 O.D.

Affi–gel 15. To determine the extent of antigen–MAb complex disruption, the dissociation steps were included between the antigen–MAb complex formation and the determination of the enzymatic activity bound to the support. The percentage of antigen–MAb complex disruption was then calculated according to equation 1.

$$\%Disruption = \frac{A_m - A_d}{A_m - A_b} \qquad (1)$$

A_m represents the enzymatic signal (absorbance) obtained without applying any disruption step, A_d represents the absorbance obtained after applying the desired disruption cycles, and A_b is the nonspecific signal obtained in absence of MAb.

Among all MAbs specifically recognizing the antigen immobilized on Affi-gel 15, only LIB–CNH 4.5 allowed its easy and quick dissociation (only 2 washing steps with 0.1 M glycine/HCl, pH 2.0, Table III). As shown in Figure 4, the I_{50} value for this MAb was around 20 nM (4 ng/mL) . Therefore, LIB–CNH 4.5 was selected as a suitable MAb for future development of biosensor systems. In this respect, the immunosorbent presented high stability in desorption conditions, since no physical or chemical changes were observed.

Table III.   Percentage of Antigen–MAb complex disruption in the indirect format performed on Affi–gel 15

	MAb (LIB–)							
Wash	CNH 3.2	CNH 3.6	CNH 3.7	CNH 4.5	CNH 8.9	CNH 10.1	CNH 10.3	CNA 3.6
1	38	45	35	80	55	40	60	40
2	60	55	60	98	70	50	75	40
3	60	65	60	100	75	70	90	55
4	65	70	70	100	80	70	95	60

CPG. Schiff–base reactions were used for glutaraldehyde–mediated coupling of the BSA–antigen conjugate to CPG. Under dissociating conditions (pH 2.0), the imine bonds that attach glutaraldehyde to both CPG and conjugate partially hydrolyzed. Therefore, a procedure was carried out to reduce the imine bonds. This immunosorbent showed high nonspecific adsorption of the secondary antibody and very low activity as well. These facts impeded the use of this support as immunosorbent for the indirect format assays.

Conclusions

Provided that a wide set of MAbs directed to the analyte of interest is available, the proposed and optimized method has been proved to be a useful screening tool in quickly selecting the best solid support, immunoreagents, format and regeneration conditions for immunosensor development. Different results were obtained depending on whether the antigen or the MAb is immobilized on the solid support. Affi–gel (–Hz or –15) was found to be a better support than CPG. In the direct format (immobilized MAb), MAbs were selected on the basis of their affinity to the analyte and their regeneration capability. In the indirect format (immobilized antigen), MAbs were mainly selected in relation to their ability to be desorbed from the antigen–MAb complex. Both selected immunoreagents immobilized on agarose gels were useful in immunosensor systems of high reusability (>15 regenerations) without signal decline and with I_{50} in the low ng/mL level. As suggested by preliminary experiments in progress, these biosensors configured as flow–through immunosensor systems are capable of achieving low detection limit for carbaryl (30 ng/L). Furthermore, the number of regenerations achievable without signal loss is 70 and 200 for direct and indirect format, respectively (unpublished data). In addition, this method could be of great value if applied for immunochromatography technique. Further studies in this field are also in progress.

Acknowledgments

This work was supported by grants (ALI92–0417) and (ALI94–0673) from the Comisión Interministerial de Ciencia y Tecnología (CICYT, Spain).

Literature Cited

1. Sherma, J. *Anal. Chem.* **1993**, *65*, 40R.
2. Sherry, J.P. *CRC in Anal. Chem.* **1993**, *23*, 217.
3. Puchades, R.; Maquieira, A.; Atienza, J.; Montoya, A. *CRC in Anal.Chem.* **1992**, *23*, 301.
4. Clarck, C.R.; Hines, K.K.; Mallia, A.K. *Biotech.Tech.* **1993**, *7*, 461.
5. Abad, A.; Montoya, A. *J. Agric. Food Chem.* **1994**, *42*, 1818.
6. Morais, S.; Lázaro, F. In *Anales de Investigación del Master en Ciencia e Ingeniería de Alimentos*; Fito, P. Universidad Politécnica de Valencia, Spain, 1994, Vol. 4; 257–270.
7. Rajkowsky, K.M.; Cittanova, N.; Desfosses, B.; Jayle, M. *Steroids.* **1977**, *29*, 701.
8. Langone, J.J; Van Vunakis, H. *Methods Enzymol.* **1982**, *84*, 628.
9. Abad, A.; Ph D Disertation, *Producción de anticuerpos monoclonales y desarrollo de inmunoensayos para el plaguicida carbaryl.* Universitat de Valencia, 1996.
10. Massom, M.; Townshend, A. *Anal. Chim. Acta.* **1984**, *166*, 111.
11. Raab, G.M. *Clin. Chem.* **1983**, *29*, 1757.
12. Lin, J.N.; Chang, I.N.; Andrade, J.D.; Herron, J.N.; Christensen, D.A. *J. Chromatogr.* **1991**, *542*, 41.

New Formats of Immunochemical Techniques

Chapter 10

Analysis of Atrazine and Its Degradation Products in Water by Tandem High-Performance Immunoaffinity Chromatography and Reversed-Phase Liquid Chromatography

David S. Hage, John G. Rollag, and David H. Thomas

Chemistry Department, University of Nebraska, Lincoln, NE 68588–0304

Atrazine is one of the most widely used herbicides in agriculture. This work examines the development of an automated chromatographic immunoassay method for water samples that provides quantitative information on both atrazine and its main degradation products (hydroxyatrazine, deethylatrazine and deisopropylatrazine). This method is based on the combined use of a high-performance immunoaffinity column, which contains immobilized antibodies capable of extracting these compounds from water, and a reversed-phase column, for separation of the various extracted solutes. The design and optimization of this method is discussed and data are presented concerning its performance in the routine analysis of environmental samples. This technique gives good correlation versus GC methods, while allowing for the direct analysis of samples in the ppb range in as little as 12 min.

Atrazine (i.e., 2-chloro-4-ethylamino-6-isopropylamino-s-triazine) is used throughout the world for the control of broadleaf weeds [1,2]. Due to its wide-spread use and occurrence in the environment [1,3,4], atrazine has received considerable attention from the Environmental Protection Agency (EPA), which currently has a maximum allowable limit of 3 parts-per-billion (ppb) for atrazine in drinking water [5]. Along with atrazine, several related degradation products can also be found in water, including hydroxyatrazine, deethylatrazine and deisopropylatrazine (see Figure 1) [6]. Deethylatrazine and deisopropylatrazine have phytoxicities that are similar to atrazine [3,6,7] and are believed to be formed through the metabolism of atrazine by fungi or bacteria [3,7]. Hydroxyatrazine, which appears to be formed by non-biological processes, has less toxicity than atrazine in animals and plants but a longer persistence than the other main degradation products of atrazine [7]. The development of new tools for analyzing these compounds is of current interest as a means for studying their effects on humans and the environment.

Current Methods for Atrazine and Related Degradation Products. Traditionally, various chromatographic methods, particularly gas chromatography (GC) and reversed-phase liquid chromatography (RPLC), have been used for the determination of atrazine and related analytes [9-13]. However, GC has difficulties in dealing with polar compounds such as hydroxyatrazine, which must be derivatized into a more volatile form prior to analysis. RPLC does not require such derivatization for polar compounds, but the relatively high detection limits for its most common detectors (i.e., UV/Vis absorbance monitors) means that sample pre-concentration is usually needed for the trace analysis of environmental agents. In both RPLC and GC, sample pre-concentration and the removal of potential interferences is generally accomplished by solid-phase extraction (SPE) or liquid-liquid extraction. However, these extraction methods also suffer from various problems, including the relatively non-selective nature of most extraction media (e.g., silica- or C_{18}-based SPE cartridges), and the increased time or cost that is associated with adding such steps to the overall analysis scheme.

Antibody-based immunoassay techniques, such as the enzyme-linked immunosorbent assay (ELISA), have been examined by a number of workers as a rapid and selective means for the determination of environmental contaminants [14-17]. One advantage of immunoassays for this type of testing is the highly specific binding that most antibodies have for a given class of analytes. Immunoassays are also easy to use and relatively cheap to perform. But conventional immunoassays can suffer from cross-reactivity between the analyte of interest and closely-related compounds in the sample (i.e., atrazine and its degradation products or related triazine herbicides) [17,18]. For this reason, most current immunoassays for environmental testing are used only for the screening of samples prior to analysis by a second technique (i.e., GC or RPLC).

This chapter will examine an alternative analytical technique for atrazine and its degradation products based on the combined use of RPLC and high-performance immunoaffinity chromatography (HPIAC) [19,20]. HPIAC is a method based on the use of immobilized antibodies in a column for the selective extraction of analytes from a sample [21-24]. This work will demonstrate how an HPIAC column can be developed for the rapid extraction of atrazine and its degradation products from water, and will show how this column can be coupled on-line with RPLC in order to separate and quantitate each of these analytes in a single run. In addition, various items that are important in the optimization and use of HPIAC/RPLC will be discussed. Finally, the performance of this method versus reference techniques will be examined and its potential advantages will be considered regarding the quantitation of atrazine and its degradation products in real-world samples.

Materials and Methods

Atrazine, deethylatrazine and deisopropylatrazine were supplied by Ciba-Geigy (Greensboro, NC). Hydroxyatrazine was obtained from Dr. James D. Carr at the Chemistry Department of the University of Nebraska (Lincoln, NE). The anti-atrazine monoclonal antibodies (cell line AM7B2) were produced at the Monoclonal Antibody Production Facility at the University of California-Berkeley (Berkeley, CA) [25]. HPLC-grade acetonitrile and methanol were obtained from Fisher

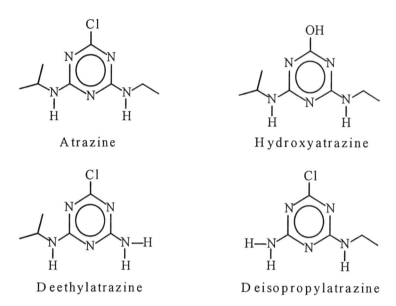

Figure 1.   Structures of atrazine and its major degradation products in water.

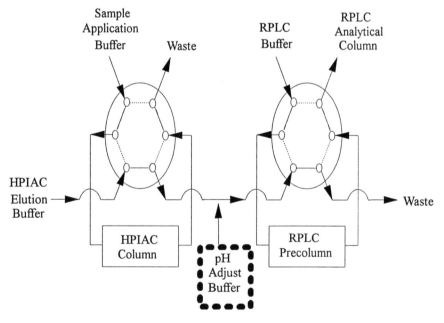

Figure 2.   Schematic of the HPIAC/RPLC system.

Scientific (Plano, TX). The Nucleosil Si-1000 silica (7 μm diameter, 1000 Å pore size) was from P. J. Cobert (St. Louis, MO). The RPLC columns were packed with a C_{18} Spherisorb support (3 μm diameter, 80 Å pore size) from Alltech (Deerfield, IL). The rabbit immunoglobulin G (IgG) used as a protein standard was from Sigma Chemical Co. (St. Louis, MO). Reagents for the bicinchoninic acid (BCA) protein assay were from Pierce (Rockfield, IL). All other chemicals were of the highest grades available. The deionized water used for solution preparation was generated by a Nanopure purification system (Barnstead, Dubuque, IA).

Instrumentation. The basic design of the HPIAC/RPLC system is shown in Figure 2. The system contained an HPIAC column for sample extraction, a small RPLC precolumn for trapping analytes as they eluted from the HPIAC column, and a RPLC analytical column for separating and quantitating the trapped analytes. The analysis was performed by first injecting the sample onto a 20 x 2.1 mm I.D. HPIAC column in the presence of a pH 7.0, 0.10 M phosphate application buffer. The retained analytes were later desorbed by passing a pH 2.5, 0.05 M phosphate elution buffer through the HPIAC column. As these compounds eluted from the HPIAC column they were trapped on a 50 x 4.6 mm I.D. C_{18} RPLC precolumn. The precolumn was then switched on-line with a 100 x 4.6 mm C_{18} I.D. RPLC analytical column and both were eluted with a mobile phase that contained a fixed amount of an organic modifier (e.g., acetonitrile or methanol) for separation of the retained analytes.

The HPIAC/RPLC system in this study consisted of two Model CM3200 pumps from LDC Analytical (Riviera Beach, FL) and one Model 350 SSI pump from Scientific Systems, Inc. (State College, PA). For degradation product determinations, an additional Model SSI 350 pump was used for adjusting the pH of the mobile phase between the RPLC precolumn and analytical column, as discussed later. Samples were applied by using a Rheodyne 7010 injection valve (Cotati, CA) operated either manually or through the use of an LDC 713 autosampler. A Rheodyne 5701 tandem switching valve equipped with a DVI actuator (Chromtech, Eden Prairie, MN) was used to control application of the mobile phases to the HPIAC and RPLC columns. The chromatographic supports were downward slurry-packed at 3500 psi using an Alltech column packer. Elution of analytes was monitored by an LDC SM3100x variable wavelength absorbance detector operated at 223 nm for atrazine or at 216 nm for its degradation products. All studies using this system were done at room temperature.

Preparation and Characterization of the HPIAC Column. The monoclonal anti-atrazine antibodies obtained from the University of California-Berkeley were purified by ammonium sulfate precipitation and immobilized onto diol-bonded Nucleosil Si-1000, as described previously [26]. After immobilization, the HPIAC support was washed several times with 2 M sodium chloride and 0.10 M phosphate buffer (pH 7.0), and stored at 4°C until further use. A portion of the silica was further washed with water, vacuum-dried at room temperature and assayed for protein content by a commercial BCA assay, using rabbit IgG as the standard and diol-bonded silica as the blank. The typical protein content of this support was about 10 mg antibody/g silica [19].

Table I. Frontal Analysis Results for the Anti-Atrazine HPIAC Column

Test Compound	Binding Capacity[a] (mol x 10^{-10})	Association Constant[a] (M^{-1} x 10^7)
Atrazine	1.8 ($\pm$ 0.2)	40 ($\pm$ 10)
Hydroxyatrazine	2.0 ($\pm$ 0.3)	6.9 ($\pm$ 0.2)
Simazine	1.7 ($\pm$ 0.2)	2.4 ($\pm$ 0.4)
Deethylatrazine	1.9 ($\pm$ 0.2)	0.9 ($\pm$ 0.1)
Deisopropylatrazine	2.0 ($\pm$ 0.3)	0.8 ($\pm$ 0.2)

[a]All numbers in parentheses represent ± 1 SD.

Frontal analysis was used to measure the binding activity of the HPIAC column for several triazine compounds. This was done by continuously applying each test solute at various concentrations to the HPIAC column at flow-rates of 0.10-0.20 mL/min and measuring the solution volume required to reach the apparent breakthrough point. These breakthrough volumes were corrected for the void volume of the system by using a diol-bonded silica column made from the same initial support as used in the preparation of the HPIAC column. From the resulting breakthrough volumes, the association equilibrium constant and binding capacity were obtained for each analyte, as described previously for other types of affinity columns [27,28].

The results of the frontal analysis work are summarized in Table I. All of the tested triazine compounds gave statistically identical binding capacities for this column (i.e., approximately 0.2 nmol). It is not surprising that consistent results were obtained for the binding capacity since monoclonal antibodies were used in preparing this column. Thus, either similar binding capacities or no measurable binding should have been seen for each solute. It was also possible from the frontal analysis experiments to estimate the association equilibrium constants for these solutes on the HPIAC column [27]. As shown in Table I, this particular column had the strongest binding for atrazine, followed by hydroxyatrazine, simazine, deethyl-atrazine and deisopropylatrazine. This trend is the same as reported for the cross-reactivity of these monoclonal antibodies when they are used in solution [17].

Determination of Atrazine by HPIAC/RPLC

Initial studies with the HPIAC/RPLC system focused on the quantitation of atrazine in water at concentrations at or near the EPA limit for this compound in drinking water (i.e., 3 ppb) [19]. Experiments performed in order to optimize the system for this application included an examination of those factors that affect the extraction of atrazine by the HPIAC column, the coupling of the HPIAC column with the RPLC precolumn, and the control of atrazine elution on the RPLC precolumn and analytical column.

Optimization of Conditions for the HPIAC Column. The degree of analyte binding to the HPIAC column was examined by injecting 250-μL of a 10 μg/L atrazine solution onto the HPIAC/RPLC system at various flow-rates. The fraction of atrazine bound at each flow-rate was determined by comparing the peak areas obtained under these conditions with those measured for direct injections of the same atrazine solution onto the RPLC analytical column. A gradual decrease in binding from 95 to 60% was seen as the flow-rate was increased from 0.1 to 4 mL/min. This behavior agrees with previous kinetic studies performed with other high-performance affinity columns [21,29]. As a compromise between assay speed and extraction efficiency, an injection flow-rate of 0.5 mL/min was selected for all further work in this section. At this flow-rate, 90% or more of the atrazine was extracted from a 250-μL sample containing 10 μg/L or less of this solute.

The conditions required to desorb atrazine from the HPIAC column were studied by injecting a 10 μg/L solution of this compound onto the HPIAC/RPLC system and desorbing the retained solute with pH 2.5, 0.05 M phosphate buffer at various flow-rates. Dissociation of bound atrazine from the HPIAC column was found to be fairly slow, with quantitative desorption requiring about 5 min, or 7.5 mL buffer applied at 1.5 mL/min. These conditions were used in all later work to minimize carryover of adsorbed solutes between samples.

Optimization of Conditions for the RPLC Precolumn and Analytical Column. The slow desorption of atrazine from the HPIAC column resulted in a peak that was too broad for direct quantitation by RPLC. However, it was found that atrazine could be reconcentrated back into a narrow plug by using the pH 2.5 desorption buffer eluting from the HPIAC column as a weak mobile phase for the application of atrazine onto a RPLC precolumn. After the atrazine was reconcentrated on the precolumn, this precolumn was then switched on-line with the RPLC analytical column and a stronger mobile phase, thus allowing the retained solutes to be eluted and separated. A similar approach has been used in the design of other HPIAC/RPLC systems [25,30,31]. Besides reconcentrating the solutes as they desorbed from the HPIAC column, this technique also helped to minimize the change in background signal that occurred when switching between the HPIAC and RPLC analytical columns and allowed the HPIAC column to be regenerated without adding any extra time to the overall analysis [19].

Experiments were performed on the RPLC columns to examine what mobile phase conditions were needed for reconcentration or the separation of atrazine from other triazine compounds. This was done by injecting these compounds onto the analytical RPLC column under isocratic conditions and in the presence of various amounts of organic modifier and pH 2.5 phosphate buffer. The capacity factors (k') for the injected solutes were then determined under each set of conditions. The results are shown in Table II. In related studies, deisopropylatrazine gave similar

Table II. Retention of Atrazine and Other Triazines on a C$_{18}$ RPLC Column

Compound	Capacity Factor (k')[a]			
	30% MeOH	40% MeOH	50% MeOH	60% MeOH
Atrazine	63	32	7.9	3.9
Simazine	25	13	4.4	2.4
Deethylatrazine	7.6	4.5	1.8	1.2
Hydroxyatrazine	3.0	1.7	0.6	0.5

[a]Data shown for various mobile phase mixtures of methanol (MeOH) and pH 2.5, 0.05 M phosphate buffer. The capacity factor was calculated as k' = (t_r - t_m)/t_m, where t_r is the compound's retention time and t_m is the column void time.

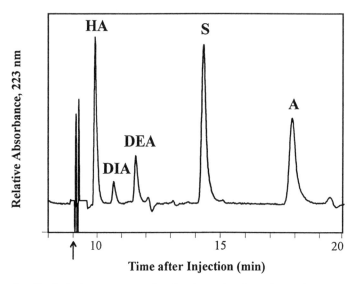

Figure 3.   Typical chromatogram for the determination of atrazine and related triazines by HPIAC/RPLC. The sample consisted of 250 µL river water spiked with 45 µg/L hydroxyatrazine (HA), 150 µg/L deisopropylatrazine (DIA), 40 µg/L deethylatrazine (DEA), 5 µg/L simazine (S) and 20 µg/L atrazine (A). The arrow indicates the time at which these components were applied onto the RPLC analytical column. (Reproduced with permission from Ref. 19)

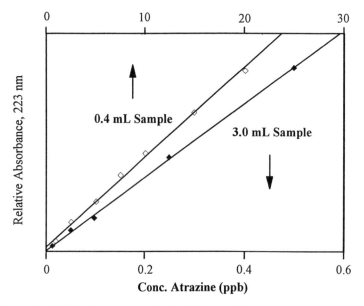

Figure 4.   Calibration curves obtained for atrazine on the HPIAC/RPLC system at various sample volumes.

behavior, with k' values that were between those for deethylatrazine and hydroxyatrazine under these conditions. Most of the test solutes showed high retention in the absence of any organic modifier (e.g., k' > 240 for atrazine), indicating that little or no elution of these compounds would occur on the RPLC precolumn when only pH 2.5 buffer was present. Table II indicates that a mobile phase containing at least 55% methanol and 45% pH 2.5 buffer was needed to resolve atrazine from other triazine compounds while allowing reasonable analysis times to be obtained (i.e., k' values of 10 or less for atrazine). A typical separation seen under these conditions is shown in Figure 3. In this case, baseline resolution was noted between atrazine, its major degradation products and simazine in less than 20 min after sample injection. However, the actual throughput was 10 min per injection since one sample could be applied to the HPIAC column while another was being separated on the RPLC analytical column. By adjusting the mobile phase further (i.e., by increasing the organic content to 60% methanol), it has been found that atrazine can be resolved from other triazines in as little as 12 min on the same HPIAC/RPLC system, with a throughput of 6 min per sample injection [19].

Performance of the HPIAC/RPLC System for Atrazine. Examples of some typical calibration curves that were obtained for atrazine on the final HPIAC/RPLC system are shown in Figure 4. When using a 250-μL sample injection, the linear range of this method (i.e., the range of analyte concentrations giving a response within ± 5% of the best-fit line) extended from 0 to 32 ppb. Under these same conditions, the lower limit of detection for atrazine at a signal-to-noise ratio of three was 0.1-0.2 ppb and the upper limit of detection extended to over 200 ppb. This range is comparable to many other analytical methods used for the routine analysis of atrazine and allows for work both above and below the EPA cut-off limit for atrazine in drinking water. The within-day precision of the HPIAC/RPLC method was evaluated by making 26 injections of a spiked river water sample, giving a value of ± 5.4% RSD at an atrazine level of 1.1 ppb. The within-day precisions at other atrazine concentrations were ± 4.1% at 0.5 ppb, ± 5.3% at 5.0 ppb and ± 7.8% at 20 μg/L (n = 3 for each level). Typical day-to-day precisions over the same concentration range were approximately 10-15%.

The accuracy of the HPIAC/RPLC method was first examined by comparing it to two reference techniques based on GC. Samples of well water, river water and surface water were analyzed in a double blind study, with one fraction being analyzed by HPIAC/RPLC and the other being measured by the reference method. For the results shown in Figure 5, a method based on GC/MS with solid-phase extraction [32] was compared with the HPIAC/RPLC system for 25 water samples containing atrazine levels between 0.2 and 9.1 ppb. The correlation coefficient between the two methods was 0.998, with a best-fit slope of 1.03 ± .02 (1 SD) and an intercept of 0.07 ± 0.20 μg/L. Similar results were obtained in a small scale study comparing HPIAC/RPLC and EPA Method 507 [8], with a correlation coefficient of 0.998 being obtained for nine water samples containing atrazine concentrations between 0.4 and 5 ppb. Accuracy was further evaluated by performing recovery studies on the HPIAC/RPLC system with spiked river water samples. Five replicate samples with an initial atrazine level of 0.7 ppb and spiked with an additional 3 ppb gave an average recovery of greater than 99%.

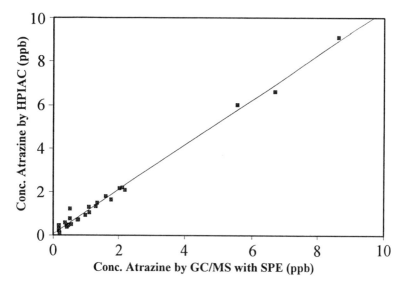

Figure 5. Correlation of HPIAC/RPLC with a GC/MS method using solid-phase extraction for the determination of atrazine in water. The parameters for the best-fit line are given in the text. (Reproduced with permission from Ref. 19)

Determination of Atrazine Degradation Products by HPIAC/RPLC

Early results with the HPIAC/RPLC system demonstrated that atrazine and related compounds, including its major degradation products, could be separated and analyzed in a single run by this approach (e.g., see Figure 3). However, the concentrations of the degradation products that were used in this earlier work were much higher than what would normally be expected in environmental samples. To address this issue, the goal of this section was to modify the HPIAC/RPLC system for the determination of these degradation products at environmentally-relevant concentrations. This involved making a number of changes in both the HPIAC and RPLC portions of the system in order to obtain the limits of detection that were required for these analytes.

Optimization of Conditions for the HPIAC Column. It was found earlier in the frontal analysis studies (see Table I) that some of the degradation products had much weaker binding than atrazine to the HPIAC column. This was of concern since it meant that these analytes might desorb during sample application, giving rise to a low extraction efficiency and poor detection limits. To study this problem, a 400 μL injection of each degradation product was made onto the HPIAC column in the application buffer followed by washing with the same buffer for various amounts of time before sample elution and quantitation by RPLC. For hydroxyatrazine and atrazine, less than a 2% decrease in the final peak area was noted when using wash times between 4 min and one hour in length. For deethylatrazine and deisopropylatrazine, less than a 5% decrease in peak area was observed over the same range of wash times. Thus, the degradation products did not have any significant desorption under the typical sample application times that were to be used in this study (i.e., 4 to 45 min).

One potential method for improving the detection limit for the degradation products was to increase the volume of sample applied to the HPIAC column, as illustrated for atrazine in Figure 4. In this case, a 7.5-fold increase in the volume of sample applied to the HPIAC column resulted in a corresponding decrease in the detection limit for atrazine from 170 to about 20 parts-per-trillion (pptr). This effect arises from the strong nature of most antibody-analyte interactions, which makes the binding of many immobilized antibody supports essentially irreversible on the time scale of a typical HPIAC run. Although using large sample volumes can increase the time required for sample injection, the increase in the total analysis time can be minimized by also increasing the flow-rate used for sample application. However, as discussed earlier for atrazine, care must be exercised in doing this since the binding efficiency may decrease as less time is allowed for the sample to interact with the HPIAC column.

Optimization of Conditions for the RPLC Precolumn and Analytical Column. The greater polarities of hydroxyatrazine, deethylatrazine and deisopropylatrazine versus atrazine meant that these solutes had less retention on the RPLC columns than their parent compound (see Table II). This creates two possible difficulties when using HPIAC/RPLC for the analysis of these solutes. First, weak retention on the RPLC precolumn could result in a low capture efficiency as these solutes elute from the HPIAC support. Second, weak retention could cause these compounds to elute near the void volume of the RPLC analytical column, making them difficult to detect if any changes in the background signal are produced as a result of the column switching event [19]. One system parameter that was altered in order to deal with these problems was the mobile phase pH used on the RPLC precolumn and analytical column (see Table III). The main analyte of concern was hydroxy-atrazine because of its low retention on the RPLC analytical column under the conditions used for the analysis of atrazine (see Table II). The results in Table III show that the retention of hydroxyatrazine on a C_{18} RPLC column increases by over two-fold between pH 2.5 and 7.0, while the other degradation products and atrazine show little or no change in retention. This is not surprising since hydroxyatrazine has a net positive charge at pH 2.5 but exists in a predominantly neutral form at pH 7.0 [33]. In contrast to this, none of the other test solutes have any significant changes in their net charge over the same pH range [12,13].

Table III. Effect of pH on Atrazine & Related Products on a RPLC Column

Compound	Capacity Factor (k')[a]			
	pH 2.5	pH 3.9	pH 5.6	pH 7.0
Atrazine	27	30	25	23
Deethylatrazine	4.4	4.6	4.6	4.6
Deisopropylatrazine	2.2	2.3	2.3	2.2
Hydroxyatrazine	1.5	1.9	3.2	3.3

[a]Data shown for mobile phase mixtures of 20% acetonitrile and 80% 0.10 M phosphate buffer at the given pH applied to a C_{18} RPLC column. The value of k' was calculated in the same fashion as shown in Table II.

Although the data in Table III indicate that pH 7.0 is preferred for the analysis of hydroxyatrazine by RPLC, this created a slight problem in work with the HPIAC/RPLC system since a pH 2.5 buffer was still required for the desorption of retained species from the HPIAC column. To deal with this, an approach was developed in which the pH 2.5 buffer coming off the HPIAC column was mixed with a second buffer stream (at a higher pH and higher molarity) prior to entry of the desorbed analytes onto the RPLC precolumn (see Figure 2). This approach made it possible to separately adjust the pH of the HPIAC elution buffer and the RPLC precolumn application solvent. In this work a pH 7.0, 0.10 M phosphate buffer was added to the pH 2.5, 0.05 M elution buffer in an equal volume ratio, providing a final pH of 6.8-7.0 for the solvent entering the RPLC precolumn.

A further increase in the retention of degradation products on the RPLC columns was made by lowering the amount of organic modifier that was used during their final separation. In addition, the modifier was changed from methanol to acetonitrile because of acetonitrile's lower background signal at the wavelength used for monitoring the degradation products. All of the degradation products gave fairly high retention (k' = 23-115) when only pH 7.0 phosphate buffer was applied to the RPLC columns, as required for the reconcentration of these solutes on the RPLC precolumn. When a 25:75 to 20:80 mixture of acetonitrile and pH 7.0 buffer was used, the capacity factors for the degradation products fell within the range of 1.5-4.7, giving both acceptable resolution and reasonable analysis times (see Figure 6).

Under the same conditions, atrazine had a capacity factor of 13.1-22.1 and could either be monitored along with the degradation products or allowed to wash from the RPLC analytical column between sample injections.

Performance of the HPIAC/RPLC System for Atrazine Degradation Products. Typical calibration curves obtained for 13-mL samples of hydroxyatrazine, deethylatrazine and deisopropylatrazine are shown in Figure 7. Under these conditions, the detection limits at a signal-to-noise ratio of three were 20-30 pptr for each analyte. The linear range extended up to about 240 pptr, and the upper limit of detection extended up to at least 1 ppb. This range of detection was sufficient for the analysis of atrazine degradation products in river water samples. When a larger sample volume of 45 mL was used, the lower limit of detection was 6 pptr for hydroxyatrazine and 10 pptr for deethylatrazine or deisopropylatrazine, making it possible to use this system in the analysis of ground water samples (see Figure 6). The same HPIAC/RPLC system could detect atrazine at pptr levels. For example, a 3-mL sample of atrazine gave a detection limit of 20 pptr, a linear range extending from 0-800 pptr and an upper detection limit of at least 2.5 ppb.

Accuracy and precision were evaluated for the HPIAC/RPLC system in the detection of atrazine degradation products in the same fashion as described earlier for the work with atrazine. The within-day precision at a level of 50 pptr in a 13-mL sample was ± 12%, ± 8.2% or ± 6.9% (1 RSD, n = 6) for hydroxyatrazine, deethylatrazine or deisopropylatrazine, respectively. The day-to-day precision over nine days of operation for the same samples was ± 11%, ± 9.5% or ± 6.2%. A series of river water samples was used in a double bind study to compare the results measured by HPIAC/RPLC with those measured by a GC/MS reference technique [34]. For deethylatrazine, the analysis of nineteen samples by both methods gave a correlation plot with a slope of 1.11 ± 0.06 (1 SD), an intercept of 30 ± 60 and a correlation coefficient of 0.9790 over sample concentrations ranging from 0-590 pptr. For deisopropylatrazine, a study using twenty-one samples gave a slope of 1.01 ± 0.05, an intercept of 10 ± 30 and a correlation coefficient of 0.9768 over concentrations ranging from 0-580 pptr. Although no reference method was available for hydroxyatrazine, similar correlation would be expected for this compound based on the results reported here and in earlier work [19]. In addition, recovery studies were performed with two sets of five replicate river water samples containing each degradation product at an initial level of 100 pptr and spiked with 500 pptr of each analyte. The average recoveries obtained were 98%, 87% and 87% for deisopropylatrazine, hydroxyatrazine and deethylatrazine, respectively.

Throughout the studies with both atrazine and its degradation products, no significant interferences were detected from other major classes of pesticides or pollutants [19]. Although some other triazine compounds also absorbed to the HPIAC column (e.g., cyanazine, simazine and propazine), these did not interfere with detection of atrazine or its degradation products since these other compounds eluted at different times on the RPLC analytical column. In addition, the HPIAC columns used in this work were found to be quite stable, with consistent performance being obtained over 700-1000 sample injections and over a 10-12 month period of time. Only a gradual decrease in column activity was noted, but this slow loss of activity was not a problem since the HPIAC columns originally

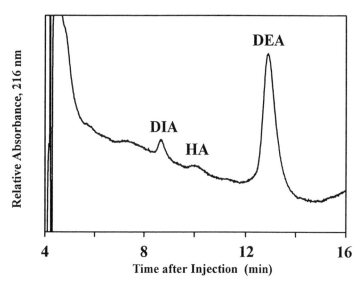

Figure 6. Typical chromatogram for atrazine degradation products in a 45-mL ground water sample. The concentrations measured for deethylatrazine (DEA), hydroxyatrazine (HA) and deisopropylatrazine (DIA) were 210 pptr, 10 pptr and 60 pptr, respectively. The mobile phase used on the RPLC analytical column was a 20:80 mixture of acetonitrile and pH 7.0, 0.10 mM phosphate buffer.

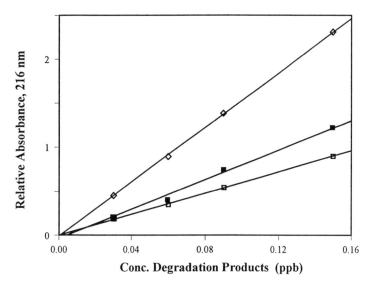

Figure 7. Calibration curves obtained with hydroxyatrazine (◇), deisopropylatrazine (■), deethylatrazine (□) for a 13-mL sample injection.

contained a large excess of binding sites relative to the amount of injected triazines and adjustments for small decreases in activity were made automatically whenever a standard curve was prepared.

Conclusions

It has been shown in this chapter how the combination of HPIAC and RPLC can provide a rapid tool for examining atrazine and its degradation products in environmental samples. One advantage of HPIAC/RPLC versus standard immunoassay methods (e.g., ELISA) is that it allows for the simultaneous and separate determination of several solutes within a given compound class. Advantages of HPIAC/RPLC versus GC methods include its ability to directly analyze solution-phase samples and its need for little or no sample pretreatment or derivatization prior to injection. HPIAC/RPLC also overcomes many of the current, general difficulties encountered during pesticide degradation studies. These include the low concentrations of analytes that must be monitored and the large difference in polarities that may be present between the degradation products and their parent compound. Furthermore, by using a different HPIAC column and modifying the RPLC elution conditions, this same approach could easily be adapted for use with other classes of pesticides or pollutants and their degradation products.

Acknowledgments

The authors wish to thank to Dr. Alex Karu at the University of California-Berkeley for the gift of the anti-atrazine antibodies. The support of Drs. James D. Carr, Roy F. Spalding and Zongwei Cai at the University of Nebraska and Candy Stock at Midwest Laboratories is also acknowledged regarding the provision of the samples and GC data that were used in this work. This work was funded by a Sect. 104 grant from the U.S. Geological Survey.

Literature Cited

1. *Agricultural Chemicals in Ground Water: Proposed Pesticide Strategy*; U. S. Environmental Protection Agency: Washington, D.C., 1987; pp. 1-150.
2. Fielding, M.; Barcelo, D.; Helweg, A.; Galassi, S.; Torstenson, L.; van Zoonen, P.; Wolter, R.; Angeletti, G. in *Pesticides in Ground and Drinking Water (Water Pollution Report 27)*; Commission of the European Communities: Brussels, 1989; pp. 16-34.
3. Somasundram, L.; Coats, J. R., Eds. *Pesticide Transformation Products: Fate and Significance in the Environment*; American Chemical Society: Washington DC, 1991; Chapter 1.
4. Wallace, L. W. *Environ. Health Perspect.* **1991**, *95*, 7-13.
5. U. S. Environmental Protection Agency, *National Survey of Pesticides in Drinking Water Wells, Phase II Report, EPA 570/9-91-020*; National Technical Information Service: Springfield, VA, 1992.
6. Cook, A. M. *FEMS Microbiol. Rev.* **1987**, *46*, 93-116.

7. Erickson, L. E.; Lee, K. H. *Crit. Rev. Environ. Control* **1989**, *19*, 1-14.
8. Barcelo, D. *J. Chromatogr.* **1993**, *643*, 117-143.
9. Lerch, R. N.; Donald, W. W.; Li, Y.-X.; Alberts, E. E. *Environ. Sci. Technol.* **1995**, *29*, 2759-2768.
10. Sabik, H.; Cooper, S.; Lafrance, P.; Fournier, J. *Talanta* **1995**, *42*, 717-724.
11. Karalangis, G.; Von Arx, R.; Ammon, H. U.; Camenzind, R. *J. Chromatogr.* **1991**, *549*, 229-236.
12. Schewes R.; Maidl, F. X.; Fischbeck, G.; von Gleissenthall, J. L.; Süss, A. *J. Chromatogr.* **1993**, *641*, 89-93.
13. Pichon, V.; Chen, L.; Guenn, S.; Hennion, M.-C. *J. Chromatogr.* **1995**, *711*, 257-267.
14. Thurman, E. M.; Meyer, M.; Perry, C.; Schwab, P. *Anal. Chem.* **1990**, *62*, 2043-2048.
15. Bushway, R. J.; Perkins, B.; Savage, S. A.; Lekousi, S. J.; Ferguson, B. S. *Bull. Environ. Contam. Toxicol.* **1988**, *40*, 647-654.
16. Dunbar, B. D.; Niswender, G. D.; Hudson, J. M. *U. S. Patent* *4,530,786*, 1985.
17. Lucas, A. D.; Jones, A. D.; Goodrow, M. H.; Saiz, S. G.; Blewett, C.; Seiber, J. N.; Hammock, B. D. *Chem. Res. Toxicol.* **1993**, *6*, 107-116.
18. Gascon, J.; Barcelo, D. *Chromatographia* **1994**, *38*, 633-636.
19. Thomas, D. H.; Beck-Westermeyer, M.; Hage, D. S. *Anal. Chem.* **1994**, *66*, 3823-3829.
20. Thomas, D. H.; Lopez-Avila, V.; Van Emon, J. *J. Chromatogr.* **1996**, *724*, 207-218.
21. De Alwis, U.; Wilson, G. S. *Anal. Chem.* **1987**, *59*, 2786-2789.
22. Hage, D. S.; Walters, R. R. *J. Chromatogr.* **1987**, *386*, 37-49.
23. Ohlson, S.; Gudmundsson, B.-M.; Wikstrom, P.; Larsson, P.-O. *Clin. Chem.* **1988**, *34*, 2039-2043.
24. Hage, D. S.; Kao, P. C. *Anal. Chem.* **1991**, *63*, 586-595.
25. Johansson, B. *J. Chromatogr.* **1986**, *381*, 107-113.
26. Walters, R. R. *Anal. Chem.* **1983**, *55*, 591-592.
27. Loun, B.; Hage, D. S. *Anal. Chem.* **1994**, *66*, 3814-3822.
28. Chaiken, I. M., Ed. *Analytical Affinity Chromatography*; CRC Press: Boca Raton, 1987.
29. Hage, D. S.; Walters, R. R. *J. Chromatogr.* **1988**, *436*, 111-135.
30. De Frutos, M.; Regnier, F. E. *Anal. Chem.* **1993**, *65*, 17A-25A.
31. Janis, L. J.; Regnier, F. E. *J. Chromatogr.* **1988**, *444*, 1-11.
32. Shepherd, T. R.; Carr, J. D.; Duncan, D.; Pederson, D. T. *J. AOAC International* **1992**, *75*, 581-583.
33. Her, N. H. Structure and Properties of Hydroxyatrazine, M. S. Thesis, University Nebraska, Lincoln, Nebraska, 1994.
34. Morris, R. T. Simultaneous Determination of the Herbicide Atrazine and Two of Its Degradation Products, Deethylatrazine and Deisopropylatrazine, in Water and Air Samples by Gas Chromatography/Mass Spectrometry, Ph. D. Thesis, University Nebraska, Lincoln, Nebraska, 1995.

Chapter 11

On-Line Immunochemical Detection System for Pesticide Residue Analysis

Development and Optimization

Petra M. Krämer[1], Renate Kast[1], Ursula Bilitewski[1], Stephan Bannierink[2], and Ulrich Brüss[2]

[1]Gesellschaft für Biotechnologische Forschung mbH, Mascheroder Weg 1, 38124 Braunschweig, Germany
[2]Meta GmbH, Oststrasse 7, 48341 Altenberge, Germany

The prospect of an automated immunochemical detection system for on-line analysis of pesticide residues has important implications for environmental monitoring and control. Over the past seven years a flow injection immunoanalysis (FIIA) system was developed and optimized in our laboratories; this led to the development of a prototype of a flow injection immunoaffinity analysis (FIIAA) device. The strategy followed during this development is described here. First results of the newest version of this system are presented. The advantages and limitations of FIIAA compared with conventional analytical techniques, such as LC and GC/MS, microtiter plate ELISA, and the commercially available test-kits are also discussed. The currently undergoing demonstration for real water samples, and the potential of this FIIAA system for water monitoring will be outlined.

Immunochemical detection methods for environmental analysis have gained acceptance over the past few years (*1*). This is true for the commercially available test-kits, such as the magnetic particle-based and microtiter plate based assays. Both formats are very useful tools for fast screening of samples at contaminated sites (*2, 3*), and also for quantitative analysis in the laboratory (*4, 5*). However, most commercial immunoassay kits are not applicable for automated analysis.

Our goal was to develop an additional format for immunochemical determination which would allow on-line pesticide monitoring in aqueous solutions, e.g. in groundwater and surface water. This technique is based on flow injection immunoanalysis (FIIA), where the antibodies are immobilized on solid supports (*6-8*).

The triazine herbicides, which are used intensively worldwide were selected as the target analytes for this study. Conventional analysis of these compounds by liquid chromatography (LC) and gas chromatography (GC) has been well established thus allowing easy confirmation of results. In addition, many sensitive immunoassays for s-triazines are available for comparison (e.g., *9-11*). Although the most widely used herbicide atrazine has been banned in some European countries, e.g. in Germany and in Denmark, its usage is still permitted in most European countries, e.g. in France, Greece, Italy, Portugal, Spain, The Netherlands, etc. It has been detected in groundwaters (*12-14*), and it is expected that atrazine and its metabolites (deethylatrazine, deisopropylatrazine) will be found in ground- and surface water supplies for many years (e.g., *12, 14*). Atrazine can also act as an indicator for the presence of other pesticide residues, such as alachlor, cyanazine, metolachlor, and metribuzin. These herbicides are also used in large quantities (*15*), and all except metribuzin are classified as probable or transient leachers (*14*).

Immobilization and Assay Strategies

A number of different assay and immobilization techniques were explored during the development of the FIIA. The following gives an account of the overriding strategy behind this development. For detailed information regarding the early research, readers are directed to the literature cited.

Immobilization of Polyclonal Antibodies on Membranes. Initial studies involved the immobilization of polyclonal antibodies on immunoaffinity membranes (Immunodyne or Fluorotrans transfer membrane, PALL Corp. Glen Cove, NY), either directly or via protein A (*6, 7*). This procedure was combined with the exchange of the membrane after each assay, and a special membrane exchanging mechanism was designed for this purpose (*7*). The main target analyte was atrazine, together with propazine and simazine, which also cross-reacted with these polyclonal antibodies. The system worked in the sequential saturation technique: analyte and enzyme-tracer were incubated sequentially with the antibodies (*16*). This implies, that the analyte and enzyme-tracer are not incubated together and therefore do not compete simultaneously for the limited binding sites of the antibodies. This method is different from the usual technique used in microtiter plate formats.

The enzyme used was horseradish peroxidase (HRP), with H_2O_2 and 3-(p-hydroxyphenyl) propionic acid (HPPA) as fluorescent substrate (*17*). The fluorescence of the product was determined downstream in a fluorometer flow through cell (λ_{ex} 320 nm, λ_{em} 404 nm). This procedure allowed the measurement of atrazine and propazine at the very low ppb level (0.1 µg/L), which is relevant for the European drinking water directive (*18*). However, this technique required large amounts of antibodies, that were about 130 times higher than in the corresponding enzyme-linked immunosorbent assay (ELISA); (*7*)), and the reproducibility of immobilization was not comparable with that achieved with the microtiter plate format. Furthermore, it was critical to exchange the membrane without the presence of the operator, which made it difficult for unattended operation. These problems needed to be overcome before the development of a prototype instrument as an automated, unattended control system was initiated.

Immobilization of Monoclonal Antibodies on Polystyrene or Glass Beads.
Following literature studies, the immobilization of antibodies on beads was explored
as an alternative to the use of membranes to obtain better performance and decrease
consumption of antibodies (*19*). Monoclonal anti-terbutryne antibodies (mab K1F4;
(*10*)) were immobilized either on polystyrene or glass beads, with diameters of about
220 μm or from 250-500 μm, respectively, and the antibodies were regenerated after
each assay with glycine/HCl (pH 2.1; (*8*)). This system offered several advantages
over the membrane method. The antibody consumption was reduced, because at least
60 regenerations of the antibodies were possible. Secondly, the unattended operation
time was extended, because no exchange mechanism was needed for approximately
one day. The reproducibility also improved as a zero standard concentration could be
determined prior to each sample analysis, providing a control value. Although the
absolute signals [mV] of the zero standards showed much variation (Figure 1), the
coefficient of variation (CV) of the corresponding %control-values for the pesticide
standards was only about 10%. The result of this %control-value was independent of
the decline in signal due to high analyte concentration of the preceding sample,
regeneration of the antibodies, or a decrease in enzyme-tracer activity.

These promising results prompted us to begin the prototype development of
this device for on-line monitoring of pesticide residues in aqueous solutions. At this
point it was noted that the use of another monoclonal antibody, namely the anti-
atrazine antibody (K4E7), caused problems.   Each antibody-hapten system has
different stability and regeneration characteristics, therefore each individual system
must be optimized, which could present difficulties to an untrained user. A new
strategy was therefore investigated with the aim of producing a universal system.

Immunoaffinity Column with Protein A or G. As an alternative to the regeneration
of the antigen-binding sites of the antibodies, a more universal system is the
regeneration of the specific binding site of protein A or G to the Fc region of IgG.
This system is very well established and utilized for affinity columns such as for
antibody purification. The binding of a specific class or subclass of IgG to protein G
or A is well characterized, with an affinity binding that is high enough (approximately
10^8 M^{-1}; (*20*)) for measurements in a flow system.

Protein G was immobilized on glass beads using the glutardialdehyde
immobilization method described earlier for antibody immobilization (*8*). Briefly,
silanized glass beads (150-212 μm) were activated with glutardialdehyde (2.5% v/v),
the beads were washed and protein G (2 mg/mL) was incubated for 4 h at room
temperature. These beads were then incubated with 1.5% bovine serum albumin
(BSA) to avoid nonspecific binding of protein. A column reactor was filled with this
material, and then anti-atrazine antibodies (mab K4E7; (*11*)) were pumped into this
protein G column, where they were incubated for 2 min. The atrazine and then the
enzyme-tracer were pumped sequentially through the column, followed by the
substrate for the enzyme-tracer.   Again, the product formation was determined
downstream in a flow through cell of a fluorometer (λ_{ex} 320 nm, λ_{em} 415 nm;
Figures 2, 3). With this system, the independence of the regeneration of the antibody-
antigen system was gained, because the antibodies were replaced after each assay, and
protein G was regenerated. It was possible to regenerate the protein G column over

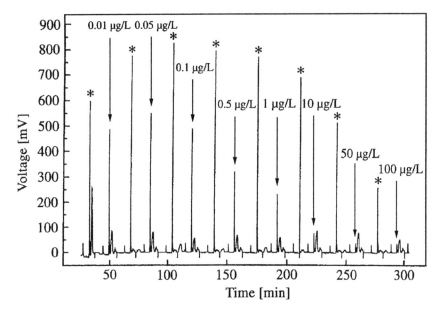

Figure 1. Flow injection immunoanalysis with regeneration of the antigen-binding site. Peak profile of a complete standard curve for terbutryne. Arrows show the peak for the corresponding concentration of terbutryne. Prior to each standard a zero concentration (✽) was determined.

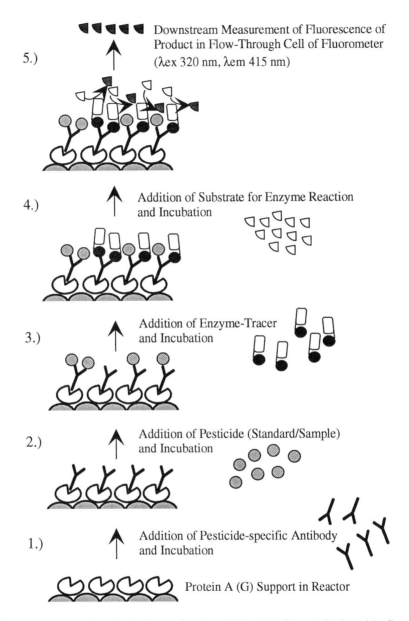

5.) Downstream Measurement of Fluorescence of Product in Flow-Through Cell of Fluorometer (λex 320 nm, λem 415 nm)

4.) Addition of Substrate for Enzyme Reaction and Incubation

3.) Addition of Enzyme-Tracer and Incubation

2.) Addition of Pesticide (Standard/Sample) and Incubation

1.) Addition of Pesticide-specific Antibody and Incubation

Protein A (G) Support in Reactor

Figure 2. Schematic illustration of sequential saturation analysis with flow injection immunoaffinity analysis (FIIAA). Steps (1-5) are performed automatically and controlled via computer. A regeneration step follows after step 5, and the protein A (G) column is then ready for the next incubation with antibodies (step 1).

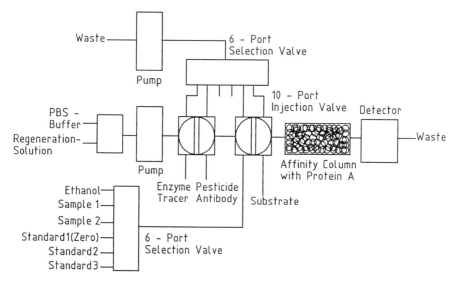

Figure 3. Most recent flow chart of the automated on-line monitoring system for one analyte, using a temperature-controlled (15°C) immunoaffinity column with protein A, where the immunochemical reaction takes place. Two sample inlets (1 and 2) are included in this system.

200 times with only little decrease in the binding capacity of the column, and with good reproducibility of the protein G regeneration. This system, with immobilized protein G (A) on the column, was designated FIIAA (flow injection immunoaffinity analysis; Figure 2).

Recently, we used a commercially available protein A affinity material. In the acidic pH range (≤ 5.5), the affinity interaction of IgG to immobilized protein A is weaker compared to immobilized protein G (20), which makes it more suitable for regeneration (antibody elution from the column). Using protein A on this affinity material, at least 1000 regenerations with 0.1 M sodium citrate (pH 2.5) are possible. High antibody and enzyme-tracer dilutions (1:20,000 to 1:80,000) can be used, which are comparable to the conventional microtiter plate ELISA. Therefore, the consumption of antibodies and enzyme-tracer is no longer a critical issue. This system was tested in an alternating mode of a zero standard and a 0.1 µg/L atrazine standard (Figure 4), which is a simulation of the on-line operation, where only minor concentration changes will usually occur. Four measurements of a zero standard were necessary before the system showed stable signals (Figure 4). After the system was stabilized, the reproducibility of the signal was very good (CV 2.4%). In addition, analysis of alternating atrazine concentrations was performed, again with a prior determination of zero concentration (Figure 5). This performance illustrates the system as in the automated laboratory version, where different concentrations will be measured in a random sequence. It should be noted though, that the concentration difference of the standards in this example was up to 500 times (0.02 µg/L was determined after 10 µg/L). Although the absolute signals (mV) of the zero concentration values and of the different atrazine concentrations showed great variations during the runs, the standard deviation (SD) was acceptable, when the atrazine standard was expressed as %control-value of the preceding zero standard (Table I).

Table I.   %Control-Values of Figure 5

Concentration [µg/L]	First Run %Control[a]	Second Run %Control[a]	Average %Control[a] ± SD
0.5	42.2	37,9	40.1 ± 3.0
0.05	77.6	79.2	78.4 ± 1.1
10	10.4	12.6	11.5 ± 1.6
0.02	83.2	79.5	81.4 ± 2.6
1	34.6	34.0	34.3 ± 0.4
0.1	74.3	67.7	71.0 ± 4.7

[a] $\%\text{Control} = \dfrac{\text{signal [mV] of standard/sample}}{\text{signal [mV] of preceding zero standard}} \times 100$

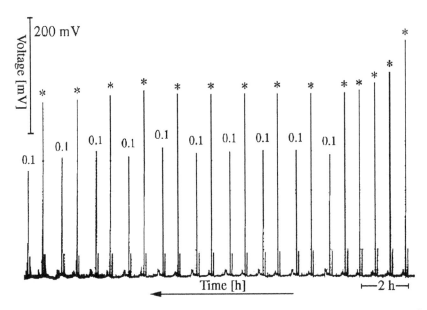

Figure 4. Scan of an original flow chart of a sequence of zero concentrations (✱) and an atrazine standard of 0.1 µg/L in flow injection immunoanalysis. A run of about 17 h is shown (right corner, bar represents 2 h). The maximal height of the zero signal corresponds to 400 mV.

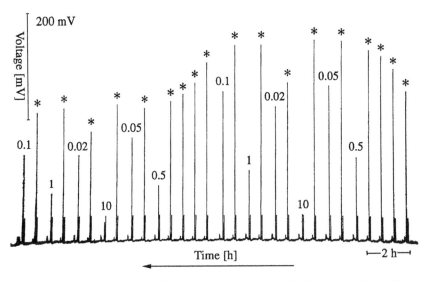

Figure 5. Scan of an original flow chart of a sequence of alternating high and low standard concentrations with a preceding zero concentration (✱). This trace shows a run of 20 h (right corner, bar represents 2 h). The maximal height of the zero signal corresponds to 400 mV.

The average CV was ca. 6%, ranging from 1.2% to 13.9%, which is within the acceptable range of the European drinking water directive, which allows a CV of 50% at the 0.1 µg/L concentration level. These results are also acceptable, if the more recently expressed aim is considered, which said, that the total error (random and systematic) should be around 20%, in order to be compatible with the directive (*14*).

An even better reproducibility was obtained, when several runs were performed with one analyte concentration. Table II gives a set of data for a standard curve for atrazine with five determinations for each concentration.

Table II. %Control-Values of Atrazine Standards[a]

Concentration [µg/L]	Average %Control ± SD[b]	CV %
0.01	94.3 ± 3.2	3.4
0.05	76.3 ± 1.6	2.1
0.1	57.2 ± 2.8	4.9
0.2	36.5 ± 2.2	6.0
0.5	21.8 ± 0.8	3.7
1.0	19.5 ± 0.6	3.1
5.0	11.7 ± 0.9	7.7
10	19.6 ± 1.2	6.1

[a]FIIAA conditions: mab K4E7 1:80,000 (100 µL, 2 min incubation), enzyme-tracer 1:80,000 (200 µL, 8 min incubation), analyte (zero standard) 4x250 µL, 4x2 min incubation), substrate (200 µL, 6 min incubation). Column was temperature controlled at 15°C.
[b]n=5

It should be mentioned that high concentrations of analyte (10 µg/L) effect the values of the succeeding zero concentration. However, this should not create a problem during the on-line measurement of real water samples, because 10 µg/L exceed the maximum allowable concentration by a factor of 100, thus causing an alarm with a subsequent attendance by the operator (e.g., rinsing of the system, exchange of the column etc.). Secondly, the background value after regeneration is very low, which is a proof for complete regeneration. There is also a signal decline over the time period measured, which is not caused by an activity loss of the protein A column, but by the aging of the immunochemical reagents (antibody and enzyme tracer). This was verified by the fact, that the signal reached its original value after repeated addition of freshly prepared immunoreagents. The acceptance criterion for the zero standard in this set-up was that it should be at least 100 mV. Below this signal, a sufficient activity of the reagents (antibodies and enzyme-tracer) could not be guaranteed.

In addition, different times and temperatures for incubation of antibody, analyte, enzyme-tracer, and substrate have been investigated, and the optimization of these parameters is ongoing. Incubation times used more recently were 2 minutes for antibody, 8 minutes each for analyte and enzyme-tracer, and 3 minutes for the substrate reaction. The results from these conditions are demonstrated in Figures 4 and 5, which show the scans of original traces.

Stabilization of Immunochemicals

As mentioned previously, the signal starts to decline after a few hours. This is mainly caused by the usage of high dilutions of antibody and enzyme-tracer solutions. These solutions have to be stabilized for measurements over a 7-14 day period. Therefore, we surveyed the literature for different stabilization methods (*21, 22*). Figure 6 gives examples of antibody stabilizations with different additives, which were tested in the microtiter plate format. Here, a 1:30,000 diluted antibody solution (K4E7) was prepared with the addition of different stabilizers. Each of these stabilizers was tested against a reference solution which was freshly prepared each day (set as 100%; Figure 6). As shown in Figure 6, an addition of 1% BSA was very useful for stabilization of the antibody solution, whereas the addition of 10% glycerol had no stabilizing effect. Addition of 0.25 M trehalose gave only a slight increase in stability.

An optimization was also carried out for the stabilization of enzyme-tracer solutions (HRP). Here, as already demonstrated by Gibson at al. (*23*), lactitol and zinc ions gave promising results (data not shown). These stabilization studies are still going on, and the results are not sufficient yet.

Lately, we started to keep the enzyme-tracer and antibody solutions at 4°C, which kept the signals stable for about one week.

Advantages and Limitations of the Flow Injection Immunoaffinity Analysis (FIIAA)

FIIAA offers a potential method for pesticide residue analysis, either on-line or as an automated laboratory instrument. FIIAA has several advantages over conventional analysis. First, increase in inherent sensitivity (up to 1000 fold) reduces the sample size requirement, from liters to a few milliliters. Second, sample pretreatment usually is not required, except for a filtration step, resulting in a quicker and simpler analysis.

Another advantage, which will gain even more importance in the future, is the enormous reduction in organic solvent consumption during the analysis. In FIIAA, only ca. 100 µL of solvent are necessary for the set-up of standards (atrazine or diuron). The rinsing of the system needed 1 mL of 50% ethanol for each sample, but recently this step was left out in FIIAA. The analysis itself is carried out in aqueous buffer solutions (40 mM phosphate buffered saline). In conventional analysis for atrazine and/or diuron, organic solvents are used for the sample preparation and for the analysis itself, and the solvent need is generally in the range of 50-100 mL per sample, depending upon the method. Organic solvents are an environmental hazard, thus, their use should be minimized.

Compared to conventional microtiter plate ELISA and test-kit formats, FIIAA offers the advantage of on-line monitoring and automation. On-line monitoring allows analysis of water samples on site with sufficient frequency to enable water supply personnel to take remedial action, thus improving drinking water protection.

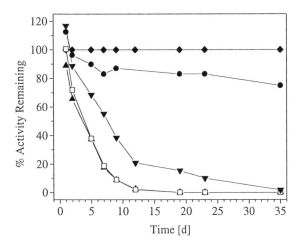

Figure 6. Stabilization of the monoclonal antibody K4E7 using different stabilizers. ◆ Reference (100%): freshly prepared K4E7 with no additives; ▲ 10% glycerol; □ K4E7 with no additives, only in phosphate buffered saline; ● 1% BSA; ▼ 0.25 M trehalose.

In addition, plastic-waste, which is still an unsolved problem for test-kit formats, is not used in FIIAA, which contributes to the environmental friendliness of this method.

All analytical methods and formats have advantages and limitations. For example, for the on-site screening of contaminants in soil commercial test-kits with their corresponding extraction kits are much better and faster than FIIAA. Furthermore, the use of conventional microtiter plate formats is recommended, when hundreds of samples have to be checked at once. At least 20 samples can be measured in about two hours in one microtiter plate, whereas FIIAA measures the samples sequentially at about 1 h per sample determination.

Conventional analytical methods, such as LC or GC/MS (mass spectrometry), can provide an exact determination of several analytes within one run, but on-line systems for LC and GC are difficult to use and very expensive. Automation and on-line control via immunoanalysis though is easier and cheaper compared to LC and GC.

On the one hand, immunochemical analysis is specific for one compound, but an additional response can be obtained from structurally related compounds. This cross-reactivity is often perceived as a disadvantage, but it also can be used to an advantage for the screening of a group of similar compounds.

Essentially the choice of format is dependent upon the analytical task. It is envisaged that FIIAA will be incorporated into the tools of the analytical laboratory, where it can offer some distinct advantages over currently available techniques.

Prospects for the Near Future

The main focus of our development strategy was atrazine and other s-triazine herbicides, because despite of the ban of atrazine in some European countries, it is predicted that groundwater contamination with atrazine residues and metabolites will still be found 5 to 10 years after its use. The increasing application of phenylurea herbicides (e.g., diuron) as substitutes for atrazine has prompted us to investigate this group of compounds as well. Currently, two different prototype FIIAA systems are being used in a European demonstration project within the program *Life**, one as an on-line system for control of water ways, the other as an automated laboratory instrument for off-line measurements of samples. The on-line device will be used for the analysis of diuron in surface water. This instrument is combined with a refrigerator unit to keep the reagents cool, and it will be located in a water control station. The second prototype (automated laboratory instrument) will be used for the analysis of atrazine in groundwater samples. This study will be carried out at different locations in The Netherlands and in Germany. Currently, a two-analyte device capable of monitoring atrazine and diuron is being developed at Meta GmbH.

Conclusion and General Note

Two prototypes of a FIIAA device for the analysis of pesticide residues in water have been developed. The instruments are computer-controlled via a specifically designed software. These systems are currently being used in a two-year demonstration project to determine the reliability and robustness of this new analytical tool in real water samples.

All immunosensor developments and immunochemical techniques such as FIIAA are dependent upon the supply with suitable antibodies. In addition, most immunosensors need an enzyme-tracer or at least supplementary haptens for the detection system. Thus, the commercial availability of antibodies and haptens for environmentally relevant compounds may lead to the development of novel biosensors and other analytical techniques suitable for environmental monitoring.

Acknowledgments

We wish to thank Dr. T. Giersch and Prof. B. Hock (TU Munich, F.R.G.) for providing the anti-triazine-antibodies K1F4 and K4E7. During this long-term project, many colleagues were involved, whom we would like to thank at this point, namely M. Dietrich, M. Stiene, Dr. M. Böcher, B. Pawletta, D. Hanisch, O. Eikenberg, and especially A. Ahlers. Financial support from the "Bundesministerium für Bildung, Wissenschaft, Forschung und Technologie" (BMBF) and the European Union (program *Life**) is gratefully acknowledged.
* *Life* - the purpose of the financial instrument *Life* is to contribute to the development and implementation of the Community environmental policy.

Literature Cited

1. Van Emon, J. M.; Gerlach, C. L. *Environ. Sci. Technol.* **1995**, *29*, 312A.
2. Meulenberg, E. P.; Stoks, P. G. *Anal. Chim. Acta.* **1995**, *311*, 407.
3. Itak, J. A.; Olson, E. G.; Fleeker, J. R.; Herzog, D. P. *Bull. Environ. Contam. Toxicol.* **1993**, *51*, 260.
4. Karu, A. E.; Schmidt, D. J.; Richman, S. J.; Cooper, C.; Tran, D.; Hsu, J. *J. Agric. Food Chem.* **1994**, *42*, 310.
5. Hill, A. S.; Skerritt, J. H.; Bushway, R. J.; Pask, W.; Larkin, K. A.; Thomas, M.; Korth, W.; Bowmer, K. *J. Agric. Food Chem.* **1994**, *42*, 2051.
6. Krämer, P.; Schmid, R. *Biosens. Bioelectron.* **1991**, *6*, 239.
7. Krämer, P. M.; Schmid, R. D. *Pestic. Sci.* **1991**, *32*, 451.
8. Dietrich, M.; Krämer, P. M. *Food Agric. Immunol.* **1995**, *7*, 203.
9. Huber, S. J.; Hock, B. In *Methods of Enzymatic Analysis*; Bergmeyer, H.U., Ed.; VCH Verlagsgesellschaft mbH: Weinheim, F.R.G., 1986, Vol XII; pp 438-451.
10. Giersch, T.; Hock, B. *Food Agric. Immunol.* **1990**, *2*, 85.
11. Giersch, T. *J. Agric. Food Chem.* **1993**, *41*, 1006.
12. Haberer, K.; Normann, S.; Schmitz, M. *Wasser + Boden.* **1988**, *5*, 258.
13. Fielding, M.; Barceló, D.; Helweg, A.; Galassi, S.; Torstensson, L.; van Zoonen, P.; Wolter, R.; Angeletti, G.   Pesticides in Ground and Drinking Water, Water Pollution Report 27, Commission of the European Communities, Brussels, Belgium, 1992, pp 1-136.
14. Barceló, D. In *Environmental Analysis: Techniques, Applications and Quality Assurance;* Barceló, D. (Ed.). Elsevier Science Publishers: Amsterdam, The Netherlands, 1993, pp 149-180.
15. Squillace, P.J.; Thurman, E.M. *Environ. Sci. Technol.* **1992**, *26*, 538.
16. Tijssen, P. In *Practice and Theory of Enzyme Immunoassays*; Burdon, R.H., van Knippenberg, P.H. (Eds.); Laboratory Techniques in Biochemistry and Molecular Biology; Elsevier Science Publishers B.V.: Amsterdam, New York, Oxford, 1985, Vol. 15, pp 139-149.
17. Zaitzu, K.; Ohkura, Y. *Anal. Biochem.* **1980**, *109*, 109.
18. Commission of the European Communities, *EEC Drinking Water Directive*, **1980**, 80/778/EEC, EEC No.L229, 11-29, EEC, Brussels, Belgium.
19. Plant, A. L.; Locascio-Brown, L.; Haller, W.; Durst, R. A. *Appl. Biochem. Biotechnol.* **1991**, *30*, 83.
20. Hermanson, G. T.; Krishna Mallia, A.; Smith, P. K. *Immobilized Affinity Ligand Techniques*. Academic Press, Inc.: San Diego, CA, 1992, pp 245-250.
21. Schein, C. H. *Bio/Technology.* **1990**, *8*, 308.
22. Manning, M. C.; Patel, K.; Borchardt, R. T. *Pharm. Res.* **1989**, *6*, 903.
23. Gibson, T. D.; Hulbert, J. N.; Woodward, J. R. *Anal. Chim. Acta.* **1993**, *279*, 185.

Chapter 12

Near-Infrared Fluorescent Immunoassays: A Novel Approach to Environmental Analysis

A. R. Swamy[1], M. I. Danesvar[1], L. Evans III[1], L. Strekowski[1], N. Narayanan[2], Ferenc Szurdoki[3], I. Wengatz[3], Bruce D. Hammock[3], and G. Patonay[1]

[1]Department of Chemistry, Georgia State University, Atlanta, GA 30303
[2]LI-COR Inc., 4421 Superior Street, Lincoln, NE 68504
[3]Departments of Entomology and Environmental Toxicology, University of California, Davis, CA 95616

The near-infrared (NIR) region of spectrum provides many advantages over conventional UV-visible region and is particularly useful when background interference is the primary concern. Many NIR fluorophores have high molar absorptivities and fluorescence quantum yields. The use of laser diodes for excitation and photodiodes for detection makes the NIR fluorophores an ideal choice as tracers for various applications. NIR fluorescence immunoassays (NIRFIAs) for disease-related antigens have been developed and optimized in our laboratory during the past few years. NIRFIA can be used in both sandwich and competitive formats. We have successfully adapted the NIRFIA as both a microtiter plate assay and a fiber optic probe. This paper is focussed on the environmental analytical applications of this novel technique. Physical properties of ideal NIR labels and engineering problems of devising remote sensors are discussed.

Conventional immunoassays have been recognized as a powerful tool in the analysis of minor components in a complex media. The use of laser induced fluorescence in the development of immunoassays has been reviewed (1). The limitations in using conventional lasers as an excitation source are size, high price, and maintenance costs. Recent advances in semiconductor laser technology have made the use of these lasers as an excitation source more practical.   The widespread use of near-infrared(NIR) emitting semiconductor laser diodes in telecommunication has made them more readily available (2).   They are inexpensive, compact and last much longer than conventional lasers.   The GaAlAs laser diode has drawn considerable interest because its emission wavelength of 750 - 870 nm is in the NIR region.

Only a small number of compounds in complex biological and environmental media are known to exhibit intrinsic fluorescence in the NIR region and cause background interference (3-4).   As a result, detection limits are dependent almost solely on the limitations of the instrumentation (e.g. detector noise).   The low interference in this spectral region makes NIR fluorophores ideal probes for both biological and environmental applications.

Heptamethine cyanines are a class of NIR fluorophores that have been used for DNA sequencing, pH and hydrophobicity determination, metal ion detection, and antibody labeling (5).   Methods for the synthesis of NIR dyes are well established (6-9).   They have high molar absorptivities (approx. $10^5 M^{-1} cm^{-1}$), relatively high quantum yields (20-40%), and short fluorescence lifetimes (500-1000 ps) (10).   The structure of the NIR dye 1 used in our studies is shown in Figure 1.   The absorption and emission spectra of dye 1 are shown in Figure 2.

Silicon photodiodes have good responsivity in the NIR region compared to conventional photomultiplier tubes (11).   A detection system comprising a silicon photodiode, emission filters and confocal optics were used to measure fluorescence from the NIR dye labeled hapten.   This detection system reduced the analysis time as well as the number of experimental steps.   Picomolar concentrations of the dye-labeled antibodies could be easily detected, which is comparable with the sensitivity of enzyme linked immunosorbent assay (ELISA).

Some of the challenges in working with the NIR dyes are problems associated with the hydrophobicity and photostability of these dyes.   Another challenge is the lack of commercial instrumentation for the near-infrared fluorescence immunoassay (NIRFIA).

Experimental

Instrumentation.   The setup used to measure NIR fluorescence from microtiter plates consisted of three main components: a proprietary scanning fluorescence

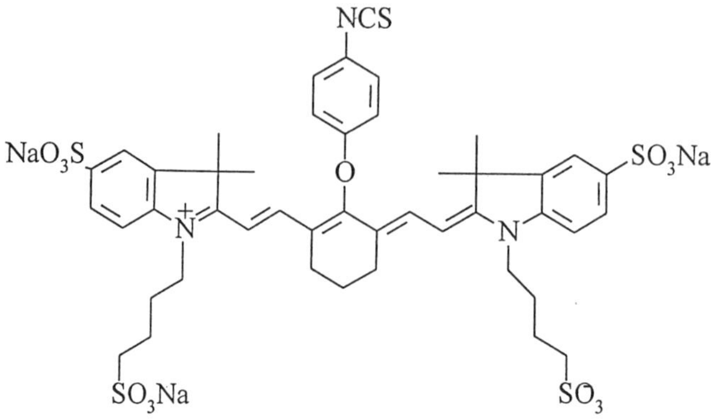

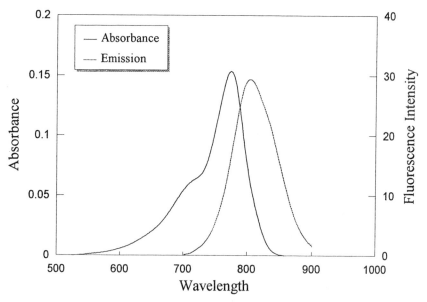

NIR Dye 1

Figure 1. Structure of NIR dye 1.

Figure 2. Absorption and emission spectra of Dye 1.

microscope (Model 4000X, LI-COR Inc, Lincoln, NE), an orthogonal scanner, an infrared analyzer and a data acquisition interface (Figure 3). The excitation source was a laser diode emitting at 785 nm(10-20 mW) and mounted on the detector microscope at an angle such that the focussed polarized radiation strikes the target at Brewsters' angle (56°) to minimize scattering. The detector in the fluorescence microscope was an avalanche photodiode cooled by a 3-stage Peltier thermoelectric cooler with detection optics. The detection optics included a 20 mm focal length aspheric objective lens, two bandpass filters (820 ± 10 nm) to eliminate scattered light from excitation source, and a focussing lens. The fluorescence microscope was mounted on a scanning platform with variable scan speeds (1.5 - 15 cm/hr). The fluorescence microscope was coupled with a orthogonal scanner programmed for nine separate scan speeds (4 - 260 cm/hr), allowing considerable flexibility in image resolution and acquisition time. The infrared analyzer had a built in lock-in amplifier and an analog to digital converter. The data acquisition interface was connected to a Pentium computer (IBM) via an IEE 488 (GPIB) cable.

The setup for the immunosensor approach comprises a laser diode for excitation, a fluorescence microscope for detection (similar to the one used for the microtiter plates), two optical fibers connected at their distal ends, and a digital voltmeter for measuring the signal (Figure 4).

A Perkin-Elmer Lambda 2 UV/VIS/NIR spectrophotometer (Norwalk, CT) was used for absorption measurements. The spectrophotometer was interfaced to a Zenith 286 computer with a proprietary software program (PECCS) from Perkin Elmer to gather data and control the spectrophotometer. Fluorescence measurements were made on an ISS K2 multi frequency phase fluorometer (Champaign, IL). The fluorometer was interfaced to a Gateway-486 computer equipped with proprietary software from ISS for data acquisition and controlling the fluorometer.

Application of the NIRFIA for Pesticide Analysis

Assay for Bromacil. Bromacil, also known as Hyvar [5-bromo-6-methyl-3-(1-methylpropyl)-2,4(1H,3H)-pyrimidinedione](**2**, Figure 5) is used for weed control worldwide (*12-14*). This herbicide is both persistent and mobile in soil depending on climate and properties of soil (*14*). There has been increasing concern about ground

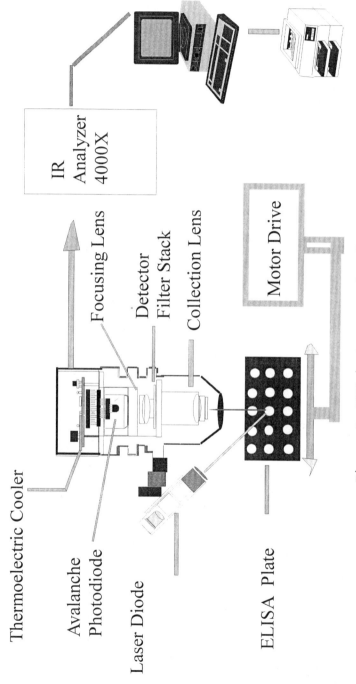

Model assembly for reading ELISA plates

IR Analyzer 4000X

Thermoelectric Cooler

Avalanche Photodiode

Laser Diode

Focusing Lens

Detector Filter Stack

Collection Lens

Motor Drive

ELISA Plate

Figure 3. NIR instrumentation for scanning ELISA plates.

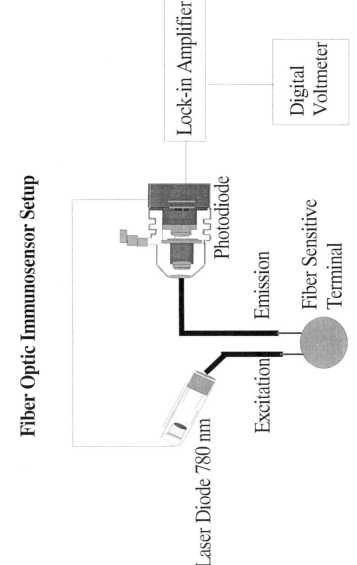

Figure 4. NIR-FFOI instrumentation setup.

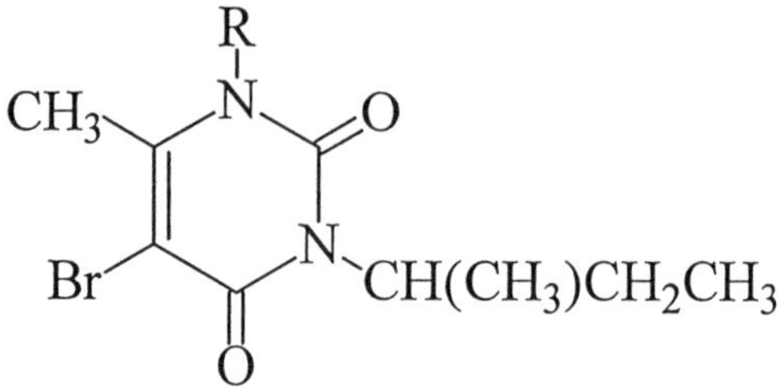

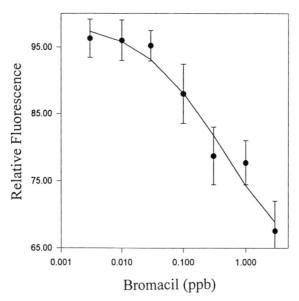

bromacil (2): R = H
bromacil hapten (3): R = $(CH_2)_5CO_2H$

Figure 5. Structure of bromacil and bromacil hapten.

Bromacil NIRFIA

Figure 6. Plot of relative fluorescence intensity versus bromacil concentration.

water contamination due to leaching of bromacil from soil (*14, 15*). A number of instrumental methods have been developed for the detection of bromacil in soil and water samples (*14, 16-20*). These analyses share drawbacks such as the need of laborious extraction and/or cleanup procedures, highly qualified analysts, and expensive instrumentation. These methods could not be adapted as field-portable sensors. Hammock and co-workers have developed immunoassays for the analysis of bromacil at trace levels which provided a low cost alternative for screening large numbers of environmental samples (*21-23*). The assay system was highly sensitive (IC_{50} of 0.25 ppb) and selective for bromacil (*21-22*). Using this assay, water samples spiked with 0.1-160 ppb levels of bromacil can be analyzed directly and 0.04-20 ppm concentrations of bromacil can be detected in soil with only a single extraction step (*22*).

A similar assay for bromacil was devised in the NIRFIA format in our laboratories(*24*) by employing the polyclonal antibodies developed for the ELISAs (*21-22*). Bromacil-hapten 5-(5-bromo-6-methyl-3-(1-methylpropyl)-2,4-(1*H*,3*H*)- pyrimidinedione-1-yl)hexanoic acid (**3,** Figure 5) was covalently linked to the amino-groups of bovine serum albumin (BSA). The resulting **3-BSA**-conjugate was then purified by extensive dialysis and treated with the NIR-dye **1** to produce the bromacil-BSA-dye-conjugate (**3-BSA-1**). The product was purified by gel filtration. The product containing about 8 dye molecules bound to the macromolecular carrier was then used as a fluorescent tracer in the NIRFIA-experiments. The different steps involved in the assay were optimized. Microtiter plates were precoated with a secondary antibody solution (goat-anti-rabbit-antibody). The plates were washed and blocked with ovalbumin solution. The plates were washed again and coated with a solution of polyclonal rabbit antiserum #2369 (*21,22*). Excess antiserum was washed off and the plates were incubated with the solution of the bromacil standard and the tracer (**3-BSA-1**). The excess tracer was washed off and the fluorescence signal from the plates was measured. A plot of the signal versus bromacil concentration is shown in Figure 6. The standard curve demonstrated that the NIRFIA can detect less than 1 ppb of bromacil and the sensitivity of the assay is comparable to that of the analogous ELISA system (*22*).

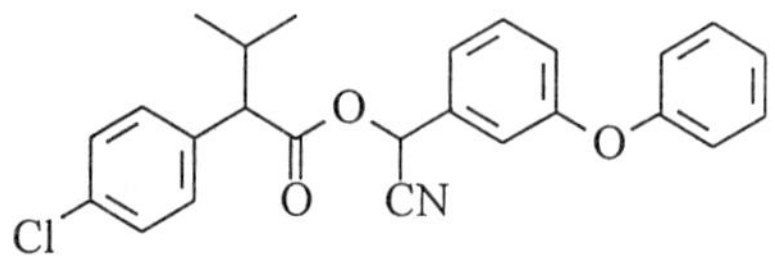

fenvalerate (4)

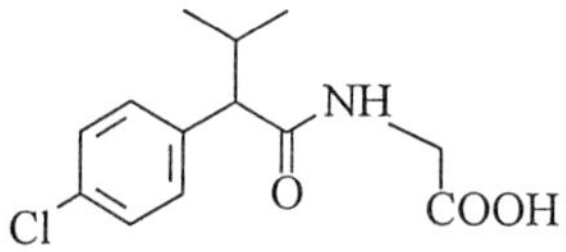

fenvalerate metabolite (5)

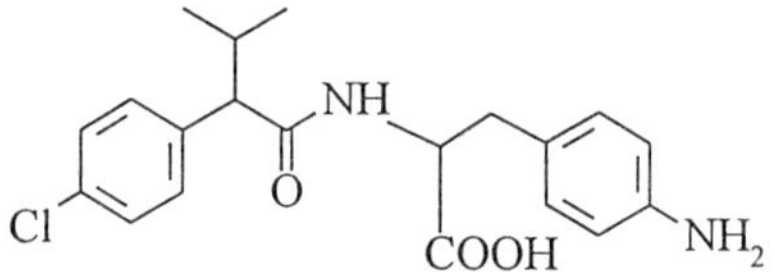

fenvalerate hapten (6)

Figure 7. Structure of fenvalerate,
fenvalerate metabolite and fenvalerate hapten

Assay for a Pyrethroid Metabolite. Fenvalerate [α-cyano-3-phenoxybenzyl 2-(4-chlorophenyl)-3-methylbutyrate] (**4**, Figure 7), a synthetic pyrethroid, is more photostable than the naturally occurring pyrethrum from which it was designed. Fenvalerate is a highly potent insecticide used predominantly in crop protection. Conventional methods for the detection of pyrethroids such as gas chromatography (GC) and high performance liquid chromatography (HPLC) involve multistep sample cleanup procedures. A simple immunoassay for the detection of the pyrethroid permethrin in meat has been reported (*25*).

Metabolites are used as biomarkers to monitor environmental and human exposure to xenobiotics (*26*). We have developed a NIRFIA and compared it to ELISA for the detection of a fenvalerate metabolite [*N*-(Carboxymethyl)-2-(4-chlorophenyl)-3-methylbutyramide] (**5**, Figure 7), in urine samples (*27*).

In order to produce a tracer for the assay the metabolite analogue *N*-[2-(4-aminophenyl)-1-carboxyethyl]-2-(4-chlorophenyl)-3-methylbutyramide (**6**, Figure 7) was covalently bound via its amino group to aldehyde groups generated on the surface of the enzyme horseradish peroxidase (HRP) using the NaIO$_4$-method (*28*). The enzyme conjugate (**6-HRP**) was purified by gel filtration. A solution of **6-HRP** in PBS was treated with slightly basic carbonate/bicarbonate buffer and an excess of the reactive dye **1**. The tracer product was purified by gel filtration using PBS as eluent. The coupling reaction was confirmed by performing a normal ELISA, which showed a successful competition with the analyte. The dye/HRP ratio of 2.8 was determined from the absorbance value of the conjugate solution at 780 nm.

In the assay, microtiter plates were first coated with a secondary antibody solution (goat-anti-rabbit-antibody) in carbonate buffer. After washing the plate, the surface of the wells were blocked with an ovalbumin solution. The plate was washed and then incubated with anti-pyrethroid antiserum (*27*). Following another washing step, analyte standard solution (0.5-500 ppb) was pipetted into the wells. Subsequently, the **6-HRP-1** tracer was added and incubated, followed by a washing step. Plates prepared this way were used either for NIRFIA or for ELISA experiments. In the ELISA substrate reaction, 3,3',5,5'-tetramethylbenzidine (TMB) and hydrogen peroxide were dissolved in an acetate buffer and added to each well. The reaction was stopped by adding sulfuric acid. The optical

Pyrethroid NIRFIA

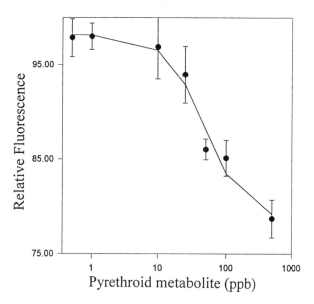

Figure 8. Standard plot of fluorescence
intensity versus pyrethroid concentration

Fiber Optic Probe

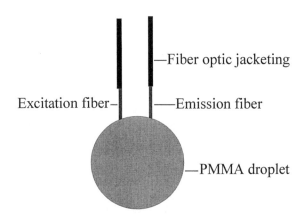

Figure 9. Schematic of the FFOI probe.
(Reproduced with permission from ref.4.
Copyright 1996 Plenum Press.)

densities of each well on the plates were read with an ELISA reader at 650 nm.

For the NIRFIA, microtiter plates were read with the NIR fluorescence reader. Images of the plates were obtained by scanning the plates with the setup shown in Figure 3. The standard curve showed that the NIRFIA assay could easily detect less than 10 ppb of pyrethroid (Figure 8).

The Immunosensor Approach. Conventional immunoassays require multiple steps. The immunosensor approach reduces the number of steps resulting in a decrease in analysis time and errors. Immunosensors are a class of biosensors which utilize the binding of an antigen or hapten to its corresponding antibody on the sensor surface. Most immunosensors are based on optical or electrical detection techniques. Optical immunosensors coupled to fiber optic have received considerable attention (*29*). NIR fluorescence fiber optic immunosensor (NIR-FFOI) offers the advantages of detection of extrinsic fluorescence with little or no background interference, making them highly sensitive (*30*). We have designed a simple, practical, sensitive and selective NIR-FFOI. Design of the NIR-FFOI involves a double stranded optical fiber. The strands of poly(methyl methacrylate) (PMMA) fiber are stripped off their jackets and their ends joined with a PMMA droplet (3g PMMA in 10 mL of dichloromethane) (Figure 9). The sensing end is activated by treatment with hydrochloric acid, which converts the methoxy carbonyl groups on the surface of the polymer to carboxylic acid groups, followed by covalent coupling of the capture antibody (goat-anti-human IgG) on the sensing tip (Figure 10). The FFOI is then exposed to the antigen (Human IgG). The antigen-coated FFOI is introduced to a second antibody (goat-antihuman IgG) labeled with NIR dye **1**. After washing off any unbound component, the fluorescence intensity from the probe is measured. The intensity of fluorescence is directly proportional to the concentration of bound antigen. For human IgG, detection limits of 10 ng/mL were achieved easily in less than 15 minutes (Figure 11).

Conclusions The use of NIR labeled antibodies and haptens for analysis of environmental contaminants has been studied. The NIR region offers considerable advantages over working in the visible region plagued with the problem of background interference. The chemistry and the instrumentation of the NIRFIA technique have been

NIRFIA Immunosensor Assay

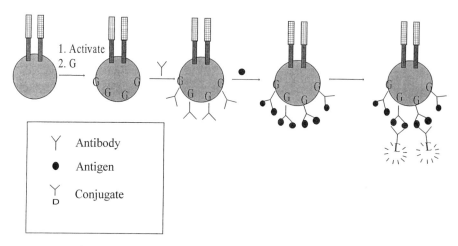

Figure 10.Antibody immobilization and
formation of fluorescent complex on FFOI.
(Reproduced with permission from ref.4.
Copyright 1996 Plenum Press.)

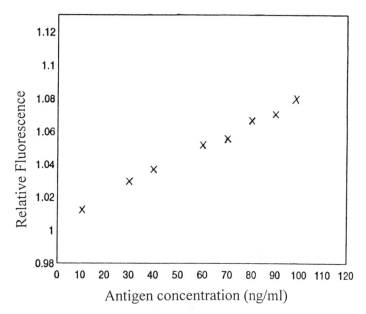

Figure 11. Plot of relative fluorescence
intensity versus antigen (IgG) concentration.
For each IgG concentration three identical
FFOIs were tested with SD= ±0.003.
(Reproduced with permission from ref.4.
Copyright 1996 Plenum Press.)

developed, but are amenable for further optimization and miniaturization. This would help to build compact instruments for analysis and to conduct fast and reliable assays at remote locations.

Protein labeling for fluorescence immunoassay calls for water soluble dyes. The dye used in our studies is water soluble, but also has hydrophobic moieties. The hydrophobicity of the dye leads to nonspecific binding in a protein rich environment. Preliminary evidence supporting this hypothesis was obtained in some of our experiments. This issue is currently being addressed in our laboratory by appropriate modifications in dye structure.

Preliminary results demonstrate the feasibility of using the NIRFIA approach for pesticide analysis and fabrication of an NIR-FFOI. Currently we are investigating a combination of the two approaches to develop a compact remote sensing FFOI for environmental analysis.

Acknowledgments

Dr. Lyle Middendorf is gratefully acknowledged for providing us with NIR dye 1 and helpful advice. The authors also wish to thank Dr. Jose Mauro Peralta for his valuable suggestions and assistance with NIRFIA experiments. Partial support was provided by the NIEHS Superfund Basic Research Program (ESO4699). U.C. Davis is a NIEHS Health Science Center (ESO5707) and an EPA Center for Environmental Health (CR819658).

Literature Cited

1. Hemmila, I. A. *Applications of Fluorescence in Immunoassays*; John Wiley and Sons Inc: New York, NY, **1991**.

2. Svelto, O. *Principles of Lasers*; Plenum press: New York, NY , **1989**.

3. Patonay, G.; Antonie, M. D. *Anal. Chem.* **1991**, *63*, 321A-327A.

4. Danesvar, M. I.; Casay, G. A.; Patonay, G. *Proc. SPIE* **1994**, *vol 2068*, 128-138.

5. Casay, G. A.; Shealy, D. B.; Patonay, G. In *Topics in Fluorescence Spectroscopy*; Lakowicz, J. R., Ed.; Plenum Press: New York, NY, Vol 4; 183-222.

6. Strekowski, L.; Lipowska, M.; Patonay, G. *J. Org. Chem.* **1992**, *57*, 4578-4580.

7. Hamer, F. M. *The Cyanine Dyes and Related Compounds;* John Wiley and Sons Inc: New York, NY **1991**.

8. Tyutulkov, N.; Fabian, J.; Melhorn, A.; Dietz, F.; Tajder, A. *Polymethine Dyes, Structure and Properties;* St. Kliment Ohridski University Press: **1991**.

9. Narayanan, N.; G. A.; Patonay, *J. Org. Chem.* **1995**, *60*, 2391-95.

10. Shealy, D. B.; Lohrmann, R.; Ruth, J.; Narayanan, N.; Casay, G. A.; Evans III, L.; Patonay, G. *Appl. Spec.* **1995**, 49, 1815-20.

11. *Topics in Fluorescence Spectroscopy;* Lakowicz J. R., Eds.; Plenum Press: New York, NY. **1983**, Vol *1*.

12. Gardiner, J. A. In *Herbicides. Chemistry, Degradation, and Mode of Action;* Kearney, P. C.; Kaufman, D. D., Eds.; Marcel Dekker Inc.: New York, NY, **1995**; pp 293-322.

13. *The Pesticide Manual. A World Compendium;* Worthing, C. R.; Hance, R. J., Eds.; The British Crop Protection Council: Farnham, Surrey, UK, **1991**.

14. James, T. K.; Lauren, D. R. *J. Agric. Food Chem.* **1995**, *43*, 684-690.

15. Tan, S.; Singh, M. *Bull. Environ. Contam. Toxicol.* **1995**, *55*, 359-365.

16. Pease H. L. *J. Agric. Food Chem.* **1966**, *14*, 94-96.

17. Goewie, C. E.; Hogendoorn, E. A. *J. Chromatography* **1987**, *410*, 211-216.

18. Wylie, P. L.; Oguchi, R. *J. Chromatogr.* **1990**, *517*, 131-42.

19. Stan, H. J.; Heil, S. *Fresenius J. Anal. Chem.* **1991**, *339*, 34-39.

20. Foster, G. D.; Foreman, W. T.; Gates, P. M. *J. Agric. Food Chem.* **1991**, *39*, 1618-1622.

21. Szurdoki, F.; Bekheit, H. K. M.; Marco, M. P.; Goodrow, M. H.; Hammock, B. D. *J. Agric. Food Chem.* **1992**, *40*, 1459-1465.

22. Bekheit, H. K. M.; Lucas, A. D.; Szurdoki, F.; Gee, S. J.; Hammock, B. D. *J. Agric. Food Chem.* **1993**, *41*, 2220-2227.

23. Szurdoki, F.; Bekheit, H. K. M.; Marco, M. P.; Goodrow, M. H.; Hammock, B. D. In *New Frontiers in*

Agrochemical Immunoassay; Kurtz, D. A.; Skerritt, J. H.; Stanker, L., Eds.; AOAC International: Arlington, VA, **1995**, pp 39-63.

24. Wengatz, I.; Szurdoki, F.; Swamy, A. R.; Evans III, L.; Patonay, G.; Stimmann, E.; Delwiche, M.; Stoutamire, D.; Gee, S.; Hammock, B. D. In *Advances in Fluorescence Sensing Technology II. Proc. SPIE;* Lakowicz, J. R.; Ed.; SPIE-The International Society of Optical Engineering: Bellingham, WA, **1995**, vol 2388; pp 408-416.

25. Stanker, L. H.; Bigbee, C.; Emon, J. V.; Watkins, B.; Jensen, R. H.; Morris, C.; Vanderlaan, M. *J. Agric. Food Chem.*, **1989**, *37*, 834-839.

26. Harris, A. S.; Wengatz, I.; Wortberg, M.; Kreissig S.; Gee, S. J.; Hammock, B. D. In *Effects of Multiple Imapacts on Ecosystems*, Cech J. J.; Wilson B. W.; Crosby, D. G.; Eds., Lewis Publishers: Chelsea, MI, accepted.

27. Wengatz, I.; Stoutamire, D. W.; Hammock, B. D. In preparation.

28. Tijssen, P. *In Practice and theory of enzyme immunoassay;* Elsevier: Amsterdam, The Netherlands,**1985**.

29. Wingard, L. B.; Ferrance, J. P. *Biosensors with optics*, Wise, D. L.; Wingard, L. B., Eds.; Humana Press: Clifton, **1991**, pp 1-27 and 111-138

30. Golden, J. P.; Shriver-Lake, L. C.; Narayanan, N.; Patonay, G.; Ligler, F.S. Proc. SPIE; Lakowicz, J. R.; Ed.; SPIE-The International Society of Optical Engineering: Bellingham, WA, **1994**, *vol 2138*; 241-250.

Chapter 13

Optimization of Electrochemiluminescence Immunoassay for Sensitive Bacteria Detection

Hao Yu

Calspan Systems Research Laboratories Corporation, Building E3549, Edgewood Research, Development and Engineering Center, Aberdeen Proving Ground, MD 21010

Electrochemiluminescence Immunoassay(ECLIA) offers high sensitivity for detection of antigens in solution. However, conventional ECLIAs developed for bacteria detection are not optimal and many factors which may affect the ECLIA results still remain unclear. In this report, ECLIA kinetics, antibody biotinylation and non-specific binding were investigated. Results of these studies showed that the system optimization in ECLIA could substantially increase detection limit by at least 10-fold compared to conventional ECLIA.

Electrogenerated chemiluminescence (ECL) as a highly sensitive detection technology has received considerable attention in chemical analysis and clinical diagnostics (1-4). Applications of ECL immuno- and nucleic acid-based assays for biological and environmental sample analysis were reported (5-7).

The principle of ORIGEN ECL (IGEN, Gaithersburg, MD) has been described previously(4). Briefly, an ECL employs a metal chelate, ruthenium (II)-trisbipyridyl, $(Ru(bpy)_3^{2+})$, in redox reaction which are conducted on the surface of an anode in the presence of electron carriers, tripropylamine (TPA). Both $(Ru(bpy)_3^{2+})$ and TPA are oxidized to $(Ru(bpy)_3^{3+}$ and TPA+, respectively. Deprotonated TPA+ spontaneously becomes TPA·, which creates a high energy state $(Ru(bpy)_3^{3+*})$. Relaxation of $(Ru(bpy)_3^{3+*})$ emits photons at 622 nm on the surface of the electrode. In the ECL Immunoassay (ECLIA), antibody-coated magnetic particles are used as primary capturing antibody and $(Ru(bpy)_3^{2+})$-conjugated antibodies, (Ru-Ab), as Tag-label to generate the ECL signal. The quantity of captured antigens is determined by measuring the ECL intensity at 622 nm. Sample media does not contribute any ECL signals. Unlike the fluorescence and chemiluminescence assays, the ECLIA have higher signal to noise (S/N) ratios and better reproducibilities (4, 6). That is because the noise contributed by natural fluorescence from biological samples remains at a minimum in ECLIA. Furthermore, unlike conventional chemiluminescence assays, the ECLIA reproducibilities are under well-controlled electric potentials and does not depend upon the substrate concentration or reaction time. Additionally, unlike Radio Immunoassay (RIA), the ECLIA is an non-radio immunoassay and is safe to use.

We have demonstrated that the ECLIA can be used for sensitive bacteria detection. Detection of biotoxoids at femtogram level. Detection of 100 spores/mL in buffer, and 2000 cells/mL of *E. coli* in environmental and food samples have been reported (5-7). This report focuses on the study of assay kinetics, non-specific binding, and varying molar incorporation ratios (biotin to protein (b/p)) in order to optimize ECLIA for the detection of biological agents, such as *B. subtilis var. niger, Salmonella typhimurium, E. coli* O157:H7 and *Yersinia pestis.*

Materials and Methods

Immunochemical Reagents. Goat anti-*B. subtilis var. niger* polyclonal antibody was obtained from the United States Naval Medical Research Institute. Polyclonal goat anti-*Salmonella sp.* and anti-*E. coli* O157 antibodies were obtained from the Kirkegaard Perry Lab.(KPL; Gaithersburg, MD). Mouse anti-*Yersinia pestis* (F1 positive) monoclonal antibody was obtained from BioDesign International (Kennebunk, ME). All antibodies were affinity purified.

Biotin-*DNP*-hydroxysuccinimide (NHS) (spacer arm 2.24 nm, Molecular Devices, Sunnyvale, CA), was used for antibody biotinylation. The initial b/p ratios used for conjugation ranged from 5 to 25. The final b/p molar ratios obtained ranged from 1.2 to 5.8 as determined using the absorbance measurement of each conjugate at 280 nm and 362 nm according to manufacturer's protocol (Molecule Device, Sunnyvale, CA). $Ru(bpy)_3^{2+}$-NHS,(Ru-NHS) was purchased from IGEN for $Ru(bpy)_3^{2+}$-antibody label. Molar ratios of Ru-antibody (Ru-Ab) conjugates were determined by absorbance at 280 and 455 nm. A final molar ratio of 4 was obtained.

Irradiated *B. subtilis var. niger* spores and anti-*Y. Pestis* F1-antigen (control #CO92) were obtained from the United States Army Medical Research Institute for Infectious Diseases (Ft. Detrick, MD). Irradiated *E. coli* O157:H7 and heat-killed *Salmonella typhimurium* were obtained from Dr. Jerry Crawford (United States Department of Agriculture) and KPL, respectively. All cells were centrifuged at 12,000 x(g) and resuspended in phosphate buffered saline (PBS) before use. Cell counts were determined by a hemacytometer. Ten-fold cell dilutions were made from a stock solution of $5x10^9$ cells/mL. Only 10 mL from each dilution was used in ECLIA. The final volume for the ECLIA was 280 mL as described (5).

Magnetic Particles. Streptavidin (SA)-coated polystyrene magnetic particles with 2.8 and 4.8 mm diameter (M-280 and M-450) were obtained from Dynal (Lake Success, NY). Super paramagnetic microbeads (#481-01) coated with SA were obtained from Miltenyi Biotec (Sunnyvale, CA). Organic polymer based, SA-coated BioMag paramagnetic particles (8-4680, size around 1 mm in diameter) were purchased from Advanced Magnetics (Cambridge, MS). SA-coated Magnetic Pore Glass (MPG) paramagnetic glass particles (MSTR0502) were given as a gift by MPG (Lincoln Park, NJ).

ECLIA, and Fluorescence Microscopy. All immunoassays described in this report used sandwich immunoassay format. Biotin-Ab and Ru-Ab from the same animal species were used for primary and secondary antigen capture, respectively. The magnetic particles along with captured antigen-antibody complex were collected on the surface of electrode by a permanent magnet. There was no wash step involved during the ECL reaction. Finally, the ECL signal (at 622 nm) was measured by a photomultiplier tube. The ECL signal intensities were proportional to the antigen concentrations. The conventional ECLIA protocols used in this report were described previously (5, 7).

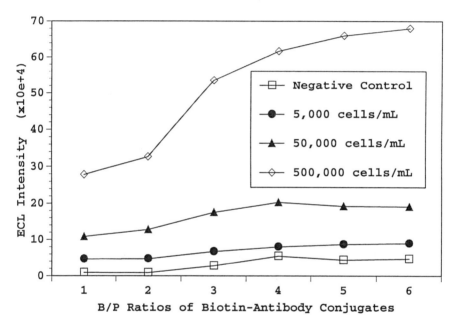

Figure 1. Determination of optimal b/p molar incorporation ratios.

A fluorescence microscopic study was performed on an Olympus BH-2 microscope for examination of the particle natural fluorescence. The excitation/emission filters (490/525 nm) were selected.

Results

The ECLIA results using different b/p molar ratios are shown in Figure 1. Results indicated that the ECLIA intensities increased as the b/p molar ratios increased. Saturation of ECLIA intensities occurred when the b/p molar ratios increased up to 4 or greater in these experiments. *B. thuringiensis* spores at 10^6 cells/mL were used as negative control (-). Negative control also increased as the b/p molar ratio increased.

Three types of SA-coated magnetic particles were examined by fluorescence microscopy. Some features of these magnetic particles were summarized (data not shown). The comparative results revealed that both SA-Dynabeads and SA-BioMag beads have minimal non-specific binding compared to SA-MPG beads under the same experimental conditions. Furthermore, fluorescence microscopy results revealed that Dynabeads (M-2.8 and M-4.5) emit broader auto-fluorescence at 520 nm. Within the visible wavelength range (such as 490 to 520 nm), the use of Dynabeads will interfere with the Fluorescein Isothiocyanate (FITC) based assays. SA-BioMag particles and SA-MPG beads, on the other hand, are good candidates for fluorescent assays because there are no natural fluorescence emitted from these beads. Dynabeads have stronger magnetization and spherical shape compared to other types of magnetic beads. Therefore, in an ECLIA, Dynabeads are good candidates for rapid capturing and rapid wash-off from the surface of the electrode. Natural fluorescence from magnetic particles is not a concern in ECLIA, therefore, Dynabeads are the best choice over others in ECLIA application.

Polyclonal anti-*E. coli* and anti-*Salmonella* sp., and monoclonal anti-*Yersinia pestis* antibodies were selected for ECLIA kinetic studies. Goat anti-*Salmonella* sp. (1000 and 2000 cells/mL), anti-*E. coli* (500 and 1000 cells/mL) were used. Negative controls (-) (without antigens in both sandwich immunoassays) were used. Results of kinetic measurement in Figure 2a shows that the ECLIA intensities are proportional to the reaction time when polyclonal Ru-Ab is added into pre-incubated Biotin-Ab and antigen complex. However, when primary and secondary antibodies were monoclonal, the ECLIA intensities were saturated after 30 minutes reaction time (Figure 2b). The increase in reaction time also slightly increased the intensities of the negative control (-) in experiments using either polyclonal or monoclonal antibodies. When using polyclonal antibodies, kinetic results suggest that the longer the reaction time between Ru-Ab and Biotin-antibody-antigen complex is, the better the S/N ratios will be. The optimal reaction time for the ECLIA for *Y. pestis* using monoclonal antibodies is about 30 minutes.

In Figures 3 and 4, the conventional ECLIA and optimized ECLIA are compared. Unlike the 2% serum used in conventional protocol, SuperBlock Blocking Buffer (Pierce, Rockford, IL) was used in optimized ECLIA for magnetic bead dilutions. An extended biotin spacer (2.24 nm) was used in current biotinylation with the b/p ratio of 4 compared to 1.35 nm biotin-Ab spacer used in the conventional assay. Reaction between the antigens and primary antibody was carried in ice for 5 minutes and incubation time with Ru-Ab was about 50 minutes using polyclonal and up to 30 minutes using monoclonal antibodies. These incubation time is significantly different compared to the time in conventional assays. All reactions were carried in polystyrene test tubes instead of glass tubes. Results of optimized ECLIA show at least 10-fold intensity increase (compared to signal to noise ≥ 3) for *B. subtilis var. niger and E. coli* detection compared to the conventional assays previously reported.

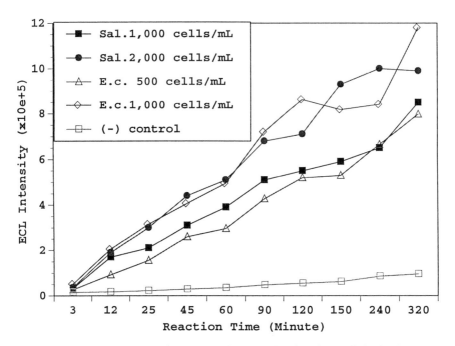

Figure 2a. ECLIA kinetic studies using polyclonal anti-E. coli O157:H7 (E. c.) and anti-Salmonella sp. (Sal.)antibodies.

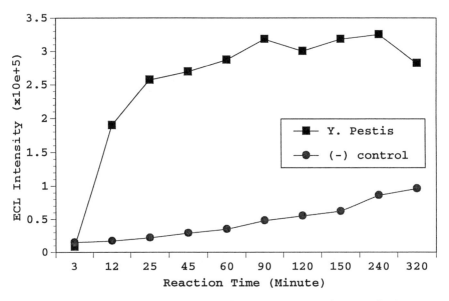

Figure 2b. ECLIA kinetic studies using monoclonal anti-Y. pestis (F1 positive) antibody.

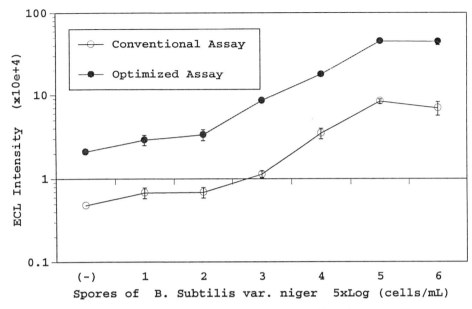

Figure 3. Results of conventional and optimized ECLIA for B. subtilis var. niger spore detection.

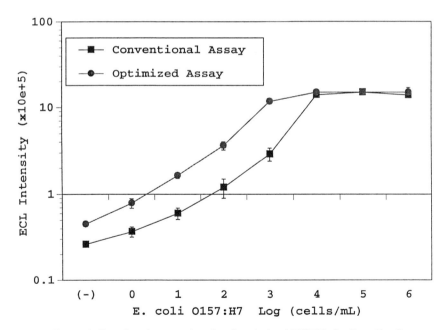

Figure 4. Results of conventional and optimized ECLIA for E. coli cell detection.

Discussion

In a sandwich immunoassay, antibodies from different species are used for antigen capturing (9). This allows the antibodies to interact with different epitopes of the target antigens. However, in this investigation the same antibodies (either goat IgG polyclonal or mouse IgG monoclonal antibodies) were used as primary and secondary antibody capture. In current ECLIA for micron-sized spores and intact bacterial cells capturing, Biotin-Ab coated on the beads are allowed to interact with antigens prior to adding the Ru-Ab. A competitive reaction between Ru-Ab and Biotin-Ab on the same antigen binding sites will begin after adding Ru-Ab. In addition to the competitive reaction, polyclonal Biotin- and Ru-Ab could also bind to different binding sites (multiple sites or epitopes) on the surface of those antigens. ECLIA results have demonstrated that the current sandwich immunoassay format can be used for bacteria cell capturing, including either using anti-*B. subtilis* spores and anti-*Salmonella* (polyclonal) or anti-*Y. pestis (monoclonal)* antibodies.

In addition, the binding sites are potentially available for both Biotin-Ab and Ru-Ab since no covalent bonds are formed between the antibody and antigen. The on/off rate of the free antigens to antibodies (Biotin-Abs and Ru-Abs) depends upon their equilibrium-constant. This constant could vary if the antibodies were modified by probe labels (in the current case, such as biotin and Ru(bpy)$_3^{2+}$)(*10*).

The b/p molar ratio seems to play an important role on the detection limit of the assay. Gretch and coworkers (*11*) reported that molar ratios of biotin to antibody which were equal to or greater than 8 had the maximum antigen recovery potential on SA-conjugated agarose matrix. Even higher b/p molar ratios were reported in the application of flow cytometry by labelling growth factor (*12*). Our ECLIA results show that b/p molar ratios between 4 and 6 were optimal in our experimental conditions. The b/p molar ratios greater than 6 did not significantly improve the ECLIA detection limit. Experiments also suggested the optimal b/p ratios have to be determined in each experiment. However, the high molar ratios of Biotin-NHS to antibody could potentially disrupt the antibody-binding capability to antigens (*12*).

Reaction times in ECLIA are critical. The longer the reaction time is, the higher the intensities are. However, over an extensive incubation, the Ru-Ab potentially could adhere to the magnetic particles. This may explain why the negative controls (-) of ECLIA results increased slightly during the longer incubation. In our lab. 2.5% to 5% serum in PBS, 3% BSA and 10% skim milk as blocking solutions were investigated (data not shown). Five percent goat serum seemed to perform well when goat polyclonal antibodies were used in ECLIA. The use of SuperBlock Blocking Buffer provides the best blocking solution, to preventing protein non-specific binding in the presence of magnetic carriers.

Optimization of ECLIA showed the enhancement of assay sensitivity (linear range) and detection limits for bacteria detection. About 10-fold increase in detection limit in optimized assay (approximately 5000 cells/mL with S/N ratio of 4.1) compared to the conventional assay (50000 cells/mL with S/N ratio of 4.9). The detection limit is approximately 10 cells/mL with S/N ratio of 3.28 in optimized assay compared to the conventional assay at 100 cells/mL (S/N ratio: 3).

ECLIA time is about 1 hour including the extended 50 minutes incubation time. In addition, the reaction between SA-beads and Biotin-Ab is almost instant because of the high affinity constant ($K_a = 10^{15}$ M^{-1}) between SA and biotin. There is no need to extend this pre-incubation time in ECLIA.

It is interesting to note that the ECLIA kinetics of polyclonal and monoclonal antibodies are different. In the polyclonal antibody case, the ECLIA responses were directly proportional to the incubation time up to 5 hours. This suggests that a competition process gradually occurred between the Biotin-Ab and Ru-Ab on the available multiple antigen binding sites. At the same time, the Ru-Ab could be adhered to magnetic beads even in the presence of blocking solution. Therefore, proportional ECLIA responses versus reaction time are expected. In using monoclonal antibody, the ECLIA responses reached a plateau around 30 minutes. Result suggest that a strong and specific antibody-antigen binding process occurred and that the mono-binding sites are occupied by either Biotin-Ab or Ru-Ab. Fifteen to 30 minutes reaction time is optimal under current ECLIA condition when monoclonal antibodies are employed.

The ECLIA is a relative new technology. Besides the major advantage of no natural fluorescence involved in detection. Some disadvantages also should be taken into account. If samples contain some chemicals, such as proline, oxalate, gentamicin, streptomycin and NADH, the ECLIA could cause false positive readings upon internal electron transfer from these chemicals (*13*). ECLIA technique has been successfully used as an alternative method in clinical and environmental applications for biological agent detection. It is believed that the ECLIA technique also could be applied in environmental monitoring for sensitive bacteria detection.

Literature Cited

1. Norffsinger, J.B.; Danielson, N.D. *Anal. Chem.* **1987**, *59*, 865-868.
2. Danielson, N.D.; He, L.; Norffsinger, J.B.; Trelli, L.; *J. Pharm. Biomed.* **1989**, *7*, 1281-1285.
3. Uchikura, K.; Kirisawa, M.; *Anal. Sci.* **1991**, *7*, 803-804.
4. Blackburn, G.F.; Shah, H.P.; Kenten, J.H.; Leland, J.; Kamin, R.A.; Link, J.; Peterman, J.; Powell, M.J.; Shah, A.; Talley, D.B.; Tyagi, S.K.; Wilkins, E.; Wu, T.; Massey, R.J.; *Clinical Chem.* **1991**, *37*, 1534-15.
5. Gatto-Menking, D.L.; Yu, H.; Bruno, J.G.; Goode, M.T.; Miller, M.; Zulich, A.W.; *Biosensors and Bioelectronics* **1995**, *10*, 501-507.
6. Yu, H.; Bruno, J.G.; Cheng, T.; Calomiris, J.J.; Goode, M.T.; Gatto-Menking, D.L.; *J. of Biolum. Chemilum.* **1995**, *10*, 239-245.
7. Yu, H.; Bruno, J.G.; *Applied and Environmental Microbiology* **1996**, *62*, 587-592.
8. Leland, J.K.; Powell, M.J.; *J. of Electrochem. Soc.* **1990**, *137*, 3127-3131.
9. *Immunological Reagents for Research catlog,* Jackson ImmunoResearch Laboratories, Inc. West Grove, Pennsylvania **1993**, 6.
10. Bredehorst, R.; Wemhoff, G.A.; Kusterbeck, A.W.; Charles, P.T.; Thompson, R.B.; Ligler, F.S.; Vogel, C.W.; *Anal. Biochem.* **1991,** *193*, 272-279.
11. Gretch, D.R.; Suter, M.; Stinski, M.F.; *Anal. Biochem.* **1987**, *163*, 270-277.
12. Jong, Marg O.D.; Rozemuller, H.; Bauman, Jan G.J.; Visser, Jan W.M.; *J. of Immunol. Methods* **1995**, *184*, 101-112.
13. Savage, M.D; Mattson, G.; Desai, S.; Nielander, G.W.; Morgensen, S.; Conklin, E. J.; Avidin-Biotin Chemistry: A Handbook, Chapter 2, Pierce Chemical Company, 1992.
14. Lee, W.Y.; Neiman, T.A.; *Anal. Chem.* **1995**, *67*, 1789-1796.

Chapter 14

Microarray-Based Immunoassays

F. W. Chu[1], P. R. Edwards[1], R. P. Ekins[1], H. Berger[2], P. Finckh[2], and F. Krause[2]

[1]Division of Molecular Endocrinology, University College London
Medical School, Mortimer Street, London W1N 8AA, United Kingdom
[2]Boehringer Mannheim GmbH, Bahnhofstrasse 9–15,
D–8132 Tutzing, Germany

Recent worldwide interest in the development of miniaturized, array-based, multianalyte binding assay methods suggests that the ligand assay field is on the brink of a technological revolution. Our own collaborative studies in this area have centered largely (but not exclusively) on antibody spot "immunoarrays" localized on "microchips" which are potentially capable of determining the amounts of hundreds of different analytes in a small sample (such as a single drop of blood). Analogous technology for genetic testing using oligonucleotide arrays is under active development both in the US and Europe. Array-based immunoassay methods are clearly likely to prove of particular importance in areas such as environmental monitoring where the concentrations of many different analytes in test samples are required to be simultaneously determined. In this presentation we review the general principles underlying this emerging technology.

Immunoassay

"Binding" or "ligand" assay methods have, in the past 35 years, been applied to the assay of a wide range of substances of biological importance. Because antibodies can be raised against many such analytes, antibody-based "immunoassay" techniques have achieved particular prominence, but the principles on which these techniques rest can be exploited using many other classes of binding agent. Of particular and increasing importance is the use - in methods relying on identical analytical concepts - of oligonucleotide probes, which bind to single chain fragments of DNA with affinities and specificities of the same order as, or greater than, those characterizing antibody-antigen and other binding reactions.

Such "binding assay" techniques were originally developed to determine the minute concentrations of hormones in biological fluids, but were subsequently exploited in many other areas of medicine in which the estimation of small amounts of biologically-important substances is required. More recently still, they have been increasingly adopted in fields such as environmental monitoring in which similar needs arise.

During the period 1960-80 the "competitive" or "saturation" assay approach - relying on the use of radiolabeled analyte markers, and typified by radioimmunoassay (RIA) - dominated the field. Radiolabeled antibody methods, usually referred to as "immunoradiometric assays" (IRMAs), were developed in the late 1960's by Wide (*1*), followed shortly by Miles and Hales (*2, 3*), and owed much to the development of immunosorbant techniques for the preparation of pure labeled antibodies. These methods were originally claimed (*3*) to be inherently more sensitive than RIA; however this claim was neither supported by rigorous theoretical analysis, nor persuasive experimental evidence, and remained controversial. Particular doubts on its validity were cast in 1973 by Rodbard's and Weiss' (*4*) detailed theoretical studies purporting to demonstrate that both labeled analyte and labeled antibody methods possess essentially equal sensitivities, IRMA methods being supposedly somewhat more sensitive for the assay of small polypeptides in which radioiodine incorporation into the analyte molecule was restricted, but less sensitive for the assay of high molecular weight analytes. These erroneous conclusions stemmed in part from the confusion surrounding the concept of sensitivity prevalent during this period; in part from the common misconception (which persists to the present day) that a crucial distinction in regard to assay performance lies between labeled antibody and labeled analyte methods *per se*. (In other words, the belief that labeling the antibody used in the system of itself affects assay sensitivity.) In reality, labeled antibody, or "immunometric", methods can be sub-classified as "competitive" and "non-competitive", immunometric assays of competitive design being essentially identical in sensitivity to competitive assays relying on labeled analyte. In short, higher assay sensitivity is not a necessary consequence of labeling the antibody used, but is a possible consequence of the adoption of a non-competitive design. Nevertheless, the low specific activity of the radionuclides conventionally used in this context implies that non-competitive IRMAs do not demonstrate significant improvements in sensitivity when compared with competitive RIAs and IRMAs and other non-isotopic competitive methods, especially when high affinity antibodies are used.

However, the theoretical demonstration that "non-competitive" assays using antibodies labeled with non-isotopic markers of high specific activity are potentially capable of greater sensitivity and may require far shorter incubation times (*5, 6*), - when coupled with the development of methods of monoclonal antibody synthesis by Köhler and Milstein in 1975 (*7*) - provided the basis for the development of a new generation of fast, "ultrasensitive", immunoassays. The first of this generation - developed by Ekins *et al* in collaboration with the instrument manufacturer Wallac Oy (*8, 9*) - relied on high specific activity lanthanide chelate labels and time-resolution techniques for the measurement of the slowly-decaying fluorescent signals such chelates emit. Subsequently other manufacturers have also introduced ultrasensitive methods based on identical principles and the use of other high specific activity, non-isotopic, labels (e.g. chemiluminescent substances, enzymes, etc), as recently reviewed by Kricka (*10*). These second generation technologies have progressively replaced the earlier isotopically-based methods during the '80s, enabling, *inter alia*, the development of the automatic immunoanalysers now widely used in clinical laboratories.

Despite the increases in sensitivity and reduction in assay performance times resulting from these developments, current immunoassay methodology remains essentially limited to single analyte measurements. However, the need for simultaneous, sensitive, determination of large numbers of analytes in small test samples is clearly apparent in many areas of clinical medicine; for example, allergy testing, the screening of transfusion blood for viral contamination, forensic investigation, etc. But the area in which such a need very clearly arises, and that has probably attracted the widest popular attention, is that of genetic analysis. Genetic testing techniques rely on the detection of specific polynucleotide sequences within a DNA strand by their binding to complementary oligonucleotide probes. Conscious of the impending completion of the Human Genome Project, US Government agencies established in 1992 a five-year collaborative study on the development of

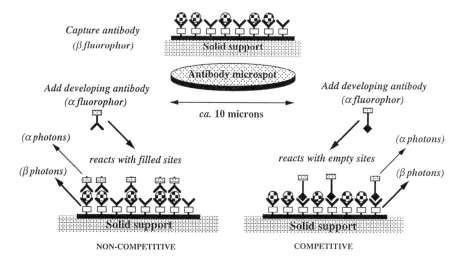

Figure 1. Following exposure of (fluorescent-labeled) capture antibodies within the microspot area to analyte-containing medium, either occupied sites (non-competitive approach) or unoccupied sites (competitive approach) are determined using an appropriate "developing" antibody labeled with a second fluorescent label. The ratio of signals yielded by the two labels reveals capture antibody occupancy.

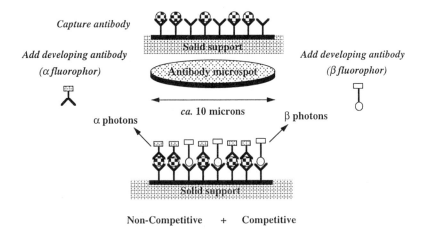

Figure 2. Combined non-competitive/competitive approach, relying on developing antibodies labeled with different fluorescent labels, and measurement of the signal ratio.

oligonucleotide-array-based analytical techniques (the 'Genosensor Project'). Albeit formally restricted to DNA analysis, this project is closely similar in its aims and approach to our own studies, and is clearly relevant to other similar multianalyte binding assays.

The current widespread interest both in the US and in Europe in the development of array-based microanalytical methods, and the related instrumentation and technology, for use in diagnostic medicine clearly has major implications with respect to other areas in which immunoassay techniques are frequently employed, such as environmental analysis.

In this presentation, we describe the concepts underlying the development of miniaturized, multianalyte, microspot array technologies, which are likely to constitute the next major revolution in the binding assay field.

Array-based Multianalyte Binding Assays: Basic Concepts

The basic concepts developed by Ekins et al (*11*) that underlie the array technologies currently under development jointly at UCL and Boehringer Mannheim GmbH have been previously reviewed on several occasions (*12, 13*) and only a brief summary is therefore necessary here. These concepts include the following:

i. The "Binding Site Occupancy" Principle of Immunoassay and Other Binding Assays. This embodies the proposition that all such assays implicitly rely on determination of the fractional occupancy by analyte of a binding agent - typically an antibody or oligonucleotide. Non-competitive assays directly determine <u>occupied</u> sites; competitive assays require observation of <u>unoccupied</u> sites, from which binding-site fractional occupancy is inferred.

ii. "Ambient Analyte" assay. This term describes assays in which the binding site concentration is so low as not to affect the analyte concentration in the medium to which the binding agent is exposed. In practice, this implies that no more than 5%, and ideally less than 1%, of the total analyte is bound. In these circumstances binding-site fractional occupancy is independent of both the amount of binding agent used in the system and the sample volume.

iii. Microspot Assays. Such assays are characterised by the use of small amounts of a binding agent localised at high surface density on a solid support in the form of a "microspot". If the system fulfils ambient analyte assay conditions, the microspot acts as an analyte "sensor", its fractional occupancy by analyte being indicative of the analyte concentration in the surrounding medium.

iv. Dual Label, Ambient Analyte "Ratiometric" Binding Assays. These assays rely on observation of the ratio of signals emitted by two labels to determine binding site fractional occupancy and hence the ambient analyte concentration. This approach reduces errors arising from variations in the amount, or surface density, of binding agent located within the microspot area. Figure 1 shows a microspot immunoassay using two distinguishable fluorescent labels. One is coupled to the solid-supported capture or "sensing" antibody, the other to a second "developing" antibody. An alternative approach locates the labels on two developing antibodies directed respectively against occupied and unoccupied sites of the sensing antibody (Figure 2). A system operating in this manner can be described as both competitive and non-competitive.

These four concepts underlie the development of antibody microspot-array technologies in which each individual microspot in an array relates to a different analyte in the medium to which it is exposed. In our initial feasibility studies, a

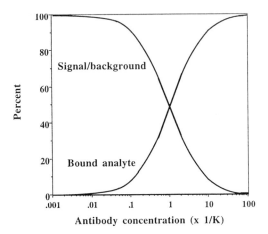

Figure 3. Increase in antibody concentration in the system (i.e. increase in microspot area) results in an increase in analyte binding, but a decrease in the surface density of captured analyte, and hence a decrease in the signal/background ratio.

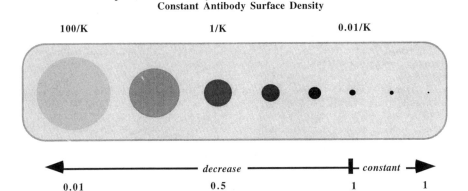

Figure 4. As microspot area increases, the signal yielded by occupied sites fades into the background. Reproduced with permission from Ekins, R.P.; Chu, F.W. In *Immunoanalysis of Agrochemicals, Emerging Technologies*; Nelson, J.O.; Karu, A.E.; Wong, R.B., Ed.; ACS Symp. Ser. 586 American Chemical Society: Washington, DC, 1995, Ch. 11. Copyright 1995 American Chemical Society.

commercially-available laser scanning confocal microscope was used to observe microspots in which were located sensor and developing antibodies labeled by a pair of conventional fluorophors - Texas Red and FITC (*12*). The latter have since been replaced by a polyfluorophor amplification system to increase assay sensitivity, but a detailed description of this modification is presently prevented by commercial constraints. This modified approach yields very high sensitivities (*13*). The use of fluorescent labels and a laser scanning confocal microscope has subsequently been adopted by several participants in the US Genosensor project, although other methods of observation of light-emitting arrays (relying, for example, on CCD-based devices) are also obviously feasible.

Theoretical Considerations: Microspot Assay Sensitivity and Speed

The proposition that microspot assays may be more sensitive and rapid than conventional systems challenges many accepted ideas in the binding assay field and often arouses skepticism. One widely-held belief is that - to determine a small amount of an analyte in a sample using a non-competitive assay strategy - it is necessary to 'capture' the majority of the analyte molecules present (*14*). Contradicting this precept, binding agent microspots sequester only an insignificant fraction of the total analyte in the sample. Another concept (deriving from the mass action laws) apparently contravened by the microspot approach is that the use of a large amount of binding agent maximizes the velocity of the binding reaction, thus yielding assays requiring shorter incubation times. These conflicts with generally accepted ideas should be briefly addressed.

Sensitivity. As indicated above, all immunoassays essentially rely on measurement of antibody binding site occupancy, either directly (non-competitive assay) or indirectly (competitive assay). A non-competitive assay generally yields higher sensitivity since it is normally preferable, in practice, to measure a small quantity directly rather than indirectly, i.e. by subtracting one large quantity from another. In certain circumstances it may be advantageous - if high sensitivity is required - to maximize binding-site occupancy; however, increasing the number of binding sites to achieve this objective is also likely to cause a concomitant increase in the background signal. The proposition that, in a non-competitive microspot assay design, sensitivity is increased by reducing the microspot area (assuming the binding agent surface density remains constant) is illustrated in Figures 3 and 4. Figure 3 shows the fall in the signal/background (s/b) ratio (expressed as a percentage of the asymptotic (s/b) limit as the antibody-coated area tends to zero) as the amount of antibody increases. Also plotted is the percentage of the total analyte bound to the antibody. The calculations underlying these figures are based on the assumptions of equilibrium and the use of a relatively low total analyte concentration ($<0.01/K$). Clearly, for a constant surface coating density, as the spot size and the amount of antibody increase the fractional occupancy, the s/b ratio, and hence - implicitly - assay sensitivity, decrease, despite an increase in total analyte bound. This phenomenon is illustrated in Figure 4. Obviously if the microspot area were progressively reduced to zero, both signal and background would likewise also tend to zero (the s/b ratio nevertheless remaining essentially constant as the antibody concentration fell below $0.01/K$). Thus, in the limit, no signal other than instrument noise would be observed, and the system would be totally insensitive. In practice, other statistical considerations come into play when the number of individual events (e.g. photons) observed by the detecting instrument is small, thus prohibiting the reduction of the sensor-antibody concentration beyond a certain lower limit. Thus as the microspot area approaches zero, the magnitude of the s^2/b ratio becomes a better indicator of assay sensitivity. The point at which reduction in the antibody-coated area causes such loss of

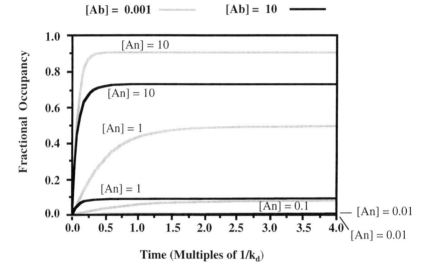

Figure 5. Antibody/analyte binding kinetics. Curves showing increase of fractional occupancy of antibody (concentrations 0.001/K and 10/K) with time for different analyte concentrations (10/K, 1/K, 0.1/K, 0.01/K). (Note the expression of all concentrations as multiples of 1/K, and of time as multiples of $1/k_d$, where K = equilibrium constant, and k_d = dissociation rate constant, implies that the curves are applicable to all antibody/antigen reactions.) Note that capture antibody fractional occupancy is greater at all times for the lower antibody concentration. Reproduced with permission from Ekins, R.P.; Chu, F.W. In *Immunoanalysis of Agrochemicals, Emerging Technologies*; Nelson, J.O.; Karu, A.E.; Wong, R.B., Ed.; ACS Symp. Ser. 586 American Chemical Society: Washington, DC, 1995, Ch. 11. Copyright 1995 American Chemical Society.

detectable signal that the precision of antibody-occupancy measurement significantly deteriorates depends on the specific activity of the labeled antibody, the higher its value, the smaller the permissible microspot size. However, an additional constraint is the number of analyte molecules captured within the microspot area which, when very small, will also cause statistical variations in the emitted signal. Nevertheless, given very high specific activity labels, the optimal capture-antibody concentration employed in a non-competitive assay design may be extremely low, although a reduction to below 0.01/K is likely to offer little or no benefit assuming thermodynamic equilibrium is reached in the system.

Non-competitive sandwich assay designs are generally inapplicable to analytes of small molecular size (such as are likely to be encountered in environmental analysis) and the competitive approach is therefore generally preferable. Although the sensitivity attainable is unlikely to be as high as in non-competitive assays, theoretical considerations (*15*) reveal that highest sensitivity is likewise achieved when a small amount of antibody located on a microspot is employed and when the fractional occupancy of antibody binding sites is therefore greatest.

Microspot Assay Kinetics. That a microspot format is capable of yielding assays that are more rapid than conventional methodologies may also cause surprise since it is widely accepted that the use of high antibody concentrations in a non-competitive assay implies faster binding of analyte, and hence attainment of equilibrium in accordance with the mass action laws. Microspot assays must, however, be viewed from a different standpoint. The more rapid of two such assays is that yielding the higher binding site fractional occupancy, and thus the highest s/b ratio, in a given incubation time. If the diffusion constraints that apply to any system in which the binding agent is linked to a solid support are, for the moment disregarded, the increase in binding site fractional occupancy with time may be readily calculated for microspots of different areas. Figure 5 demonstrates that occupancy is at all times greater using the lower antibody concentration. Thus the s/b ratio is always greater in the case of the smaller spot despite the larger spot's more rapid attainment of equilibrium.

However, diffusion constraints on the velocities of binding reactions are exacerbated when capture binding agent molecules are linked to a solid support. Constraints on analyte migration to and from the support reduce the effective association and dissociation reaction rates observed prior to the attainment of equilibrium, albeit the final equilibrium state is unaffected. Analysis of the diffusion of analyte molecules (*16*) towards a circular absorber or "sink" demonstrates that diffusion constraints are reduced for smaller microspots. Such analysis reveals that the diffusion flux (molecules/sec) is given by 4Dr[An], where D = diffusion coefficient (cm^2/sec), r = radius (cm) and [An] is the analyte concentration (molecules/cm^3). Assuming uniformity of the antibody surface density (d_{Ab} (molecules/cm^2)) within the microspot area, the migration rate of analyte molecules to a circular microspot of radius r_m is thus proportional to r_m; meanwhile the number of analyte molecules that must be bound to reach a specified occupancy level of antibody located within the microspot is proportional to the microspot area, i.e. to πr_m^2, and thus to r_m^2 (Figure 6). Hence - assuming the reaction is diffusion controlled - the rate of antibody occupancy within the microspot is proportional to $1/r_m$.

More detailed consideration of the rate at which analyte molecules migrate towards, and bind to, an antibody microspot reveals that the initial antibody occupancy rate (OR) is given by:

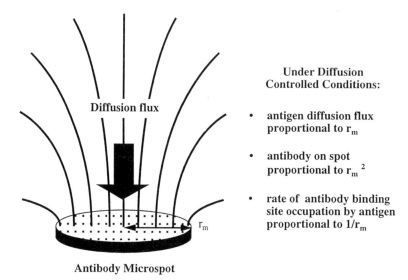

Under Diffusion Controlled Conditions:

- antigen diffusion flux proportional to r_m

- antibody on spot proportional to $r_m{}^2$

- rate of antibody binding site occupation by antigen proportional to $1/r_m$

Antibody Microspot

Figure 6.   The rate of antibody occupancy by analyte molecules is inversely proportional to microspot radius (assuming constant capture antibody density). Reproduced with permission from Ekins, R.P.; Chu, F.W. In *Immunoanalysis of Agrochemicals, Emerging Technologies*; Nelson, J.O.; Karu, A.E.; Wong, R.B., Ed.; ACS Symp. Ser. 586 American Chemical Society: Washington, DC, 1995, Ch. 11. Copyright 1995 American Chemical Society.

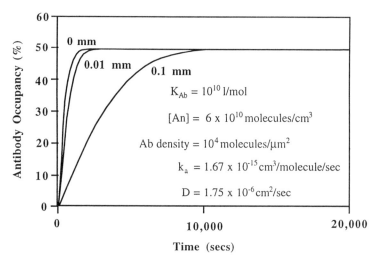

Figure 7.   Calculated antibody occupancy as a function of time for micro-spots of different radius (0.1 and 0.01 mm) compared with kinetics of antibody binding in a homogeneous medium (assuming parameter values shown). Diffusion constraints are reduced in the case of the smaller spot.

$$OR = \frac{4r_m k_a D[An]d_{Ab}}{\left(\pi r_m^2 k_a d_{Ab} + 4Dr_m\right)} \quad \text{molecules/sec/cm}^2$$

where k_a = association rate constant (cm^3/sec/molecule). As r_m tends to zero, the term $\pi r_m^2 k_a d_{Ab}$ becomes small compared to $4Dr_m$, implying that OR approximates $k_a[An]d_{Ab}$. In other words, the kinetics of the reaction increase consequent upon reduction in r_m, ultimately approximating those seen in a homogeneous solution.

We have also developed computer models revealing the full sequence of events following introduction of antibody microspots of different diameters into an analyte-containing solution, these embracing the kinetics of the antibody-analyte reaction *per se,* the establishment of analyte concentration gradients in the solution, etc. These have confirmed that the smaller the microspot, the lower the diffusion constraints on the rate of analyte binding to antibody, and the more closely the reaction kinetics approximate those characterising a homogeneous liquid-phase reaction system (see Figure 7).

Such considerations indicate that higher signal/background ratios are attained in a shorter time using a microspot format, implying that microspot assays are likely to prove at least as rapid as assays of conventional macroscopic design. However, use of relatively small amounts of binding agent implies that equilibrium will not be reached during short incubation times, increasing the possibility of assay drift. In practice, this phenomenon is unlikely to constitute a major problem when using automatic equipment, and highly standardised conditions of incubation and sample processing.

Practical Implementation

Construction of Antibody and Oligonucleotide Microspot Arrays.   Our preliminary studies in this field relied on crude techniques involving brief (*ca* 1 sec) exposure of suitable plastic surfaces to minute droplets of antibody-containing solutions followed rapidly by conventional washing and protein-blocking procedures. Such manual techniques are clearly inappropriate to large scale industrial production. The array construction technology currently being developed by ourselves relies on small disposable polystyrene carriers onto which microspots are deposited using an "inkjet" technology. Arrays of as many as 200 spots may readily be deposited in this manner on the flat bottom (ca 2.25 mm radius) of the carrier wells, each droplet of reagent (antibody or antigen) solution possessing a volume of less than 1 nl and yielding a spot of diameter approximating 80-100 μm. Spot size, and array pattern and shape, are controlled automatically by a digital imaging feedback device resulting in a high precision of microspot application. An alternative approach that is applicable to the construction of oligonucleotide arrays has been adopted by Fodor et al (*17*) using combinatorial synthetic techniques. In this approach, selected areas on a solid substrate are unmasked by a photolithographic process and exposed to a particular nucleoside. Chemical reactions occurring at these areas result in attachment of the corresponding nucleotide. Repeated masking, unmasking and exposure of different areas in a series of cycles, leads to the creation of an array of oligonucleotides, each differing in their nucleotide sequence. By this means, the set of all possible oligonucleotides (4^n) of length n can be produced in 4n cycles. Thus a set of 65,536 (8-mer) oligonucleotides can be created in only 32 cycles. However, such techniques are clearly not applicable to the construction of antibody arrays, nor to arrays of other molecules which are not of a relatively simple single-stranded form.

Oligonucleotide arrays are clearly of potential importance for DNA analysis , but we have also used such arrays as array templates, or "proto-arrays", with which individual researchers and other small users can construct antibody arrays, using their own antibodies to which complementary oligonucleotide sequences have been linked.

Although a detailed description of the techniques we have used to construct such "proto-arrays" is inappropriate in this presentation, a brief outline will suffice to give an indication of their simplicity.

Oligonucleotides with 5'-phosphorothioate modification are conjugated to antibodies using the following procedure:

Antibody preparations are activated with a hetero-bifunctional reagent such as {sulfosuccinimidyl 6-[3'-(2-pyridyldithio)-propionamido] hexanoate}. The activated-antibody preparations are purified on disposable PD10 Sephadex columns supplied by Pharmacia, concentrated, and then reacted with a selected 5'-phosphorothioate modified oligonucleotide possessing a unique sequence. The oligonucleotide-conjugated antibody preparations are finally purified on Sephadex G200 columns and again concentrated. This entire operation may be completed in approximately 24 hours. Subsequently the mixture of oligonucleotide-tagged antibodies is exposed to the template array comprising oligonucleotide spots complementary to the oligo-nucleotide moieties of the oligonucleotide/antibody conjugates and hybridized, the conjugates hybridizing only to spots containing the complementary oligonucleotide (see Figure 8). The arrays are finally washed and dried, and stored for future use.

The multi-microspot array approach offers other possible advantages, some of which are likely to be of particular importance in the context of environmental monitoring. For example, it offers for the first time a means of recognizing, and correcting for, the effects of cross-reacting sample components by inclusion of microspots on each carrier which flag specific interferences, or that preferentially recognize the different known components of a complex mixture. By appropriate computations, the ambient concentrations of each of the individual components may subsequently be determined (although the accuracy of such determinations will clearly deteriorate as the number of cross-reacting components and/or their structural similarity increases). The use of multiple spots for the same analyte leads to higher precision and added confidence in test results; moreover the use of spots of differing antibody affinity or density for an individual analyte implicitly increases the dynamic range of the assay (in the latter case by obviating saturation effects in the optical system used), thereby obviating additional sample dilutions.

Instrumentation. We have developed a prototype analyzer which permits the use of a wide variety of reaction conditions, all components including the incubators and washers, having been specifically designed for dealing with small reagent volumes.

For a typical assay procedure, up to 20 µl of sample are used. Following sample incubation and a first washing step, a mixture of analyte-specific reagents is introduced into the sample carrier followed by a second washing step. After incubation of developing antibodies and final washing, fluorescent signals generated from individual spots in the array are measured in the detector. Total assay time is about 20-30 minutes.

The detection system comprises a dedicated confocal scanning unit claimed to possess a sensitivity sufficient to permit detection of 1 labeled reagent molecule and possessing a dynamic range of 10^5. A polyfluorophore label is used to provide fluorescent signal amplification. The array is initially scanned very rapidly to confirm exact spot positions; subsequently the fluorescence signal emitted by each individual spot, and the background fluorescence emanating from outside the spot area, are determined. Scanning of the entire array is completed within 12 seconds. This procedure reveals sample specific backgrounds, thereby facilitating interpretation of results if and when interfering substances are present.

Results

We have developed a variety of non-competitive and competitive Microspot assay systems, including sandwich assays, labeled analyte back-titration assays for analytes of low molecular weight, and solid phase capture-antigen assays for the determination of serum antibodies. Such assays have related to analytes within the fields of endocrinology, allergy and infectious disease; similar techniques have also been employed for the screening of therapeutic drugs. TSH assays have been of high sensitivity (< 0.01 μU/ml), and good correlation with results yielded by other comparable analyzers has been obtained. Likewise good correlation with the results obtained using the Pharmacia CAP test kits has been demonstrated for a number of allergens (birch, cat epithelia, house dust mite, α-amylase, bee venom and total IgE), assay precision and sensitivity being superior to those of the Pharmacia tests. For total IgE a lower detection limit of < 0.01 IU/ml has been obtained. Similarly Microspot assays relating to a number of infectious diseases (e.g. HIV, HBsAg, anti-HBC, rubella) have been developed and are currently being evaluated.

We have also successfully developed combined competitive/non-competitive assays using two labeled developing binders as described above to extend the assay working range. This effect may be illustrated by plotting the precision profiles of an ambient analyte Microspot TSH assay using the combined method (Figure 9).

Discussion

Although still at a relatively early stage in their development, miniaturized array technologies are now attracting widespread scientific and industrial interest and are likely to bring about the next major revolution in the microanalytical field. Our own studies in this area have been specifically targeted towards diagnostic assays used in medicine, in which field high assay sensitivity, precision and sample throughput are generally vital. The instrumentation required in this context is therefore likely to remain sophisticated and relatively bulky, notwithstanding miniaturization of the assay process *per se*. However in areas of application in which less rigorous requirements apply, it is likely that smaller and simpler instruments will ultimately be developed. It should be noted that the signal-measurement technology employed in this context is comparable with that relied on in compact disk players, the main instrumental complication arising from the additional need for fluid sample handling and treatment under carefully controlled temperature conditions in many diagnostic applications. Nevertheless more simple instruments are likely to be generally adequate for environmental monitoring.

As indicated above, many substances of interest in an environmental context are of small molecular size, and may therefore not be measurable by so-called "sandwich" or "two-site" immunoassays (albeit conventional single-site, non-competitive, assays of the type developed by Miles and Hales (*1, 2*) may, in principle, be developed for such analytes). For this reason, these are generally measured in assays of competitive design, using either a labeled form of the analyte or a labeled anti-idiotypic antibody to determine unoccupied (anti-analyte) antibody binding sites. The lower limits of detection of competitive assays may be shown to be restricted, in theory and in practice, to analyte concentrations in the order of 10^7 molecules/ml; moreover such assays also require considerably longer incubation times than those of non-competitive design if these detection limits are to be achieved. Although these fundamental disadvantages are likewise inherent in Microspot assays of competitive design, it should nevertheless be noted that - contrary to generally accepted views - highest assay sensitivity and shortest incubation times are achieved in such assays by using a vanishingly small amount of antibody (provided non-specific binding of the

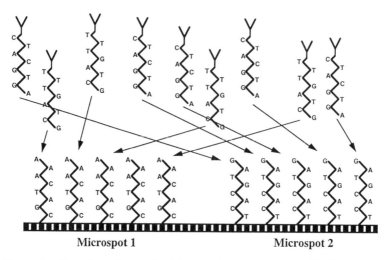

Figure 8.   Use of an oligonucleotide array to construct an antibody array.

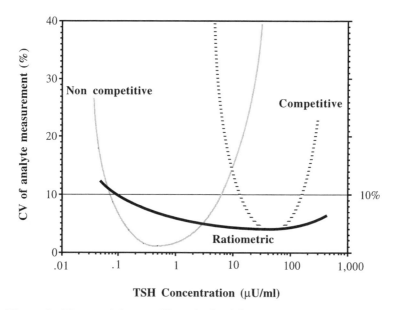

Figure 9. The precision profiles obtained by separate analysis of the competitive and non-competitive data are compared with that obtained using the ratio of the signals emitted by both labeled developing antibodies (ie the combined competitive/non-competitive ratiometric method).

labeled material is low). In other words, a Microspot format exploits to the full the sensitivity potential of competitive assay designs.

In order to overcome the sensitivity limitations of competitive assays, certain workers (*18, 19*) have adopted a novel, albeit somewhat complicated, approach, i.e. that of exposing the initial incubation mixture (containing unlabeled antibody-analyte complex and residual anti-analyte antibody) to either an unlabeled anti-idiotypic antibody or an analyte-protein complex reacting with, and thus effectively blocking, residual, unoccupied, antibody binding sites. The incubation mixture is then further incubated with a second, labelled, anti-idiotypic antibody capable of binding to the analyte-antibody complex, but sterically prevented from reacting with blocked sensor-antibody binding sites. The resulting "idiometric" assay system thus possesses a superficial resemblance to a two-site assay, and can legitimately be described as of non-competitive design. It is claimed to yield very high sensitivity; however the labelled anti-idiotypic antibody is blind to the nature of molecules occupying anti-analyte-antibody binding sites, so that, in regard to specificity, the system relies solely on this antibody's characteristics, and thus functions essentially as a single-site assay.

In principle, such an approach may be exploited using the Microspot format; however it requires additional incubation steps, and for this and other reasons we have not as yet examined it in detail.

Another question occasionally asked in regard to the Microspot technology is its vulnerabilty to the presence of fluorescent substances in test samples. This constitutes a potential hazard in regard to samples encountered in both biomedical and environmental monitoring contexts. This problem is very largely obviated by the sequential nature of incubations carried out in our current procedures; thus the test sample itself is washed away following the initial incubation prior to exposure of the sample carrier to developing antibodies. Moreover since, in the final laser scanning of the sample holder, areas <u>surrounding</u> antibody spots (as well as the spots themselves) are examined, any fluorescent material non-specifically bound to the sample carrier is immediately apparent and its interfering effects - if not overwhelmingly large - corrected for. Finally, of course, fluorescence of analyte molecules themselves may occur, but the resulting effects would inevitably be small in comparison with signals emitted by labeled antibodies, and would in any case implicitly corrected for by the standardisation procedure. To summarise, we have not experienced problems, in practice, due to the presence of fluorescent substances in biological samples, and would not expect them to arise in samples encountered in an environmental setting except in extremely exceptional circumstances.

Other multianalyte immunoassay techniques have, of course, been known for many years. These have generally depended on the use of different labels to monitor the individual reactions, though some have relied on their spatial separation, for example on paper strips. The former are clearly restricted by the number of different labels that can be readily distinguished; the latter have been somewhat cumbersome, not readily automated, and have required relatively large sample volumes. The Microspot technology represents the ultimate miniaturized form of this approach, reflecting the realisation that the use of vanishingly small concentrations of antibody yields assays that are faster and more sensitive than any other.

Immunosensor techniques and other similar sensing methods have frequently mentioned as candidate technologies for use in both the biomedical and environmental monitoring fields, and a great deal of time and money has been devoted by manufacturers on their development over many years. However the development of transducer-based sensing devices of adequate sensitivity and specificity in itself presents formidable difficulties (some of a fundamental nature), putting aside the obvious desirability of multiparameter testing in environmental monitoring and many other potential areas of application. For these reasons, we believe such sensors lie far in the future, and that the development of array technologies of the kind described in this presentation offers a far greater prospect of success in the foreseeable future.

Acknowledgments

We acknowledge the meticulous and dedicated technical work of Antony Burt of the Department of Molecular Endocrinology, University College London in the feasibility and other preliminary studies on which the project has been based. Later studies were possible as a result of the varied development work of the extensive Boehringer Mannheim Microspot project team. The generous financial assistance of the Wolfson Foundation to the Department is also acknowledged with gratitude.

References

1. Wide, L.; Bennich, H.; Johansson, S.G.O. *Lancet* **1967**, *2*, 1105-1107.
2. Miles, L.E.H.; Hales, C.N. *Nature* **1968**, *219*, 186-189.
3. Miles, L.E.H.; Hales, C.N. In *Protein and Polypeptide Hormones;* Margoulies, M., Ed.; Excerpta Medica: Amsterdam, Holland, 1968, Part 1; pp 61-70.
4. Rodbard, D.; Weiss, G.H. *Anal. Biochem.* **1973**, *52*, 10-44.
5. Ekins, R.P. In *Radioimmunoassay and Related Procedures in Medicine*; International Atomic Energy Agency Vienna: Vienna, Austria, 1977, Vol 1; pp 241-268.
6. Jackson, T.M.; Marshall, N.J.; Ekins, R.P. In *Immunoassays for Clinical Chemistry*; Hunter, W.M.; Corrie, J.E.T., Eds.; Churchill Livingstone: Edinburgh, UK, 1983, pp 557-575.
7. Köhler, G.; Milstein C. *Nature* **1975**, *256*, 495-497.
8. Marshall, N.J.; Dakubu, S.; Jackson, T.; Ekins, R.P. In *Monoclonal Antibodies and Developments in Immunoassay*; Albertini, A.; Ekins, R.P., Eds.; Elsevier: North Holland, Amsterdam, Holland, 1981, pp 101-108.
9. Soini, E.; Lövgren, T. *CRC Critical Reviews in Analytical Chemistry* **1987**, *18*, 105-154.
10. Kricka, L.J. *J. Clin. Immunoassay* **1993**, *16*, 267-271.
11. Ekins, R.P.; Chu, F.; Micallef, J. *Journal of Bioluminescence and Chemiluminescence* **1989**, *4*, 59-78.
12. Ekins, R.P.; Chu, F.; Biggart, E. *Analytica Chimica Acta* **1990**, *227*, 73-96.
13. Ekins, R.P; Chu, F. *Tibtech.* **1994**, *12*, 89-94.
14. Hay, I.D.; Bayer, M.F.; Kaplan, M.M.; Klee, G.G.; Larsen, P.R.; Spencer, C.A. *Clin. Chem.* **1991**, *37*, 2002-2008.
15. Ekins, R.P.; Newman, B; O'Riordan, J.L.H. Theoretical Aspects of 'Saturation' and Radioimmunoassay. In *Radioisotopes in Medicine: In Vitro Studies,* Hayes, R.L.; Goswitz, F.A.; Murphy, B.E.P., Eds.; Oak Ridge Symposia, USAEC: Oak Ridge, Tennessee, 1968, pp 59-100.
16. Crank, J. In *The Mathematics of Diffusion, 2nd edition*; Oxford U P: Oxford, UK, **1975**.
17. Fodor, S.P.A.; Read, J.L.; Pirrung, M.C.; Stryer, L.; Lu, A.T.; Solas, D. *Science* **1991**, *251*, 767-773.
18. Barnard, G; Kohen, F. *Clin. Chem.* **1990**, *36*, 1945-1950 .
19. Self, C. *Determination Method, Use and Components.* International patent application: PCT 01033.

Chapter 15

Development and Applications of a New Immunoassay Method: Latex Piezoelectric Immunoassay

S. Kurosawa[1], M. Muratsugu[2], H. O. Ghourchian[3], and N. Kamo[3]

[1]National Institute of Materials and Chemical Research,
1–1 Higashi, Tsukuba, Ibaraki 305, Japan
[2]Osaka Prefectural College of Health Sciences,
Habikino, Osaka 585, Japan
[3]Faculty of Pharmaceutical Sciences, Hokkaido University,
Sapporo, Hokkaido 060, Japan

Latex piezoelectric immunoassay (LPEIA) is a new immunoassay method that requires no immobilization of antigen or antibody on the electrode surface, in contrast to earlier immunoassays in which a piezoelectric crystal is used as a microbalance and immobilization is essential. The mechanism of the frequency change observed during the aggregation of antibody- or antigen-coated latex particles is discussed here. LPEIA was used for the detection of C-reactive protein, antistreptolysin O antibody and rheumatoid factor, and is sufficiently sensitive for clinical applications. This method is adaptable and ideal for environmental analysis because of its portability and cost-effectiveness.

An increase in the level of some proteins including some hormones, nonprotein hormones, leukocytes, tissues and viruses in human blood or urine is an indicator of disease or infection. To monitor these levels, various good diagnostic techniques have been developed, among the most sensitive and selective of which involve immunoreactions. For clinical tests, the latex agglutination test is usually employed. When an antigen is added to a suspension of antibody-coated latex particles, immunoreaction causes agglutination of the particles (Figure 1). These reactions are measured in terms of the degree of light scattering or magnitude of change in absorbance of light. The method for optical measurement of the agglutination is called latex photometric immunoassay (LPIA) (*1, 2*). Many kits including antibody-coated or antigen-coated latex particles are commercially available for LPIA. Table I shows the principle behind some latex reagent kits for the determination of various substances such as C-reactive protein (CRP), antistreptolysin O (ASO), rheumatoid factor (RF), ferritin, immunoglobulin, α-fetoprotein and β_2-microglobulin. LPIA is sensitive, precise, gives reproducible results, and is widely used in clinical laboratories for serum testing (*3*).

Table I. Principles Behind some Commercially Available Immunologcal Latex Reagent Kits

Molecule Immobilized on Latex[a]	Target / Aim
C-reactive protein (CRP) ab[b]	detection of CRP[d] in serum
streptolysin O (SO)	detection of antistreptolysin O (ASO)[d] in serum
human immunoglobulin	detection of rheumatoid factor (RF)[d] in urine
human immunodeficiency virus (HIV) ab	detection of AIDS virus in serum
hepatitis B surface antigen (HBs) ab	detection of HBs virus in serum
ferritin ab	detection of ferritin in serum
human growth hormone (hGH) ab	detection of hGH in serum
human chorionic gonadotropin (hCG) ab	detection of hCG in urine / pregnancy diagnosis
estrogen, estrogen ab	detection of estrogen in serum
fibrinogen ab	detection of fibrin and fibrinogen degradation product (FDP)
β_2-microglobulin ab	detection of β_2-microglobulin / kidney dysfunction
human serum albumin (HSA) ab	detection of HSA in urine / kidney dysfunction
α-fetoprotein ab	detection of α-fetoprotein / cancer diagnosis
hIgE ab	allergic contact dermatitis inspection
atrazine[c] ab	detection of atrazine in water and soil / environmental analysis

[a]Applications of LPIA are very wide ranging, because the physically adsorbed antibody or antigen is easily substituted; [b]ab denotes antibody; [c]atrazine: 2-chloro-6-ethylamino-4-isopropylamino-s-triazine; [d]Latex piezoelectric immunoassay was used.

A piezoelectric quartz crystal functions as a microbalance: adsorption on the surface of the crystal changes the oscillation frequency. The adsorption of about 1 ng of material decreases the frequency by 1 Hz when a 9 MHz of AT-cut quartz crystal is used (*4*). The detection limit can be as low as 10 Hz ($\approx$ 10 ng) using a relatively simple apparatus. The advantages of using piezoelectric quartz crystal for analysis is its use is high sensitivity and portability. To be useful in liquid matrices, we must know the properties of the quartz crystal in the respective solutions (*5*). When the crystal is immersed into a solution, the oscillation frequency of the crystal changes in proportion to the product of the density (ρ) and viscosity (η) of the solution, as was shown in equation **1** by Kanazawa and Gordon (*6*) and Bruckenstein and Shay (*7*):

$$\Delta F = -K \, (\rho\eta)^{1/2} \qquad (1)$$

where ΔF represents the frequency change in solution, K a proportionality coefficient. Many attempts have been made to use such a highly sensitive, convenient and portable apparatus for immunoassay. In almost all studies, a thin film was formed on the quartz crystal surface to which antibodies or antigens were fixed, usually by chemical treatment (*8*). In contrast, we developed a new immunoassay method, combining antibody or antigen-coated latex particles and a piezoelectric quartz crystal (*9, 10, 11, 12*). This method does not require the formation of films for the fixation of the antibody or antigen. We named this method latex piezoelectric immunoassay (LPEIA).

Because of the portability of the piezoelectric immunoassay system this technique may be adapted for environmental monitoring and on-site analysis. The equipment of LPEIA consists of a universal counter, a small electronic circuit with quartz crystal and the measuring cell, whole size of which is less than 15×15×15 inches and 5 kg without computer. In addition, a piezoelectric quartz crystal currently costs the equivalent of only $1 in Japan. The cost of an entire experimental LPEIA system (hardware) is only about $3,000. The portability and the low cost of this method makes it suitable for analysis in field studies of pollutants. In this paper, we will describe the principles of LPEIA and its possible application to environmental analysis.

Materials and Methods

Ultrapure water with a specific resistance of more than 18 MΩ cm^{-1} was prepared using Milli-Q (Millipore Ltd., Tokyo, Japan) or NANO pure II (Barnstead Co., Newton MA, USA). Antigen- or antibody-coated latex particles such as ones coated with anti-CRP antibody (Seratestam CRP-H), streptolysin O (Seratestam ASO-E) or rheumatoid factor (Seratestam RF) were gifts from Hitachi Chemical Co., Ltd (Tokyo, Japan). The latex to which the antigen or antibody was adsorbed was made of polystyrene with particle size of approximately 0.1 μm. CRP standard serum (Seratestam S CRP, 5.1 mg dL^{-1}), ASO standard serum (Seratestam S ASO, 204 IU mL^{-1}) and the serum of high ASO titer (1040 IU mL^{-1}), or rheumatoid factor serum (Seratestam S RF, 72 IU mL^{-1}) were gifts from Hitachi Chemical Co., Ltd (Tokyo). Human serum albumin (HSA) and bovine serum albumin (BSA) were purchased from Sigma Chemical Co. (St. Louis, MO, USA). Other chemicals used were of guaranteed reagent grade and purchased from Wako Pure Chemical Co. (Osaka, Japan) or Nakarai Tesque Inc. (Kyoto, Japan). Normal human serum containing no CRP (with total protein concentration of 5.2 g dL^{-1}) was obtained from I.I.C. Japan and diluted 6-fold before use with 10 mM phosphate-buffered saline (PBS), pH 6.5, containing 5 % BSA, 16.5 mM NaN$_3$ and 135 mM NaCl.

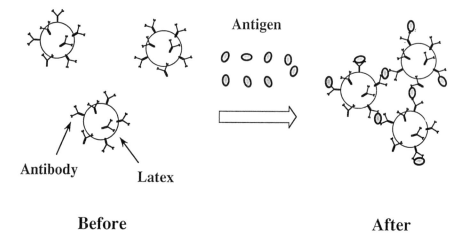

Figure 1. Latex agglutination upon addition of antigen to antibody-coated latex particles.

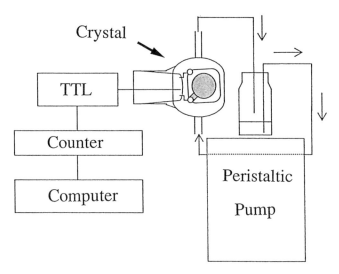

Figure 2. Scheme showing the experimental setup of the latex piezoelectric immunoassay (LPEIA).

AT-cut piezoelectric quartz crystals (9 MHz resonance frequency, 8×8 mm) were purchased from Yakumo Tsushin Co. (Tokyo, Japan) and a silver electrode was deposited on each surface. In order to achieve stable oscillation in saline, one side of the crystal was sealed with silicon sealant (Sealant-45, Shin-Etsu Kagaku, Tokyo) as described previously (5), except in a rheumatoid factor experiment. An electronic circuit was made from TTL gates (SN7400). The resistance and capacitance of this circuit were the same as those of circuit II described in a previous work (5). The signal was fed to a universal counter (Model 7202, Iwatsu, Tokyo) and the data generated from changes in oscillation frequency were stored in a microcomputer (NEC PC9801, Tokyo).

Results and Discussion

Prototype LPEIA: detection of C-reactive protein. Figure 2 shows a schematic illustration of the experimental setup of a prototype LPEIA, in which a flow-cell is used (9). The entire apparatus (except the microcomputer) was accommodated in an air chamber maintained at 37±1 °C, unless otherwise noted. The latex particles were suspended in PBS with the concentration being 5.56×10^{-3} % (1 mL of the stock latex suspension was added to 8 mL of PBS). The suspension (9.0 mL) was circulated over the crystal by means of a peristaltic pump with the rate of circulation being 5.7 mL min^{-1}. The frequency was stabilized in about 10 to 30 min and the stabilized oscillation frequencies differed from one crystal to another. This variation came from the one-side sealing procedure: the variation of amounts of silicon sealant that work as a foreign mass on the crystal changes the frequency. It is noted that the deflection of frequency by addition of samples was taken as response (see below). Figure 3 shows typical data on the frequency decrease caused by the addition of CRP. At t=0, 0.1 mL of CRP stock solution (final concentration, 56 μg dL^{-1}) was added and a reduction in frequency (130 Hz) was observed. About 1 hr after the addition, the frequency became constant and this frequency change (ΔF) was taken as the magnitude of response. The reproducibility was examined; ΔF upon addition of CRP (final concentration of 56 μg dL^{-1} and final latex concentration of 5.56×10^{-3} %) was measured for nine independent crystals. The mean value and the standard deviation were 105 and 20 Hz, respectively, and the coefficient of variation was 19.3 %. As described above, sealing one side of the crystal was necessary for this prototype, and it seemed probable that this sealing altered the sensitivity, which varied from one crystal to another. This might be one reason for the relatively large coefficient of variation observed. Thus, we determined the sensitivity of each sealed crystal, assuming that the frequency change, ΔF, upon immersion in a solution follows the relationship (5, 9):

$$\Delta F = -K (\rho\eta)^{1/2} \qquad (2)$$

Theory describes that the thin liquid layer adjacent to the crystal surface should be forced to move due to the oscillation of the surface and this weight of the layer is accounted as a foreign mass to change the frequency. K values were used as sensitivity coefficients and obtained using 5 or 10 % sucrose solutions, because those physicochemical properties are well known and thus the frequency change will not be dependent on species present in solutions but only on the $\rho\eta$ product. The responses were expressed in terms of $-\Delta F/K$. Adopting this calibrated response resulted in a significant reduction in the coefficient of variation, from 19.3 % to 9.3 %, suggesting

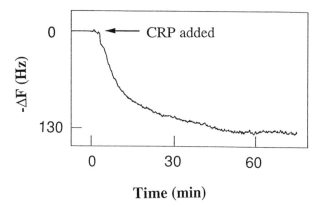

Figure 3. Typical data showing frequency decrease when CRP to a final concentration of 56 μg dL^{-1} was added.

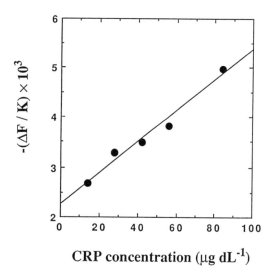

Figure 4. Plot of magnitude of frequency decrease against CRP concentration. The ordinate is expressed as -ΔF/K. For detail see text.

that the sensitivity of the crystal depends on the amounts of the sealant. Figure 4 shows the dependence of the frequency decrease on the CRP-concentration, expressed in terms of $-\Delta F/K$. A linear relationship was observed up to 90 µg dL[-1] final concentration of CRP.

Next, we investigated the specificity of frequency change. A frequency change of only 20 Hz ($-\Delta F/K$ of 1.60×10^{-3}) was detected upon addition of 0.1 mL of normal human serum (essentially free from CRP). Upon addition of 0.2 mL of normal human serum, only 25 Hz was observed. In another control experiment we replaced the latex particles with ASO latex (with the concentration being constant), and observed a change in $-\Delta F/K$ of 1.0×10^{-3} upon addition of 56 µg dL[-1] CRP ($\Delta F/K = 3.8 \times 10^{-3}$ if proper latex particles were used). These results show that the present system using anti-CRP antibody-coated latex particles is specific for CRP. Another control experiment was performed without any latex particles. Addition of CRP or normal human serum (0.1 mL) resulted in a change of 250 to 300 Hz. Comparison of this value with 20 Hz observed in the presence of the latex particles suggests that a large part of proteins added adsorbs to the latex particles. The adsorption of proteins to latex particles, however, does not promote the frequency change but the large changes in frequency are observed only for the proper combination of antigen and antibody. In other words, latex particles prevent the nonspecific adsorption of proteins to the crystal. In clinical analysis, serum CRP levels of 1,000 to 10,000 µg dL[-1] usually needs to be monitored. Thus, the addition of 90 µL of the patient's serum to the present system should be adequate to give a working range of 10 - 100 µg CRP per dL. Figure 5 shows the dependence of the magnitude of the response on the latex concentration: the magnitude of the response increases with increasing latex concentration.

Although the prototype LPEIA was successfully used to measure the concentration of the antigen, several modifications may be done, such as reduction of assay volume, reduction of assay time, and increase in ΔF. Figure 5 shows the magnitude of the increase in response with respect to latex concentration. Although these experiments were done at a relatively low concentration of the latex to minimize consumption of the expensive antibody-immobilized latex, this figure shows clearly that a higher concentration of latex would be more effective. Therefore, some means of reducing the assay volume are necessary to reduce the amounts of latex required. About 1 hr was required for testing this system. It will be ideal to shorten this assay time.

Improvement of assay time and volume: initial rate method with a batch-cell.
The ASO antibody test is the most valuable serologic test used to detect infection by group A β-hemolytic streptococci. Streptolysin O produced by most strains of the streptococci stimulates the production of a specific antibody, ASO (*13*). We modified the original LPEIA to reduce the assay time and volume using streptolysin O-coated latex particles (*10*). The assay volume was reduced to 1.2 mL, and a one-side-sealed quartz crystal was placed in a small cuvette containing a small stirring bar. Circulation with a peristaltic pump was replaced by gentle stirring. The entire apparatus except the microcomputer was accommodated in an air chamber maintained at 25 ± 1 °C. The experimental procedure was essentially the same as the previous one. Calibration using K values was necessary because a one-side-sealed crystal was used. To reduce the assay time, we employed the rate method; the time-dependent frequency change was differentiated with time using computer software and the constant rate of frequency change was obtained. Figure 6 shows the method

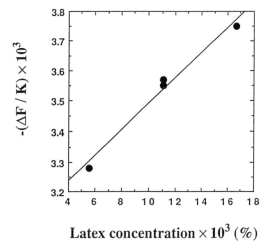

Figure 5. Relationship between latex concentration and magnitude of frequency decrease due to agglutination. The latex concentrations used were 5.56×10^{-3}, 11.12×10^{-3}, and 16.68×10^{-3} %, and the CRP concentration was maintained at $28 \ \mu g \ dL^{-1}$.

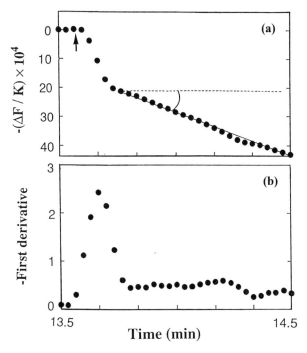

Figure 6. Determination of the initial rate of frequency change. (a) Smoothing of some of the original data (the frequency data were analyzed on computer by the method of least squares with convolution of data points; 5-point quadratic smoothing does not excessively smooth the raw data). The arrow indicates the time of addition of serum to the latex suspension. The slope of the solid line is the initial rate (V). (b) First derivatives of the smoothed data in (a). The ordinate shows the frequency change (ΔF) divided by the proportionality coefficient, K.

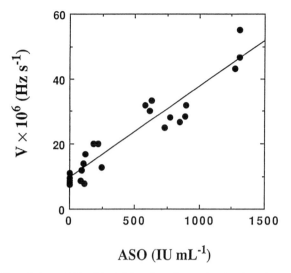

Figure 7. Correlation of V with ASO value determined using the conventional turbidimetric agglutination method. The correlation coefficient (r) was 0.950 (P < 0.01, n = 24).

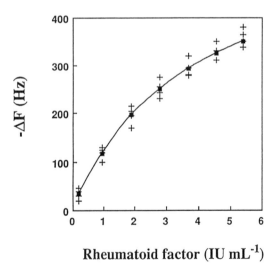

Figure 8. The pre-treatment of the crystal surface made a calibration curve fallen a single curve. Standard rheumatoid factor (72 IU mL^{-1}) of 1 to 30 μL was added to the 360 μL of buffer solution (pH 7.4) containing 10 μL of antibody-coated latex particle. Temperature was 25 °C. The symbols (+) designate the standard deviation from the mean value (denoted by square symbols) of four independent measurements.

used to determine the initial rate of frequency change. A one-side-sealed crystal was dipped in 1.2 mL of suspension containing SO-coated latex to which 0.02 mL of ASO serum (1040 IU mL^{-1}) was added. After addition of the serum, the first derivative changed significantly and then became almost constant (Figure 6b). We determined the initial rate of gradual frequency change (V = -ΔF/K/time) as the slope of the solid line in Figure 6a. After addition of serum, steady values of V were obtained within 2 to 3 min, which represents a great reduction of the assay time.

To investigate the possibility of clinical applications of LPEIA, we used clinical specimens for measurement of ASO. After each serum from 24 individuals was added to the SO-latex suspension, V was determined and then compared with ASO values determined using the conventional turbidimetric agglutination method (Figure 7). The coefficient of correlation between V and the ASO value was 0.950 (P < 0.01), indicating that the values obtained using LPEIA correlated well with those obtained using the conventional method.

Constancy of sensitivity between crystals: detection of rheumatoid factor. In order to achieve stable oscillation in PBS, one side of the crystal was sealed with silicon sealant. However, this treatment is time-consuming and results in differences in sensitivity between crystals. The sensitivity of the crystal was changed after its use, presumably due to alteration of its surface (9, 10). Efforts were made to keep the sensitivity constant for all crystals. Towards this end, a detection cell that did not require sealing was designed.

The newly designed cell had a capacity of 400 µL, and the assay volume was reduced to less than 100 µL when necessary (11). The crystal was washed carefully with water followed by ethanol. It was then placed in a petri dish and heated at approximately 200 °C for 5 min and then cooled to room temperature. Care was taken so that the crystal surface did not come in contact with the petri dish during the heating process. The cooled crystal was allowed to stand overnight at room temperature and was then heated again for 5 min before use. The calibration curves for crystals subjected to this treatment were almost identical. Figure 8 shows an example of pre-treatment of the crystal surface where -ΔF was plotted against rheumatoid factor concentration. The experiments were done for 4 independent crystals. It is noteworthy that -ΔF itself obtained using 4 independent crystals follows almost a single curve, which should be compared with the large variation obtained in the previous experiments using one-side sealing crystal. It is also noted that judging from the measurable rheumatoid concentration range this is clinically applicable.

Mechanism of the frequency change: clues for increasing sensitivity. As described above, two factors can affect the oscillation frequency in solutions: one is $\rho\eta$ change of the solution, and the other is adsorption of molecules from the solution onto the crystal surface. The agglutination may induce the magnitude of the $\rho\eta$ change, and was thought to be the main reason for the frequency change; however, the following observations have ruled this out. After stabilization of the frequency change due to agglutination (-ΔF = 350 Hz), the assay solution was gently removed from the cell and fresh PBS was added. If the $\rho\eta$ change due to the agglutination was the main cause of the frequency change, this replacement is anticipated to restore the frequency value before the agglutination. This was not the case however: only a small frequency change of 20 Hz was observed. Thus, the binding of agglutinated latex particles is an important factor in the change. Also, the frequency change in LPEIA depended mostly on properties of the electrode surface. In addition, LPEIA responses were roughly proportional to the frequency change by the addition of latex particles before agglutination. These suggest that the surface properties that facilitate the binding increase the response (*12*). We previously found that the amount of antibody adsorbed is increased 4-fold when the substrate is coated with a plasma-polymerized allylamine (*14, 15*). Such coating may increase the LPEIA response, and other surface modifications should be tried.

Comments on further use of LPEIA. LPEIA is not limited to clinical use; it can also be used in other fields such as in the analysis of compounds of environmental interest. For the determination of trace amounts of pesticides in water, soil, plant and food samples, a GC/MS or HPLC/MS method is generally used. Compared to these methods, LPEIA has some distinct advantages, such as allowing rapid screening of samples, portability, and simplicity of the instrumentation required.

Triazine herbicides are major agricultural chemicals used for the control of broadleaf plants. One such herbicide, atrazine, is one of the most widely used agrochemicals in the world (*16*). Due to the importance of this compound, a number of analytical methods have been developed to screen for and quantitate this agent in water and other environmental samples (*17*). Examples of these methods include immunoassays using anti-atrazine antibodies immobilized on either polystyrene or glass beads (*18*). Therefore, we are currently testing the feasibility of using anti-atrazine antibody-coated latex for LPEIA to detect atrazine.

References

(1)   Suzuki, T.; Meneki-Kessei Kensa; Kanai, M., Ed.; Rinsyou Kensa Teiyo, 29th ; Kinbara Publ. Co: Tokyo, 1983, pp 1153-1349.
(2)   Tsuda, Y.; Hirose, S.; *Rinsyou Byouri*. **1986**, *67*, 109-117.
(3)   Cambiaso, C. L.; Leek, A. E.; Steenwinkel, F. D.; Billen, J.; Masson, P. L. *J. Immunol. Methods*. **1977**, *18*, 33-41.
(4)   Ward, M.D.; Buttry, D. A. *Science*. **1990**, *249*, 1000-1007.
(5)   Kurosawa, S.; Tawara, E.; Kamo, N.; Kobatake, Y. *Anal. Chim. Acta*. **1990**, *230*, 41-49.
(6)   Kanazawa, K. K.; Gordon, J. G. II. *Anal. Chem.* **1985**, *57*, 1771-1772.
(7)   Bruckenstein, S.; Shay, M. *Electrochim. Acta*. **1985**, *30*, 1295-1300.
(8)   Thompson, M.; Kipling, A. L.; Duncan-Hewitt, W. C.; Rajakovic, L. V.; Cavic-Vlasak, B. A. *Analyst*. **1991**, *116*, 881-894.
(9)   Kurosawa, S.; Tawara, E.; Kamo, N.; Ohta, F.; Hosokawa, T. *Chem. Pharm. Bull*. **1990**, *38*, 1117-1120.
(10) Muratsugu, M.; Kurosawa, S.; Kamo, N. *Anal. Chem*. **1992**, *64*, 2483-2487.
(11) Ghourchian, H., O.; Kamo, N.; Hosokawa, T.; Akitaya, T. *Talanta*. **1994**, *41*, 401-406.
(12) Ghourchian, H., O.; Kamo, N. *Anal. Chim. Acta*. **1995**, *300*, 99-105.
(13) Bach, G. L.; Wiatr, R. A.; Anderson, T. O.; Cheatle, E. *Am. J. Clin. Pathol*. **1969**, *52*, 126-128.
(14) Kurosawa, S.; Kamo, N.; Arimura, T.; Sekiya, A.; Muratsugu, M. *Jpn. J. Appl. Phys*. **1995**, *34*, 3925-3929.
(15) Kurosawa, S.; Kamo, N.; Arimura, T.; Sekiya, A; Muratsugu, M. *Mol. Cryst. Liq. Cryst*. **1995**, *267*, 447-452.
(16) Van Emon, J. M.; Gerlach, C. L. *Environ. Sci. Technol*. **1995**, *29*, 312A-317A.
(17) Thurman, E. M.; Aga, D. S.; Zimmerman, L. R.; Goolsby, D. A. *Preprint of 211th ACS National Meeting*. **1996**, *36*, 74-76.
(18) Dietrich, M.; Krämer, P. M. *Food & Agricultural Immunology*. **1995**, *7*, 203-220.

IMMUNOASSAYS IN ENVIRONMENTAL STUDIES AND MONITORING

Chapter 16

Transition from Laboratory to On-Site Environmental Monitoring of 2,4,6-Trinitrotoluene Using a Portable Fiber Optic Biosensor

Brian L. Donner[1], Lisa C. Shriver-Lake[2], Alfonza McCollum Jr.[2], and Frances S. Ligler[2,3]

[1]Geo-Centers Inc., 10903 Indian Head Highway, Fort Washington, MD 20744
[2]Naval Research Laboratory, Code 6910, U.S. Department of the Navy, 4555 Overlook Avenue, Southwest, Washington, DC 20375

The detection of the explosive TNT using a fiber optic biosensor has successfully undergone the transition from laboratory analysis to on-site, real time sample monitoring due to advances in miniaturization and portability. A larger "breadboard" system capable of monitoring a single probe was replaced by a lightweight, portable sensor that can monitor 4 optical probes simultaneously. This new fiber optic biosensor was taken to two contaminated military bases for on-site testing for TNT in groundwater. Prior to on-site testing, assay variables, including buffers and fluorescence analog concentration, were optimized to permit TNT detection in the 5-20 µg/L range. In addition, several cross reactivity studies were performed demonstrating interference from TNB, a degradation product of TNT but not the explosive RDX or its degradation product, HMX.

A fiber optic biosensor, which employs antibodies as the recognition element, has been developed (1,2). Antibodies have been immobilized on the surface of a fiber optic probe in order to detect various analytes (proteins, explosives, bacteria). A competitive immunoassay has been developed for the explosive 2,4,6-trinitrotoluene (TNT)(3). Due to problems from munition storage and/or ordnance demilitarization, high levels of TNT have been found in soil at several military bases. The mobility of TNT in soil has permitted these explosives in the soil to migrate into the groundwater (4). As TNT and several of its degradation products are toxic, clean-up and monitoring are essential, especially in these days of base closure (5). On-site monitoring would greatly reduce time and cost involved with remediation relative to the current method which involves analysis off-site in a laboratory. Transitioning the fiber optic biosensor from the laboratory to the field was envisioned.

[3]Corresponding author

The TNT assay was initially performed using a fiber optic biosensor which weighed 150 pounds with dimensions 2' x 3', and for all intents and purposes, limited the operator to the confines of a laboratory (Figure 1, top). This sensor monitored a single 200 μm fused silica optical fiber laden with antibodies to detect the analyte. A miniaturized, portable version of this sensor, called Analyte 2000, was developed by Research International (Woodinville, WA) in collaboration with Naval Research Laboratory (Golden, J., et al., A portable multichannel fiber optic biosensor for field detection, submitted). This new sensor is based on the same principles of detection as its predecessor, but has been reduced to 6.5" x 4.5" x 3.5", and weighs 2.5 pounds (Figure 1 bottom). It can monitor four 600 μm fused silica optical fibers probes simultaneously. In addition, the sensor and the laptop computer required to operate it are both fully battery operable, resulting in a system that is highly portable.

The on-site detection capabilities of this sensor were tested at two military bases in the Northwest for the detection of TNT in groundwater. The sites, Umatilla Army Weapons Depot (Hermiston, OR) and Naval Submarine Base Bangor (Bangor, WA), are but two of the more than 50 sites on the US Environmental Protection Agency's (EPA) Superfund list slated for hazardous waste characterization and subsequent remediation. The field trials were sponsored to test alternative methods of site characterization in order to minimize the cost and duration of the characterization/remediation project. The current method of characterization is high-performance liquid chromatography (HPLC) (6) at an off-site laboratory, which takes 2-4 weeks and costs hundreds of dollars per sample. Logically, an inexpensive on-site means of sample analysis would address both the cost and time issues. The portable fiber optic biosensor and 3 commercially available test kits were tested for their correlation with the standard HPLC analysis, and were compared by their efficacy, speed, cost, accuracy, and simplicity. This article will concentrate on the fiber optic biosensor.

Use of this new fiber optic biosensor for the detection of TNT required adapting the protocols and techniques of the TNT assay from the previous system. In addition, new variables were identified for on-site field analysis and testing of real water samples. Several questions were examined, including: a) what conditions of the TNT assay on the breadboard system still hold for the portable system?, b) how do real groundwater samples affect an assay compared to laboratory buffer?, c) what is the ideal buffer for detecting TNT with this system?, d) what is the ideal concentration of our competitor compound for this competitive immunoassay?, and e) to what extent do related compounds cross-react with TNT for binding to the antibody?

Portable Fiber Optic Sensor System

The Fiber Optic Biosensor. The Analyte 2000 (Figure 1, bottom) fiber optic biosensor, developed as a compact, portable update to the breadboard device, consists of four identical computer cards, each supporting a diode laser (635) and a photodiode detector (*1,Golden, J., et al., submitted*). Signals are transmitted to and

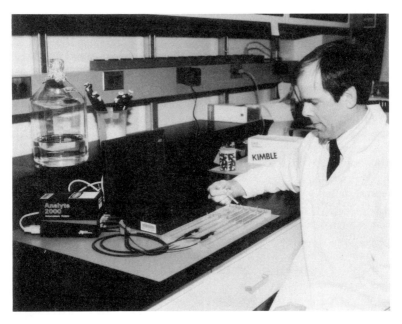

Figure 1. Versions of the fiber optic biosensor. The top picture is the original breadboard FOB which monitored 1 optical probe. The bottom picture is the portable Analyte 2000 FOB which monitors 4 independent optic probes simultaneously.

from the fiber via a specially designed jumper (*7*). The jumper consists of a fiber bundle, with a single central fiber providing the fluorescence excitation and the surrounding fibers in the bundle collecting the fluorescence emission from the fiber probe. Data from the Analyte 2000 is read in pico-Amperes (pA) and displayed graphically in real time on a laptop computer screen.

The principle of detection in both the Analyte 2000 and the breadboard fiber optic biosensor is the same (*1,Golden, J., et al., submitted, 8*). A region of energy known as the evanescent wave extends approximately 100 nm from the surface of the fiber when light is propagated along the fiber via total internal reflection. While, by Snell's law, light striking the core/cladding or core/solution interface reflects within the core, an electromagnetic component of the light traverses the core forming the evanescent wave. The power of this wave decays exponentially with distance from the core, resulting in a sensing region virtually unaffected by fluorescing molecules outside the confines of the evanescent wave. By immobilizing antibodies onto the core surface, one can perform an immunoassay using either a fluorescently tagged antigen or antibody. By using a light source above 600 nm and a fluorescent label that absorbs at this wavelength, one greatly reduces the effect of background fluorescence found in many natural samples (*9*).

Fiber Tapering and Chemistry. The antibody immobilization chemistry used on the fiber probes was first described for glass cover slips (*10,11*). It was then adapted to prepare 200 µm fibers in conjunction with a tapering step (*12*). Preparing 600 µm fibers for use in the portable fiber optic biosensor required only slight modifications from the 200 µm fiber technique. Briefly, the distal 12.5 cm portions of fused silica optical fibers are stripped of their protective cladding and tapered in hydrofluoric acid. The continuously tapered geometry has been shown to help overcome losses of the return signal from the sensing (declad) region back into the transmitting (clad) region of the fiber without loss of excitation light (*13*). The unclad region of the fibers are then silanized, resulting in exposed thiol groups which are then bound to the maleimide reactive site of a heterobifunctional crosslinker. The other functionality of the crosslinker (succinimide) binds to an amine group on the antibody.

Competitive Immunoassays for TNT. Detection of TNT in this assay was based on the competition between unlabeled TNT and a fluorescently labeled analog of TNT for a limited number of antibody binding sites (*3*). A predetermined concentration of the labeled analog, referred to as the Reference or 100% solution, is introduced to the sensing region and allowed to bind. After 4 minutes the fiber is exposed to laser light, exciting any fluorescently labeled antigen within the evanescent wave, which then emits higher wavelength light that couples back into the fiber core. The bound TNT can be removed, thus "regenerating" occupied binding sites, by 1-minute exposure to 50% ethanol (*3*). Then, the same Reference solution is introduced along with the sample containing unlabeled TNT. A reduction in signal proportional to the concentration of TNT is observed (Figure 2), since both compounds are competing for the same number of antibody binding

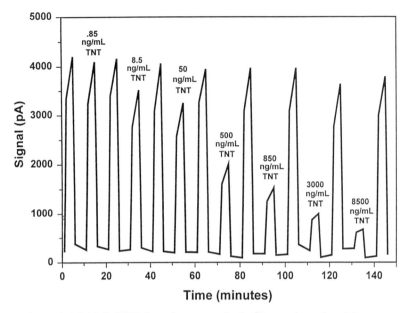

Figure 2. Multiple TNT detections on a single fiber optic probe. Measurement of reference signals and 7 concentrations of TNT were performed on a single fiber.

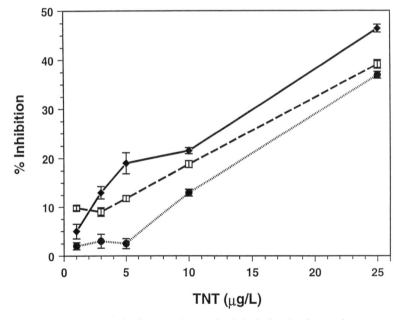

Figure 3. TNT standard curves (± standard deviations) using various concentrations of Cy5-EDA-TNB (♦ - 3 μg/L, □ - 5 μg/L, ● - 10 μg/L).

sites. This signal reduction, referred to as the % Inhibition, is calculated by using the Reference trials before *and* after the TNT sample trials. The mean of these two Reference trials is used to account for any reduction of antibody activity due to repeated tests and regenerations. The percent inhibition is calculated by:

$$\% \ Inhibition \ = \left| 1 - \frac{Test \ signal}{(\dfrac{Reference_{pre} + Reference_{post}}{2})} \right| *100 \qquad (1)$$

Using a fixed concentration of the fluorescently labeled analog along with known concentrations of TNT, one can establish standard curves which can be used to quantitate unknowns.

Optimization of TNT Assay for On-Site Monitoring

The use of the portable fiber optic biosensor in the field required both adapting the assay for TNT from the breadboard device (*3*) to the portable device, as well optimizing several variables one could expect to encounter in the field. The following experiments were performed prior to on-site monitoring.

Labeled Analog of TNT. The compound used to compete with TNT in these assays was a labeled version of trinitrobenzene (TNB) (*3*). This compound, cyanine 5-ethylendiamine-trinitrobenzene (Cy5-EDA-TNB), was chosen not only because of the structural similarity between the TNT and TNB, but also because the monoclonal antibodies used in this assay (NRL clone 11B3)(*14*) were raised against TNB. The fluorescent label Cy5 was chosen because it absorbs at 650 nm, virtually eliminating any interference from naturally fluorescing compounds in real samples.

Several concentrations of Cy5-EDA-TNB were tested to determine the optimal concentration for a competitive assay with TNT on the portable unit by determining the lowest limit of detection of TNT, the best signal-to-noise ratio, and the magnitude of the maximum signal. For each concentration of Cy5-EDA-TNB, a competitive immunoassay using increasing concentrations of TNT (described above) was performed. A minimum of 10% inhibition was arbitrarily picked as the cut-off for the detection of TNT in a sample as this level of inhibition was just above 2 standard deviations from the mean blank samples values. The inhibition curves using 3 ng/mL, 5 ng/mL and 10 ng/mL as the concentration of Cy5-EDA-TNB in the Reference solution are shown in Figure 3.

As evidenced from Figure 3, a Reference solution containing 10 ng/mL Cy5-EDA-TNB detected TNT at a concentration of 10 ng/ml or greater. While both 3 ng/mL and 5 ng/mL Cy5-EDA-TNB appeared to detect TNT at 5 ng/mL, the maximum fluorescence signal when using 5 ng/mL Cy5-EDA-TNB was, logically, greater than that for 3 ng/mL Cy5-EDA-TNB. Low fluorescence signals from the Reference solution are undesirable because small inhibitions are not easily observed. For these reasons, 5 ng/mL was chosen as the concentration of Cy5-EDA-TNB in the Reference solution and the TNT standards and unknown samples tested.

Buffer Components. Several buffer components were tested to determine the optimal buffer for this assay using field samples. Variables tested consisted of the exclusion of salts or the inclusion of detergents, blocking proteins, or organic solvents into the phosphate buffer. The need for an organic solvent directly relates to the hydrophobicity of TNT and similar compounds. The solubility of TNT is 0.02 g/100 g in water, whereas it is 1.5 g/100 g in ethanol, and as high as 132 g/100 g in acetone, at 25° C (5). These solubility characteristics are also why a 50% ethanol solution was previously found to effectively dissociate the antibody-antigen interaction and regenerate the fiber in 1 minute (3). Detergents and blocking proteins were included to minimize nonspecific adsorption.

The buffer components tested included acetone (1%, 5%, and 10%), methanol (5%), ethanol (10%), bovine serum albumin (BSA) (2 mg/ml), Triton X-100 (0.1%), Tween 20 (0.1%), and phosphate buffered saline (PBS). Several Reference buffers were made using combinations of the above components along with 5 ng/mL Cy5-EDA-TNB, with the remaining volume of the samples occupied by reverse osmosis (RO) deionized water. These Reference solutions were compared by their relative signal stability when the RO water was replaced with real water samples that contained no TNT or related compounds. After testing well water, bilge water, salt water, and Potomac River Water, two conclusions were made. The first is that the most stable buffer was composed of PBS (1x), BSA (2 mg/mL), acetone (5%), and Tween 20 (0.1%), which is referred to as PBAT. The second is that all of the real waters samples tested affected the signal to a noticeable extent. Thus Reference solutions were made to include an uncontaminated water blank at the same concentration as the samples. The water blank was obtained from water near the the contaminated site.

Protocol for Sample Analysis. The analysis of unknowns or standards begins with a 2-minute incubation of the buffer PBAT void of any Cy5-EDA-TNB. After the incubation, the fibers are exposed to laser light and a reading is taken. This value is considered the background noise, and is subtracted from all the data. Following the background determination, either a Reference, standard or unknown sample is introduced to the fiber. A reading is taken after a 4-minute incubation period, and the background is subtracted. The Reference, standard and unknown samples all have the same concentration of Cy5-EDA-TNB (5 ng/mL), the same buffer components (PBAT), and the same volumetric concentration of either deionized water (for Reference and standard samples) or blank uncontaminated groundwater (for References and samples). These solutions should differ only by the amount of (unlabeled) explosives and/or explosive degradation products they contain. To regenerate the fiber, a 50% ethanol solution in PBS is incubated with the fiber for 1 minute. Fibers are typically regenerated over 20 times. The average total time for sample analysis, including two Reference solutions, is 18 minutes for four fibers run in parallel. As long as the samples and Reference solutions can be analyzed in an alternating mode, a new set of 4 samples can be tested every 12 min.

Field Trials. Two field trials were conducted at the Umatilla Army Weapons Depot (Hermiston, OR) and Naval Submarine Base Bangor (Bangor, WA) during the summer of 1995. Both sites are on the EPA's Superfund list. In the case of Umatilla and Bangor, the environmental pollution is from contamination of soil and groundwater by explosives, a result of storage, recycling, disposal, and manufacturing of munitions. While a detection protocol exists (*6*), a thorough characterization of a site must be performed so that only contaminated sections are remediated. The current EPA-approved method for this characterization is SW-846 Method 8330, which is an HPLC analysis that takes place off-site at a certified laboratory. While accurate, the method is time consuming (2-4 weeks/sample) and expensive ($250-$1000/sample). In addition, the groundwater is remediated by filtration through granulated activated charcoal (GAC). Saturation of the filters needs to be determined with minimal turnaround time, and an on-site method would be better suited for this application than analysis in a central laboratory. In search of an on-site method of sample analysis that addresses both the cost and time concerns, the EPA sponsored field trials of on-site methods using both commercially available test kits as well as biosensors under development.

Over 100 fibers were prepared at the NRL for the field trials at Umatilla and Bangor using the protocol described above. The antibody-coated fibers were each then secured in a 200 µL glass capillary tube with a "T" connector at each end, which serve as an input and output for the samples. Fibers were stored either in PBS at 4°C, or dried down in a 100 mM trehalose solution and stored at -20°C until the field trials.

The protocol for sample analysis described above was used for all standard and unknown samples during the field trials at both sites. At Umatilla, all assays were performed on-site in an indoor storage area. Assays at Bangor were performed in a water treatment facility on the site partially open to the outdoors. While the purpose of the field trials was to demonstrate a fully portable sensor, electricity, available at both sites, was used for power.

At both Umatilla and Bangor, all samples, standards and their dilutions were provided by Black and Veatch Waste Science, and were stored cool in amber glass bottles. Several standards containing 2,4,6-trinitrotoluene (TNT), hexahydro-1,3,5-trinitro-1,3,5-triazine (RDX) 1,3,5-trinitrobenzene (TNB), octahydro-1,3,5,7-tetranitro-1,3,5,7-tetrazocine (HMX), or a combination thereof were tested at concentrations ranging from 1-10,000 µg/L for standard curves and cross-reactivity information. A blank water sample void of explosives, but representative of the water sample, was provided at each site and was used in the Reference solution, occupying a volume identical to that of the sample being tested in the unknown.

Nineteen unknown groundwater samples and twenty-three unknown groundwater and leachate samples were analyzed with the portable fiber optic biosensor at Umatilla and Bangor, respectively. All samples and standards were analyzed with 85% of the total volume being groundwater (blank or test), using a minimum of three fibers per sample. Many samples and standards from both sites were later reanalyzed at the NRL with the fiber optic biosensor and HPLC.

The limit of detection of TNT using the portable fiber optic biosensor was 20 μg/L for the field samples, compared to a limit of detection of 5 μg/L in spiked laboratory buffer previously reported (3). The limit of TNT detection for the HPLC method SW-846 Method 8330 using a direct injection protocol was 20 μg/L. Qualitative comparison between the fiber optic biosensor and SW-846 Method 8330 using linear regression gave a slope of 1.23 and an r^2 of 0.91. This indicates that on the whole the fiber optic biosensor assay produced slightly higher TNT values than Method 8330. This may be due to cross-reactivity of the antibody with trinitrobenzene since this same trend is seen with other immunoassays for TNT. There were 11.4% false positive and 2.9 % false negative identifications. A sampling of quantitative data from the trials is shown in Table I.

Table I. TNT Detection with Method 8330 and Fiber Optic Biosensor

Sample	Method 8330 (μg/L)	Fiber Optic Biosensor (μg/L)
Umatilla Well 47-4	<20	<20
Umatilla Well 009	2043	3674
Umatilla Well 4-P2	3155	3401
Umatilla Well 4-1-2	198	208
Umatilla Well 4-18-2	299	366
Bangor BEW-2	30	30
Bangor BEW-3	65	185
Bangor LCH-2	<20	<20
Bangor LCH-1PF	2025	1566
Bangor LCH-3PF	170	190
Bangor LCH-4PF	70	148

A full synopsis of the field trial results is in preparation. Overall, the portable fiber optic biosensor was able to detect TNT in both groundwater and leachate samples at both military sites in a quantitative manner. The results of the sensor were similiar to the approved U.S. EPA Method 8330. Limited data was collected on degradation products and other compounds that might cross-react with the antibody to give false results. It is known that the groundwater at Umatilla Army Depot is high in nitrates due to the extensive agricultural farming in the area. These nitrates did not seem to effect the TNT assay.

Cross-reactivity Studies

The parent compounds of the contamination at both sites were TNT and RDX. However, degradation of these compounds results in a whole family of co-contaminants. Many of these degradation products resemble the parent compound, such that they too may bind to antibodies raised to the parent compound. Information about cross-reactivity with 11B3 antibody was gleaned from both the field trial standards tested, as well as additional studies in the laboratory.

Field Standards Cross-reactivity. Of the explosives standards tested in the field, only TNB exhibited a significant cross-reactivity with the 11B3 antibody. Only at the highest concentrations was there any inhibition of signal observed from the RDX and HMX standards, indicating little cross-reactivity of the 11B3 antibody. On the other hand, TNB standards had larger inhibitions than TNT at concentrations greater than 10 µg/L . It should be noted that this monoclonal antibody was raised against a TNB conjugate, accounting for the cross-reactivity and seemingly higher affinity for TNB. Cross-reactivity with TNB is not necessarily a detriment since it is one of TNT's degradation products. Therefore, the presence of TNB indicates the presence of TNT and in addition, while TNB is not the initial source of contamination, it is still a toxic compound and needs remediation as well.

Laboratory Cross-reactivity Studies. RDX and several degradation products of TNT were tested at the NRL to determine the degree with which they bind to the 11B3 antibody and thus interfere with the binding of Cy5-EDA-TNB. Varying concentrations of RDX, TNB, HMX , 2,4-dinitrotoluene (2,4-DNT), 2,6-dinitrotoluene (2,6-DNT), 2,4-dinitrophenol (2,4-DNP), 2-amino-4,6 dinitrotoluene (2-amino-4,6-DNT), and 4-amino-2,6-dinitrotoluene (4-amino-2,6-DNT), all from Ultra Scientific were tested for their interference with the 11B3 antibodies' binding to 5 ng/mL Cy5-EDA-TNB using the competitive immunoassay format described above. Distilled water was used to solvate both the Reference solution and sample solutions for these studies.

Table II shows the concentration needed to get 50% inhibition of the Reference signal (IC_{50}). As mentioned earlier, there is a large amount of cross-reactivity with TNB as would be expected. Two of the compounds, RDX and 2 amino-4,6 DNT leveled off prior to reaching 50% inhibition. Another compound 2,6 DNT reached 50% inhibition at 50 ng/ml but no additional inhibition was observed when higher concentrations of 2,6 DNT was assayed. The compounds that do have some degree of cross-reactivity are degradation products of TNT and would indicate the presence of TNT.

Table II. Cross-reactivity with Fiber Optic Biosensor TNT Assay

Explosive	IC_{50} ($\mu g/L$)
Trinitrotoluene (TNT)	440
1,3,5 Trinitrobenzene (TNB)	30
2,4 Dinitrotoluene (2,4 DNT)	750
2,6 Dinitrotoluene (2,6 DNT)	50*
2-amino-4,6 Dinitrotoluene (2-amino-4,6 DNT)	>1000*
4-amino-2,6 Dinitrotoluene (4-amino-2,6 DNT)	>1000
2,4 Dinitrophenol (DNP)	>1000
hexahydro-1,3,5-trinitro-1,3,5-triazine (RDX)	>1000*
octhydro-1,3,5,7-tetranitro-1,3,5,7-tetrazocine (HMX)	>1000

*These explosives leveled off at or prior to the IC_{50}.

Conclusion

The fiber optic biosensor has been successfully transitioned from a laboratory breadboard system to a field portable instrument. Optimization of the competitive fluorescent immunoassay for TNT reduced variability due to differences in different groundwater samples. Field portability and sensitvity of the biosensor was demonstrated in two successful field trials. In addition, further characterization of possible cross-reactants have shown that only explosives that are degradation products of TNT seem to have any effect on the assay. The portable fiber optic biosensor is able to detect TNT on-site which will result in a reduction in time and cost over the time frame of site characterization and remediation.

Acknowledgements

The authors thank Anne Kusterbeck, Linda Judd, John Bart, George Anderson and Joel Golden for their assistance and suggestions in preparing for the field demonstrations. This work was funded by Office of the Undersecretary of Defense Environmental Security Technology Certification Program (ESTCP). The views expressed here are those of the authors and do not represent those of the U.S. Navy, Department of Defense or the U.S. Goverment.

Literature Cited

1.	Ligler, F.S.; Golden, J.P.; Shriver-Lake, L.C.; Ogert, R.A.; Wijesuria, D.J.; Anderson, G.P. *Immunomethods.***1993**, *3(2)*, 122.
2.	Anderson, G.P.; Golden, J.P.; Cao, L.K.; Wijesuriya, D.; Shriver-Lake, L.C.; Ligler, F.S. *IEEE Engineering in Medicine and Biology.* **1994**, *13(3)*, 358.
3.	Shriver-Lake, L.C.; Breslin, K.A.; Charles, P.T.; Conrad, D.W.; Golden, J.P.; Ligler, F.S. *Anal Chem.* **1995**, *67(14)*, 2431.
4.	Jenkins, T.F.;Walsh, M.E. *Talanta.* **1992**, *39(4)*, 419.
5.	Yinon, J. and Zitrin, S. *Modern Methods and Applications in Analysis of Explosives*, John Wiley & Sons: Chichester England, **1993**.
6.	U.S. Environmental Protection Agency , SW-846 Method 8330.
7.	Saaski, E.; Bizak, M.; Yeatts, J. *SPIE Procceedings.* **1995**, *2574*, 56.
8.	Golden, J.P.; Shriver-Lake, L.C.; Anderson, G.P.; Thompson, R.B.; Ligler, F.S. *Opt Eng.* **1992**, *31(7)*, 1458.
9.	R.P. Haugland, in *Biosensors with Fiber Optics*; D. Wise, L. Wingard, (Eds.); Humana Press: Clifton, N.J. 1991; 85-110.
10.	Bhatia, S.K.; Shriver-Lake, L.C.; Georger, J.; Calvert, J.; Prior, K.; Bredehorst, R.; Liger, F.S. *Anal Biochem.* **1989**, *178*, 408.
11.	Ligler (misspelled Eigler), F.S.; Georger, J.H.; Bhatia, S.K.; Calvert, J.; Shriver-Lake, L.C.; Bredehorst, R. U.S. Patent 5,077,210, December 31, 1991.
12.	Shriver-Lake, L.C., Anderson, G.P., Golden, J.P., Ligler, F.S. *Anal Letters.* **1992**, *25(7)*, 1183.
13.	Anderson, G.A.; Golden, J.P.; Ligler, F.S. *Biosensors & Bioelectronics.* **1993**, *8*, 249.
14.	Whelan, J.P.;Kusterbeck, A.W.; Wemhoff, G.A.; Bredehorst, R.; Ligler, F.S. *Anal Chem.* **1993**, *65*, 3561.

Chapter 17

Detection and Quantitation of the Explosives 2,4,6-Trinitrotoluene and Hexahydro-1,3,5-trinitro-1,3,5-triazine in Groundwater Using a Continuous Flow Immunosensor

John C. Bart[1], Linda L. Judd[2], Karen E. Hoffman[1], Angela M. Wilkins[1], Paul T. Charles[1], and Anne W. Kusterbeck[1]

[1]Naval Research Laboratory, Code 6910, U.S. Department of the Navy, 4555 Overlook Avenue, Southwest, Washington, DC 20375
[2]Geo-Centers Inc., 10903 Indian Head Highway, Fort Washington, MD 20744

Detection of contaminants in a fast, accurate, and cost-effective manner is the first step in the process of environmental clean-up of hazardous wastes. Processes which utilize the high degree of selectivity for a single analyte afforded by antibodies are called immunoassays, and are currently becoming accepted as alternatives to more traditional instrumental analytical techniques in environmental testing. One such immunoassay is described here, and data from two on-site field tests of the sensor are compared to the results obtained by high-performance liquid chromatography analysis. The accuracy and reliability of this sensor is clearly demonstrated.

The U.S. Department of Defense currently owns over 50 military sites which are on the U.S. Environmental Protection Agency's (EPA) Superfund list, many of which are contaminated with explosive compounds. In recent years, immunoassays have become accepted as serious alternatives to more traditional instrumental methods of analysis in the field of environmental monitoring and remediation because they offer advantages in portability, cost per sample, and sample processing time (1). In the specific realm of monitoring explosives in contaminated water or soil samples, the EPA-approved technique (SW-846 Method 8330) is high-performance liquid chromatography (HPLC) coupled with ultraviolet (254 nm) detection (2). Recently, the EPA has been investigating various immunoassays as alternatives to HPLC for the on-site analysis of samples. Hand-held test kits or portable instruments are potentially superior to the standard HPLC test in that they can be used by non-technical personnel on-site to get accurate results in minutes at a fraction of the cost of off-site analysis. In the scenario where a large number of samples must be tested (to determine the extent of contamination at a polluted site, for example), an immunoassay performed on-site could easily afford savings of one order of magnitude in both time and money. During the summer of 1995, two separate tests of commercially available kits (both immunoassays and non-immunoassays) and prototypes of explosives detectors were performed at EPA-monitored sites to determine the accuracy of such on-site analyzers compared to HPLC.

Site Characteristics for the Field Trials

Field tests of the assay kits and prototypes were conducted at two military bases containing sites currently undergoing EPA-monitored remediation of groundwater and soil contamination by various explosive compounds. The health advisory values for 2,4,6-trinitrotoluene (TNT) and hexahydro-1,3,5-trinitro-1,3,5-triazine (RDX) as determined by the EPA are 2 ppb (3,4), though most analytical techniques currently have limits of detection about an order of magnitude higher. TNT levels at these sites had a mean concentration of 915 ppb with a maximum of 3630 ppb. RDX levels had a mean concentration of 1132 ppb and a maximum value of 4580 ppb. Umatilla Army Weapons Depot located outside of Hermiston, Oregon was the first site where analyses were performed. This base is situated in the arid northeast section of the state, where the soil is very sandy and dry during the summer months and contains relatively high levels (1-32 ppm) of inorganic nitrate salts. Operations at this location, dating back to the Second World War, included the dismantling of outdated or damaged munitions. Explosive charges were removed from the various weapons and collected for mass burnings on the open desert floor. This practice led to acres of soil becoming highly contaminated; currently these regions are undergoing composting treatment to remove the explosives from the soil. The empty shell casings were steam cleaned so that the metal could be recycled; a procedure which produced large quantities of explosive-contaminated water which was dumped into shallow, unlined waste lagoons. Over the past five decades, the toxic waste in these lagoons has not remained confined to the topsoil, but has leached into the groundwater aquifer which lies about 90 feet below the surface. While several explosives and their decomposition products are found at this site, the contamination is predominately caused by TNT and RDX. RDX is the more water soluble of the two explosives and can be found in a 45 acre plume in the groundwater. TNT, on the other hand, is less water soluble and is thus confined to a somewhat smaller contamination zone. Since the depot is bordered by several large farms which rely on wells for crop irrigation, remediation of the groundwater aquifer is of the utmost importance. Many monitoring wells have already been drilled throughout the contaminated area and it was from these wells that the samples which were analyzed by the field-portable sensors and test kits were taken.

   The second site where the sensors were tested was Naval Submarine Base Bangor, another former weapons demilitarization site which is located outside of Seattle, Washington. In contrast to the rather sparse terrain at Umatilla, SUBASE Bangor is

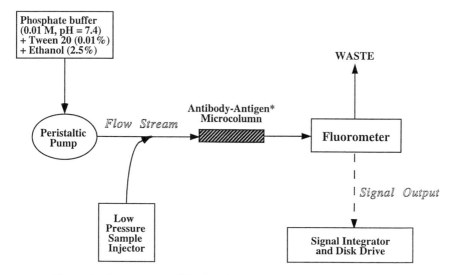

Figure 1. Components of the Continuous Flow Immunosensor (CFI).

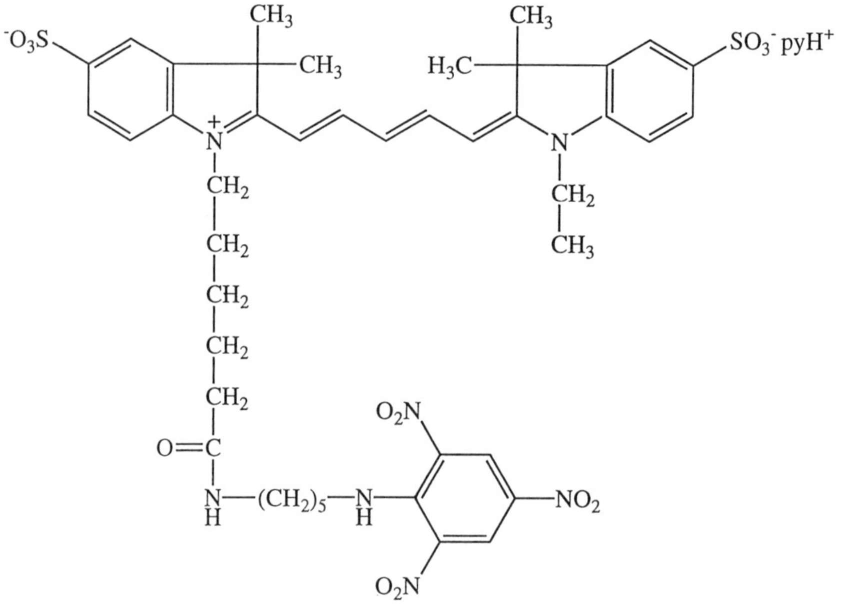

Figure 2. Structure of the sulfoindocyanine dye-labeled TNT analog.

situated in the forests of northwest Washington. This location provided the opportunity to test the devices' capabilities to accurately measure the amount of explosives in aqueous samples which contained significant quantities of humic material. Samples from Bangor also tended to have significantly higher concentrations of 1,3,5-trinitrobenzene (TNB), a photodecomposition product of TNT, and octahydro-1,3,5,7-tetranitro-1,3,5,7-tetrazocine (HMX), a by-product of the synthesis of RDX and an explosive which is occasionally mixed with TNT or RDX to form plastic explosives (5). Contaminated soil has been isolated in plastic-lined pits and is currently being flushed of explosives with large quantities of water. This soil leachate solution is then processed in granular activated carbon (GAC) units. Groundwater which contains explosives is treated by pumping the water up from the aquifer, processing it in a GAC unit, and then returning it to the aquifer. Samples analyzed during this trial included groundwater, soil leachates, and post-treatment water samples.

Operation of the Continuous Flow Immunosensor

Work at the Naval Research Laboratory (NRL) has focused on trying to develop an automated sensor which couples the inherent high degree of selectivity provided by antibodies with the excellent sensitivity of fluorescence spectroscopy. The result is the Continuous Flow Immunosensor (CFI), a device which is capable of detecting explosives at low parts per billion (ppb) levels in aqueous samples (6). A pictorial representation of the CFI is shown in Figure 1. Although the sensor which was used to collect all of the data presented here was the laboratory device (occupying approx. 0.5 m^2 of tabletop space) and not the more portable sensor currently under development, it could be completely unpacked and assembled in less than one hour. The microcolumn contains 100 µL of polyacrylamide beads (Emphaze AB1, 3M) to which monoclonal antibodies against the explosive of interest have been covalently attached. The beads average 60 µm in diameter and contain an azlactone group which is capable of forming a stable amide bond with any primary amine group on the antibody. The 11B3 strain of monoclonal anti-TNT antibody was generated specifically for use at NRL (6) and is not commercially available. The 50518 strain of monoclonal anti-RDX antibody was purchased from Strategic Diagnostics, Inc. (Newark, DE). Screening assays using standard ELISA (enzyme-linked immunosorbant assay) techniques were performed to determine the relative binding affinities of the monoclonal antibodies for labeled and unlabeled antigens prior to assay development. Crossreactivities were determined using the CFI to test a panel of explosives.

Once the antibodies are immobilized on the surface of the support, a solution of dye-labeled antigen (which we synthesized) is added and allowed to fill the available binding sites on the antibodies. An example of the dye-labeled antigen for TNT is shown in Figure 2. The particular sulfoindocyanine dye used to label the antigens was chosen because of its adventitious physical properties. Its multiple charges make for excellent solubility in aqueous solutions, and its large (2.5×10^5 M^{-1} cm^{-1}) extinction coefficient and high (>0.28) quantum yield ensure that even an extremely low concentration of the dye can be detected by fluorescence spectroscopy. Most importantly, this compound has an absorbance maximum at 650 nm and fluoresces at 667 nm, a region of the visible spectrum where there is virtually no interference from naturally-occurring species which might be present in environmental samples. This provides the low fluorescence background needed to get a limit of detection in the low ppb concentration range.

In order to prevent the relatively hydrophobic explosive compounds from aggregating or adsorbing to the walls of the tubing in the sensor, the solvent used in the CFI is a pH 7.4 phosphate buffer (0.01 M) to which has been added an organic cosolvent (2.5% v/v ethanol) and a surfactant (0.01% v/v Tween 20, Aldrich). As the buffer sweeps the sample into the column, the analyte of interest can come in close proximity to the antibody binding sites and there is the possibility that a dye-labeled antigen molecule will be displaced in favor of binding the native explosive. Thus the effluent from the column is a mixture of non-fluorescent analyte which did not bind to the antibodies within the column and fluorescent antigen which was replaced by the analyte present in the sample. Unlike other immunoassays, the best antibody for use in the CFI is not the one with the highest affinity for its antigen, rather it is the one which has the optimal binding coefficient. If the dissociation constant, k_d, is too large, the dye-labeled antigen can dissociate before being replaced by the analyte, thus increasing the background fluorescence level and, consequently, the limit of detection while shortening the column's useful lifetime. On the other hand, if k_d is too small, the analyte will have difficulty displacing the fluorescent antigen and the signal from the fluorometer will therefore be quite small. This property can be controlled two ways: via the strain of antibody used, but also to a lesser extent by the structure of the dye-labeled antigen. We have found that varying the length of the alkyl chain spacer between the fluorescent dye and the explosive analog has a significant effect on the affinity of the antibody for this compound.

Detection of the dye-labeled species is achieved using a fluorometer (Jasco 821-FP, equipped with a 16 μL flow cell) which is located downstream from the column. The species in the flow cell are excited at 632 nm while the fluorescence is recorded at 663 nm. When the analyte of interest is not present in the sample, the signal from the fluorometer vs. time is a flat line, indicating only the background fluorescence was detected. On the other hand, if the explosive being monitored is present in the sample aliquot, a peak in the fluorescence vs. time trace will be observed. There is a direct correlation between the peak area and the amount of analyte present in the sample. Each run was then stored on a 3.5" floppy disk for future reference. Unlike other commercially-available assays which achieve ppb sensitivity only via the concentration of liters of sample down to a few milliliters, the CFI is able to analyze the samples directly, which saves processing time and reduces the amount of sample required for analysis from several liters to less than 1 mL.

Sample Handling and Analysis. All groundwater samples were analyzed on-site shortly (<7 days) after they were obtained from the monitoring wells. In order to prevent small changes in the background signal due to the difference in fluorescence between the buffer and the analyzed solution, a concentrated mixture of buffer, ethanol, and surfactant was added to the sample before it was analyzed. This procedure caused only a minor (<5%) dilution of the sample, so little sensitivity is lost. Because the amount of dye-labeled antigen in the column decreases as more explosive-containing samples are injected, the amount of analyte cannot be measured via a calibration curve. Rather, a standard of known concentration is injected after the sample for direct comparison. The sample was diluted so that the concentration of the analyte was within the linear portion (generally 20-600 ppb) of the dose-response curve and a standard was chosen which was within 20% of the amount of explosive in the sample. Since the concentration of the explosives was unknown prior to analysis, the samples were always diluted at least 30:1 before injection and then reanalyzed at a lesser dilution if necessary. The sample volume injected into the CFI was only 100 μL, thus a 1 mL aliquot contained more than enough sample to analyze for both RDX and TNT in triplicate. In order to prevent loss of analyte, no filtration of the samples was performed, although some samples contained visible particulate material which was allowed to settle before the analyzed aliquot was removed.

Safety Considerations. Because all of the samples which were analyzed contained 0-10 ppm explosives in aqueous solutions, no special steps were taken above the usual safety precautions used in handling organic species (i.e. wearing safety glasses, labcoats, and gloves). Similar precautions were observed when working with antibody solutions. All standard solutions of explosives or their degradation products were purchased as dilute solutions dissolved in either acetonitrile or methanol (Ultra Scientific), thus neat samples of these dangerous materials were never handled. Special caution should be exercised if pure explosives (or soil samples which are suspected of containing significant traces of neat material) are to be used.

Analysis for TNT and RDX by the CFI

Standard Solutions. Before analyzing the samples obtained at the site, standard solutions of explosives in distilled water were injected to verify that the sensor was operating properly. These standards were prepared by an analytical laboratory which had no other function in the field trials. It should be noted that the 100 ppb RDX standard appears to have been improperly prepared, as our results indicated a concentration of 118 ± 17 ppb (average of six injections) and were consistently above 100 ppb. After this issue was raised, the standard was later tested at the U.S. Army's Cold Regions Research and Engineering Laboratory (CRREL) via EPA Method 8330 and was found to contain 164 ppb RDX (an average of two injections). The data for the analysis of some of the standards containing TNT and RDX are shown in Table I. Since the limit of detection of the CFI for TNT is about 50 ppb, it is not surprising that the 10 ppb TNT standard was found not to contain a detectable amount of explosive. Likewise, the limit of detection for RDX is currently about 20 ppb, so the 10 ppb RDX standard would also be expected to be a non-detect. The limit of detection is determined by the amount of background fluorescence (which decreases with increasing antibody/dye-labeled antigen binding affinity) and the sensitivity of the fluorometer. For the more concentrated samples, the sensor was able to predict the concentration of the explosive generally within 10% and with little variation between injections. Once convinced that the sensor was performing accurately with the explosive standards in distilled water, the field samples were then analyzed for TNT and RDX.

Table I. Concentration of TNT and RDX Found in the Standard Solutions

Explosive Concentration	TNT found, in ppb[a]	RDX found, in ppb
10 ppb TNT	N.D.[b]	N.D.
100 ppb TNT	125 ± 17	N.D.
1.00×10^3 ppb TNT	$(1.11 \pm 0.26) \times 10^3$	N.D.
1.00×10^4 ppb TNT	$(9.24 \pm 0.75) \times 10^3$	N.D.
10 ppb RDX	N.D.	N.D.
100 ppb RDX	N.D.	118 ± 17
1.00×10^3 ppb RDX	N.D.	$(9.71 \pm 0.79) \times 10^2$
1.00×10^4 ppb RDX	N.D.[c]	$(9.31 \pm 0.49) \times 10^3$

[a]All concentrations of explosives measured by the CFI are reported as the mean value $\pm$ one standard of deviation.
[b]Not detected.
[c]Of three injections, only one produced a detectable signal (equal to 60 ppb).

Analysis of the Field Samples for TNT. The CFI proved to be an accurate and reliable device for the detection of TNT in the field. As shown in Table II, the samples often contained a complex mixture of explosives, including the degradation product TNB (which crossreacts with the anti-TNT antibody used in this study when present at high concentrations) and high levels of RDX and HMX. As with the case of the explosive standards, samples which contained concentrations of TNT below the detection limit of the CFI (such as 4-1-4) did not give a measurable response and were thus classified as non-detects. On the other hand, even samples which had TNT levels just above the detection limit (such as BEW-3) could be analyzed with excellent accuracy, even in the presence of larger amounts of RDX and TNB. In fact, even groundwater which was contaminated with RDX and HMX at parts per million (ppm) concentrations could be tested for TNT and reasonable results obtained. Table II accurately displays a general trend that was found upon comparison of the HPLC results to those acquired using the CFI, namely that the CFI numbers tended to be higher. Water from well 4-P1 is an extreme example of this trend, where the CFI found almost 2000 ppb more TNT than reported by HPLC. It should be noted that the three other TNT assays operating during these field trials found TNT in concentrations between 3245 and 6600 ppb in this particular sample. This finding will be discussed further below. Figure 3 shows the contour plots obtained for the analysis of well water at the Umatilla Army Weapons Depot. With the exception of the differences in TNT concentration mentioned above, the plots are very nearly identical.

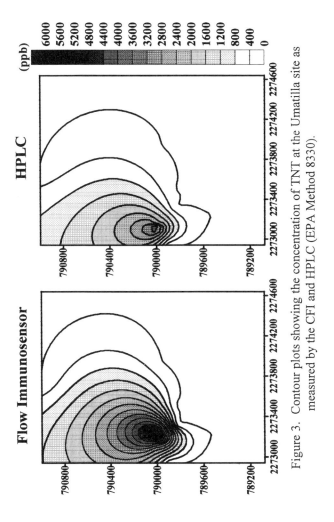

Figure 3. Contour plots showing the concentration of TNT at the Umatilla site as measured by the CFI and HPLC (EPA Method 8330).

Table II. Concentration of TNT Found in Selected Field Samples

Sample	TNT, ppb via HPLC	TNT, ppb via the CFI[a]	TNB, ppb via HPLC	RDX, ppb via HPLC	HMX, ppb via HPLC
4-1-4	25	N.D.[b]	N.D.	32	N.D.
BEW-3	67	62	101	333	N.D.
4-18-2	299	368	21	188	180
4-P1	2396	4124	218	2550	1450
4-P2	3155	3522	285	2790	1610

[a]All concentrations of explosives measured by the CFI are reported as the mean value of 2-3 injections. The overall average percent error for analysis of TNT by the CFI was $\pm$ 16%.
[b]Not detected.

Analysis of the Field Samples for RDX. All groundwater samples that were tested for TNT were also analyzed by the CFI to determine the amount of RDX that was present. This is one of the advantages that the CFI possesses, as only three of the five devices or analysis kits tested at these field trials were able to measure the concentration of both explosives. A small sampling of the results from these site tests is shown in Table III. Overall, the assay for RDX was more accurate than the corresponding analysis for TNT, as manifested by the fact that the standard of deviation in the CFI's RDX assay was smaller than that for the TNT assay. There are several reasons for this trend, and all will be presented below. As with the TNT data, contour plots (not shown) mapping the

Table III. Concentration of RDX Found in Selected Field Samples

Sample	RDX, ppb via HPLC	RDX, ppb via the CFI[a]	HMX, ppb via HPLC	TNT, ppb via HPLC	TNB, ppb via HPLC
BEW-5	82	66	N.D.[b]	N.D.	N.D.
PRE 456	175	140	N.D.	N.D.	N.D.
4-1-1	740	711	335	535	45
Well 009	3270	3527	1770	2043	449
4-P4	4500	4439	81	N.D.	41

[a]All concentrations of explosives measured by the CFI are reported as the mean value of 3 injections. The overall average percent error for analysis of RDX by the CFI was $\pm$ 19%.
[b]Not detected.

concentration of RDX over the entire Umatilla site revealed that the CFI was quite capable of generating the same results as HPLC, but at a fraction of the cost and time required by the conventional method.

Discussion of the Field Trial Results

The CFI performed quite well at the task of quantifying TNT and RDX in samples of groundwater which often contained a complex mixture of explosives in addition to naturally-occurring species such as humic material (which could interact with the antibodies in a deleterious fashion). As shown in Figure 3, the CFI excels at defining the regions of high, low, and non-existent contamination by explosives. As such, it provides on-site analytical data in a manner which is much faster and more cost-effective than HPLC with little sacrifice in accuracy. If necessary, HPLC could be used to confirm the most crucial results, such as defining the boundary between wells which contain explosives at concentrations above the remediation level and those which do not. Differences in the measured TNT levels between the CFI and HPLC displayed an *apparent* tendency of the CFI to overestimate the amount of TNT in the samples. This is most probably due to the problems encountered in acquiring accurate independent HPLC analyses; the first analytical lab (contracted by the company running the trials for the EPA to measure the amount of explosives and degradation products in the field samples) produced unusable data. Errors included: misuse of the low-level salting-out method of explosives extraction prior to HPLC analysis (extraction only needed for parts per trillion sensitivity, thus the high-level direct injection method should have been employed), incomplete extraction of the analytes (44-77% recoveries), holding times for extraction and analysis missed for a majority (52 of 90) of the samples, surrogate recoveries outside of QC limits, sample fortification errors, samples outside of the precision relative percent difference (RPD) requirement ($\pm 50\%$), and positive detections in the method (i.e. solvent) blanks. The samples were reanalyzed via HPLC at a different lab (CRREL), but only after the acceptable stability period of nitroaromatic explosives in groundwater (40-60 days if refrigerated) had been exceeded (7). Many samples were found to contain substantial quantities of aminoaromatics, the products of the microbial degradation of the nitroaromatics TNT and TNB (8). Thus the HPLC data came from samples which truly did contain significantly less TNT and TNB than when the CFI data was collected several months earlier. Since RDX and HMX are known to be much more stable over the time period involved here (7), it is clear that the major reason that the CFI's TNT data were consistently high (but the RDX data were much more accurate) was due to the microbial degradation of the TNT during the time between when the field analyses and the HPLC analyses were performed. Thus, while data for both TNT and RDX are presented above, no quantitative conclusions can be drawn from the comparison of the CFI data to that of HPLC for the TNT assay. The instability of nitroaromatic explosives in aqueous solutions only further underscores the need for immediate analysis of groundwater samples on-site to achieve the most accurate measurement of the concentration of these compounds.

Acknowledgments. J.C.B. graciously acknowledges the American Society for Engineering Education for providing his postdoctoral fellowship. Thanks to the Black & Veatch Special Projects Corp. for organizing all of the data collected by the various on-site tests and the analytical labs. This work was supported in part by a grant from the Strategic Environmental Research and Development Program.

Literature Cited.

1. Van Emon, J. M.; Lopez-Avila, V. *Anal. Chem.* **1992,** *64,* 79A.
2. U.S. Environmental Protection Agency *Nitroaromatics and nitramines by high-performance liquid chromatography;* Draft Method 8330, SW-846; U.S. EPA Office of Solid Waste and Emergency Response: Washington, DC, 1992.
3. U.S. Environmental Protection Agency *Trinitrotoluene health advisory;* U.S. EPA Office of Drinking Water: Washington, DC, 1989.
4. U.S. Environmental Protection Agency *Health advisory for RDX;* U.S. EPA Criteria and Standards Division, Office of Drinking Water: Washington, DC, 1988.
5. U.S. Army *Military explosives;* Department of the Army Technical Manual TM9-1300-214: Washington, DC, 1984.
6. Whelan, J. P.; Kusterbeck, A. W.; Wemhoff, G. A.; Bredehorst, R.; Ligler, F. S. *Anal. Chem.* **1993,** *65,* 3561.
7. Golden, S. M.; Grant, C. L.; Jenkins, T. F. *Evaluation of pre-extraction analytical holding times for nitroaromatic and nitramine explosives in water;* U.S. Army Cold Regions Research and Engineering Laboratory; Special Report 93-24: 1993.
8. Walsh, M. E.; Jenkins, T. F.; Schnitker, P. S.; Elwell, J. W.; Stutz, M. H. *Evaluation of SW-846 Method 8330 for characterization of sites contaminated with residues of high explosives;* U.S. Army Cold Regions Research and Engineering Laboratory; Special Report 93-5: 1993.

Chapter 18

Evaluation of Immunoassay for the Determination of Pesticides at a Large-Scale Groundwater Contamination Site

T. R. Dombrowski[1], E. M. Thurman[1], and G. B. Mohrman[2]

[1]U.S. Geological Survey, 4821 Quail Crest Place, Lawrence, KS 66049
[2]Rocky Mountain Arsenal, Headquarters Building,
Commerce City, CO 80022

Pesticide concentrations in ground water at Rocky Mountain Arsenal (RMA) near Denver, Colorado, were determined using solid-phase extraction (SPE) gas chromatography/mass spectrometry (GC/MS) procedures and enzyme-linked immunosorbent assay (ELISA) for cyclodiene insecticides and triazine herbicides. Matrix interferences resulted in inconclusive results for some GC/MS analyses due to baseline disturbances and co-elution, but ELISA analyses consistently gave definitive results in a minimum amount of time. ELISA was used initially as a screening method, and pesticide concentrations and plume extents identified by ELISA were confirmed by SPE-GC/MS. A high degree of correlation was seen between results from GC/MS and ELISA methods for the triazine herbicides (correlation coefficient (R^2) = 0.99). All areas with high pesticide concentrations were found to be within the boundaries of RMA.

Enzyme-linked immunosorbent assay (ELISA) has proven to be an increasingly important technique for the analysis of an expanding list of environmentally significant compounds. One major example of this growing field of research is the application of immunoassay techniques to the evaluation of pesticide concentration and transport in ground and surface water in the central United States *(1,2,3)*. Within these evaluations, ELISA has been demonstrated to perform in a consistent, reliable manner in the characterization of agricultural pesticide residues in water and soil. The major part of the contamination described in these evaluations, however, usually is confined to the natural water systems present in agricultural areas, and the potential contaminants likely to be encountered (pesticides and fertilizers) are fairly well known. Matrix interferences usually are due to naturally occurring dissolved organic matter (such as humic and fulvic acids) and pesticide degradation products and metabolites, which have been characterized to a broad extent *(4)*. The use of ELISA to characterize extensively contaminated ground water has not been as well documented, and the performance of the technique in these situations is not as well understood.

Rocky Mountain Arsenal (RMA), a 70-km^2 tract of land 14.5 km northeast of Denver, Colorado (Figure 1), has been the site of several intensive chemical research and production projects. From the 1940's to the 1960's the RMA was the principal center for the production of chemical agents and their constituents and for emptying

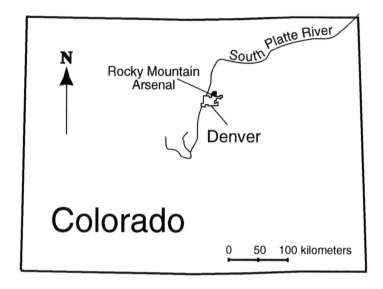

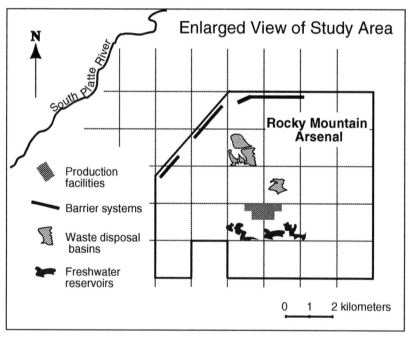

Figure 1. Location of the study area and significant features at the Rocky Mountain Arsenal near Denver, Colorado.

canisters of unused mustard gas by the U.S. Department of the Army *(5)*. In the late 1940's, production facilities at RMA were leased to private industry and were used to manufacture agricultural pesticides.

Waste and other industrial byproducts of the manufacturing processes from both the U.S. Department of the Army and private industry were disposed of in accordance with the accepted protocols of that time period. Since the first contaminant problems were recognized in the mid-1950's, innovative containment and cleanup of these wastes have been undertaken on a massive scale by all organizations associated with RMA property *(6)*. Extensive barrier systems have been installed on the downgradient boundaries of RMA (Figure 1). These systems consist of an impermeable, physical barrier (bentonite) to the ground water flowing off post to the north and northwest of RMA, combined with an activated-carbon filtration system that intercepts ground water. The ground water is pumped through a series of activated-carbon filters and then recharged back into the subsurface on the opposite (downgradient) side of the barrier system. The emplacement of these and other related measures has resulted in the reduction of contaminants moving offsite to concentrations less than the maximum acceptable levels established by the U.S. Environmental Protection Agency (EPA) and the Colorado Department of Health *(7)*.

Target analytes and "fingerprint" compounds (materials known to be unique to RMA) were identified early in characterization and remediation activities to allow the specific contaminant-plume boundaries to be positively identified and mapped. Aggressive monitoring policies have resulted in the establishment of a network of more than 1,200 wells that extends throughout RMA property and into the surrounding areas. The volume of analytical work required to support the monitoring network activities combined with sampling constraints due to well volume, hazard and accessibility, and the complex chemical matrix present in many areas of RMA, present a major challenge in time, expense, and overall complexity of analysis.

The application of ELISA as a rapid screening technique for the identification of pesticide contaminants at RMA was examined. ELISA was selected to be evaluated as a screening technique at RMA because several characteristics of the immunoassay may serve to minimize the constraints listed above. ELISA techniques have several advantages over conventional instrumentation methods currently in use at RMA in that ELISA has very small sample-volume requirements, is fast, field-portable, and is highly specific. The small sample volumes required by ELISA (usually 300 μL or less) are a benefit where sample volumes are limited and where several analyses may need to be done on a single sample. The small sample-volume requirement is also beneficial when the samples are highly contaminated with toxic materials. Small volumes minimize the risk of exposure to these compounds during handling, transport, and storage of the samples.

The specificity of the technique comes from the inherent physiologically selective requirements for antibodies produced in a living system. Immunoassay utilizes the specific binding sites of antibodies to recognize and bind to a single compound or members of a class of compounds. This "targeted" binding can be used in conjunction with a labeled enzyme conjugate of the target analyte (competitive binding assay) or a separate, labeled antibody specific to the structural conformation of the antibody-analyte complex (noncompetitive sandwich assay) to provide information on the presence and concentration of target analytes in a sample.

Because of the highly specific nature of the ELISA technique, many other contaminants present in the same sample can be effectively "screened out," and the analyte(s) of interest reproducibly determined *(8)*. The specificity of the binding site is based on chemical structure and conformation, however, and allows ELISA to exhibit some cross reactivity to compounds with a chemical conformation similar to the target analyte. The extent of this cross reactivity, therefore, needs to be adequately characterized to identify the configuration of the compounds to which the ELISA kit will respond, particularly when dealing with a highly complex, uncharacterized sample

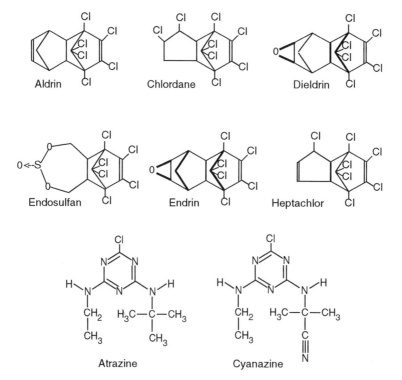

Figure 2.   Chemical structures of target cyclodiene and triazine pesticides. Note: For dieldrin and endrin, heavy lines indicate bond planes above the plane of the illustration.   Narrow lines indicate bond planes below the plane of the illustration.

matrix. A separate analytical method for validation also needs to be used, both in the initial evaluation of the kit performance and in routine analyses after the kit response has been fully characterized. Positive identification methods such as gas chromatography/mass spectrometry (GC/MS) or liquid chromatography/mass spectrometry (LC/MS) often are used for this purpose. Initially, all samples are usually analyzed by both the ELISA and the validation methods. After routine analysis procedures and responses have been established, a designated fraction or percentage of the samples are analyzed by the validation method for quality-assurance purposes. In several areas of RMA, a highly complex chemical matrix, including pesticides and a variety of complex organic and inorganic salts, was present in ground water. Some of the organic compounds were present at parts-per-thousand concentration levels *(5)*. As it was possible that compounds with structural similarities to the target analytes were present, it was unknown how the ELISA would perform in this complex matrix and what possible interferences might be encountered.

The ELISA kits selected for the research described herein have been used extensively in evaluations of herbicides in the water of the midwestern United States *(1,2,3)*. Concentration ranges for the triazine and cyanazine ELISA kits used in these evaluations were found to be adequate for the concentration levels resulting from the agricultural application of atrazine and cyanazine. These and other evaluations have shown a high degree of correlation between the data obtained from both the triazine and cyanazine ELISA analyses and that obtained from concurrent GC/MS analyses *(9)*. This previous research, combined with the ease of operation and small sample-volume requirements of the ELISA kits, was a major factor in selecting ELISA as a screening technique for current research.

This study was conducted by the U.S. Geological Survey in cooperation with the U.S. Department of the Army from August 1994 through October 1995. Specific compounds of interest to this research were the cyclodiene insecticides--aldrin, chlordane, dieldrin, endosulfan, endrin, and heptachlor, and the triazine herbicides, atrazine and cyanazine (Figure 2). The specific objectives of this study were: (1) The identification of pesticide contamination (both current location and potential path) at RMA using ELISA and (2) the evaluation of ELISA as a rapid screening technique for a large-scale, ground-water contamination site with a highly complex chemical matrix. This paper focuses mainly on the second objective.

Experimental Methods

All wells involved in this study were sampled from May 1994 through October 1995. The sampled wells were selected from either the confined and unconfined aquifer systems (with the majority in the unconfined system) and represented a geographical distribution that included a range of contaminant concentrations from generally uncontaminated to very contaminated water. Duplicate samples were collected at a frequency of 1 in 10. Sample and trip blanks were collected with the same frequency. Well depths, screened intervals, pumping methods, and water levels were noted at the time of sample collection. Samples were collected in clean, baked, 125-mL amber glass jars with Teflon[1]-lined lids, stored at or below 4 °C, shipped within 48 hours of sampling, and analyzed (ELISA) or extracted (solid-phase extraction (SPE) GC/MS) upon receipt at the U.S. Geological Survey laboratory in Lawrence, Kansas.

ELISA Analyses. All ELISA analyses were performed according to the manufacturer's instructions (Millipore Corporation, Bedford, MA, for the cyclodiene and triazine kits, and Ohmicron Corporation, Newton, PA, for the cyanazine kit) using reagents included with the kits. Cross-reactivity and sensitivity studies were carried out using standards obtained through National Institute of Standards and Technology (NIST) traceable sources. The separate standard solutions required for each kit were prepared from neat standard materials. Appropriate amounts of each of these materials

---

[1]NOTE: Please see Acknowledgments page 232.

were weighed and dissolved in methanol (HPLC grade, Fisher Scientific, Pittsburgh, PA). The standard solutions then were diluted to working concentrations using distilled water generated through activated charcoal filtration and deionization with a high purity, mixed-bed resin, followed by a second activated charcoal filtration step, and finally distillation in an automatic still.   Stock standard concentrations were calculated to ensure that dilute standards would contain less than 0.5% organic solvent to minimize interference with the antibodies in the immunoassay. Cyclodiene ELISA analyses were performed using chlordane as the calibration standard.   Cyanazine ELISA analyses used cyanazine as the calibration standard. Triazine ELISA analyses used atrazine as the calibration standard. Standard solutions and negative controls were analyzed with all sample sets.   All standards and negative controls were analyzed in duplicate; samples were analyzed in duplicate or triplicate.

Calibration curves were generated for each sample set. Plots of the B/B_0 values (B/B_0 values represent the optical density of the sample concentration (B) divided by the optical density of the blank solution (B_0)) versus log of the corresponding standard concentration were used to identify the linear concentration range of each standard and to identify IC_{50} (concentration of the compound required to give a B/B_0 value of 50%) and LDD (least-detectable dose, defined as B/B_0 of 0.90) values.

SPE-GC/MS Analyses.   Validation of the data obtained from the triazine and cyanazine ELISA analyses was obtained through a SPE-GC/MS method established for the determination of pesticide residues in water samples *(10-13)*. A brief overview of this method follows.

Pesticide-grade methanol and ethyl acetate (Fisher Scientific, Pittsburgh, PA) were used.   Triazine herbicide (atrazine, cyanazine, propazine, simazine) standard materials were obtained from Supelco (Bellefonte, PA); terbuthylazine (the surrogate recovery standard) was obtained from EPA Pesticide Chemical Repository (Research Triangle Park, NC); phenanthrene-d_{10} (the internal standard) was obtained from Ultra-Scientific (North Kingstown, RI); deethylatrazine and deisopropylatrazine (triazine metabolites) and cyanazine amide (a cyanazine metabolite) were obtained from Ciba-Geigy Agricultural Division (Greensboro, NC). All standard materials were obtained at >97% purity for standard preparation. Concentrated stock and spiking solutions were prepared in methanol, except for phenanthrene-d_{10} which was prepared in ethyl acetate. Distilled water was generated as discussed previously.

As the isolation and analysis procedures used have been published previously *(10-13)*, only a brief summary of the method is outlined below.   Herbicides (triazine class and cyanazine) and metabolites were isolated from water samples using C-18 SPE cartridges (Sep-Pak plus, Waters, Milford, MA).   Prior to extraction, each water sample was spiked with terbuthylazine as a surrogate standard. A 100-mL volume of each sample then was pumped at 20 mL/min through a SPE cartridge that had been conditioned by a Millilab 1A Workstation (Waters-Millipore, Milford, MA). The SPE cartridge then was eluted with 2.5 mL ethyl acetate, and the eluate spiked with internal standard, phenanthrene-d_{10}. The ethyl acetate eluate then was evaporated to a volume of approximately 100 μL under a nitrogen stream and transferred to a 200-μL glass-lined polystyrene crimp-top vial. All sample extracts were stored at -10 °C until analysis by GC/MS.

GC/MS analyses were performed on a Hewlett-Packard 5890A GC with a 5970A mass selective detector (MSD) (Palo Alto, CA).   Chromatographic separation was accomplished by a Hewlett Packard (Palo Alto, CA), 12-m x 0.2-mm-i.d., HP-1 or Ultra-1 capillary column with 0.33-μm methylsilicone film. Quantification for each analyte was based on the internal standard, phenanthrene-d_{10}. Identification of target analytes was based on the presence and relative response ratios of the base peak and one or two confirming ions (molecular ion wherever possible) and with a retention time match of $\pm$ 0.2% relative to the phenanthrene-d_{10}. The quantitation limit for all compounds analyzed by this method was 0.05 µg/L.

Results and Discussion

Quality-control samples analyzed included trip and sample blanks, duplicate samples (samples taken at the same site and time) and replicate analyses (repeated analyses of the remaining sample volume from a single sample). For the ELISA analyses, all method blanks agreed to within 10% (coefficient of variance) of the negative control, and duplicate samples agreed within 20% of the mean value. Replicate analyses agreed within 12% of the mean. All initial analyses that gave contaminant concentrations that exceeded the linear range of the ELISA kits were diluted with distilled water and reanalyzed. The data obtained from the cross-reactivity studies showed the same overall trends as noted in product information supplied by the manufacturer.

Cyclodiene Results. The cyclodiene ELISA microtiter plate kit initially used in this study was rated by the manufacturer as having a quantitation range of 5.0 to 100 µg/L (as chlordane). The kit showed the highest sensitivity toward endrin and endosulfan (100% and 50%, respectively) The 56 ground-water samples analyzed showed a wide concentration range. Cyclodiene concentrations obtained by ELISA analyses ranged from <15 to 2,200 µg/L. Most (87.5%) of the samples showed cyclodiene contamination. Highly contaminated areas were localized inside the RMA boundaries.

The method quantitation limit (10 times the standard deviation of the blank) for the ELISA analyses was set at 15 µg/L, resulting in a linear quantitation range of 15 to 100 µg/L. Historical GC data available in unpublished reports from RMA for wells used in this research showed cyclodiene concentrations that ranged from nondetectable to about 100 µg/L (composite values from the four cyclodienes that were evaluated separately--aldrin, chlordane, dieldrin, and endrin) with most of the samples showing concentrations less than 10 µg/L. The quantitation range of the cyclodiene ELISA kit, therefore, approximated the concentration range expected in RMA samples.

Although the relative concentration trends were similar, the cyclodiene ELISA results displayed a positive bias when correlated with the historical GC data for the same wells. This may indicate that an interfering compound(s) is present in the ground water. Another potential explaination may be that the historical GC data used in the correlation only identified concentration levels for aldrin, chlordane, dieldrin, and endrin. The ELISA kit is sensitive to several other compounds in this same family (heptachlor, endosulfan, etc.) and the presence of these other compounds at high concentrations may have resulted in the positive bias displayed in the correlation curve.

Because low-concentration samples may benefit from preconcentration methods, a SPE procedure is being developed for these samples. Due to the highly complex ground-water matrix, the standard SPE procedures previously used are being modified to increase selectivity in retained compounds. The development of a validation procedure for the cyclodiene insecticides by GC/MS is currently in progress.

Triazine Results. The triazine ELISA microtiter plate kit initially used in this study was rated by the manufacturer as having a quantitation range of 0.1 to 2.0 µg/L (as atrazine). For the 56 wells sampled, triazine concentrations ranged from <0.1 to 23 µg/L by ELISA. GC/MS atrazine concentrations ranged from <0.05 to 5.32 µg/L, indicating either cross-reactivity from related compounds, or interference from non-specific binding of other matrix components, or both. No false positives were observed for the triazine ELISA. Of the wells sampled, more than 50% showed nondetectable concentrations (<0.1 µg/L) of triazine compounds. A small percentage (15.4%) of the wells sampled showed triazine concentrations >1.0 µg/L, and 9.2% showed concentrations >3.0 µg/L, the EPA Maximum Contaminant Level for atrazine in drinking water (*14*). All of the samples that were above 3.0 µg/L cyanazine were from wells inside the RMA boundaries.

Triazine concentrations >5 µg/L (analyzed by ELISA) in samples from four

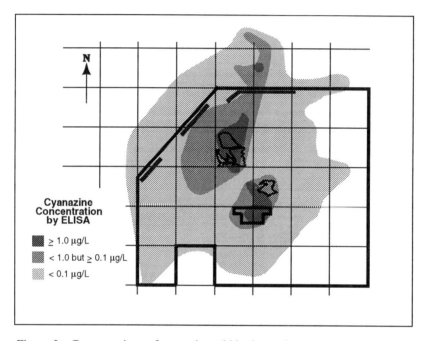

Figure 3. Concentrations of cyanazine within the study area as identified by ELISA analyses. Note: Concentration intervals were extrapolated by hand from existing well data for visualization purposes only and do not represent a statistical calculation of probable cyanazine concentration for nonsampled wells falling within the concentration interval.

wells prompted evaluation by GC/MS techniques and led to the identification of cyanazine as a contaminant at RMA. Samples exhibiting high triazine concentrations by ELISA were extracted and analyzed by the standard GC/MS herbicide method outlined previously to confirm the specific atrazine concentrations present. The results showed atrazine concentrations much higher than those commonly encountered in samples in agricultural areas but substantially less than the concentrations indicated by the triazine ELISA. Cyanazine concentrations in these samples, however, were found to exceed 5.00 µg/L in all cases (average = 34.8 µg/L, maximum = 97.3 µg/L). Atrazine concentrations averaged 3.70 µg/L (maximum = 5.32 µg/L) for samples from the same four wells.

Because of the similarity in structural conformation of atrazine and cyanazine (Figure 2), the triazine ELISA displayed some sensitivity to both compounds; the response to cyanazine was 2.6% of the response to atrazine. It is theorized that the sensitivity of the triazine ELISA toward atrazine, when combined with the cross reactivity of this ELISA toward cyanazine when both atrazine and cyanazine were present at high concentrations, resulted in the increased triazine concentrations reported by the ELISA. These findings point out the potential benefits of a broad or class-sensitive ELISA in which cross reactivity may lead to the identification of a previously unsuspected contaminant. Whereas the total response of this ELISA kit was due primarily to the presence of atrazine and cyanazine, GC/MS analysis of these samples showed that there were other cross-reacting triazine compounds present that also contributed to the overall response such as propazine, simazine, prometryn, and prometon.

Cyanazine Results. With the confirmation of cyanazine by GC/MS methods in samples from 4 of the initial 56 wells sampled, a much larger scale ELISA screening program involving 353 wells was established to characterize the extent of cyanazine contamination at RMA, and cyanazine subsequently was evaluated separately for all RMA samples. The cyanazine ELISA kit used in this study had a magnetic particle format, rated by the manufacturer as having a quantitation range of 0.1 to 3.0 µg/L. Cross reactivity of this ELISA toward compounds other than cyanazine was extremely low (with no significant response to atrazine), making it nearly compound specific.

Cyanazine concentrations ranged from <0.1 to 180 µg/L by ELISA. GC/MS cyanazine concentrations ranged from <0.05 to 97.3 µg/L. A total of 15 false positives were observed for ELISA (reported at <0.05 by GC/MS). These samples exhibited a highly complex chemical matrix. The presence of interfering and co-eluting peaks on the GC/MS analyses casts doubt on the cyanazine values reported as <0.05 µg/L for these samples. For these samples, the method of standard addition was applied to verify the ELISA results. The samples were re-analyzed by ELISA concurrently with sample aliquots that had been spiked with a similar concentration of cyanazine. All standard addition analyses by ELISA agreed with the expected value (original sample concentration + the spiked concentration) to within ± 10%. Method development is presently underway to modify the SPE method used so that it will remove a greater proportion of the interfering compounds and allow a more reliable identification of cyanazine concentrations in these highly contaminated areas. Most of the area (91.3% of the sampled wells) showed cyanazine concentrations <0.1 µg/L (Figure 3). A small percentage (6.4%) of the sampled wells showed cyanazine concentrations of >0.1 to 1.0 µg/L, and 2.3% showed levels above 1.0 µg/L, the EPA Health Advisory Level (HAL) (*14*). All areas exhibiting high cyanazine concentrations were located inside the RMA boundaries. A single well outside of the barrier system exhibited cyanazine contamination exactly at the 1.0 µg/L level (equal to but not exceeding the HAL). Historical data available for this well record the presence of high levels of other target contaminants before the installation of the barrier systems. Contaminant levels have not increased following the installation of these systems and have decreased in the immediate vicinity of the barrier systems for a number of target compounds. Transport of cyanazine through the barrier system, therefore, is not believed to be the cause of

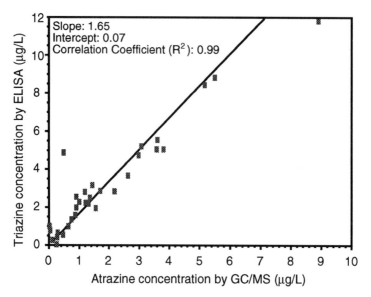

Figure 4. Triazine concentrations by ELISA compared to atrazine concentrations by GC/MS in 52 ground-water samples from the Rocky Mountain Arsenal. (To better display the curve fit, this plot does not include data from the four samples known to contain high levels of cyanazine.)

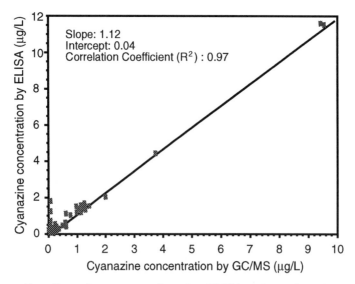

Figure 5. Cyanazine concentrations by ELISA compared to cyanazine concentrations by GC/MS in 353 ground-water samples from the Rocky Mountain Arsenal. (To better display the curve fit, this plot does not include data from the four samples known to contain high levels of cyanazine.)

cyanazine contamination in the area of this well. Rather, the transport of cyanazine as a part of an older contaminant plume, predating the barrier system, seems a more likely explanation.

The ability to respond immediately to the question of contaminant distribution and concentration generated by the initial identification of cyanazine at RMA is perhaps the most significant benefit of using ELISA as a screening technique. Fortunately, the identification of cyanazine was made during an established sampling sequence, so no special sampling had to be accomplished. Within 2 days, 353 samples were analyzed by cyanazine ELISA, and most of the wells were found to contain no significant contamination. Because of the short turnaround time and small sample-volume requirements of the technique, an area-wide screen was possible, the scope of which fell well within the sampling period. Had significant contamination been observed, the overall sampling plan could have been modified, and the samples collected with greater frequency in the affected areas. Cyanazine was shown instead, by both the initial ELISA screen and a later GC/MS validation, to be only a minor contaminant at RMA. Future analyses may include cyanazine, but the initial question of distribution and contaminant concentration was answered quickly and efficiently and at reasonable cost.

GC/MS correlations for atrazine (Figure 4) and cyanazine (Figure 5) showed good agreement. Both the triazine and cyanazine ELISA results displayed a positive bias when correlated with concurrent GC/MS data for the same samples. This is most likely explained by the cross reactivity of the ELISA kits. The positive bias for the triazine ELISA kit (correlation curve slope = 1.65) is much greater than that for the cyanazine ELISA kit (correlation curve slope = 1.12). A completely non-biased correlation would have a slope of 1.0. The triazine ELISA kit (Figure 4) displays a higher relative sensitivity to compounds other than atrazine (the compound used for the GC/MS correlation) and, therefore, has a much larger probability of responding to the presence of these other triazine compounds than the more selective cyanazine ELISA kit. This response to both atrazine and non-atrazine compounds in the same sample increases the slope of the correlation plot to greater than 1.0. The cyanazine ELISA kit (Figure 5), however, is much more specific to cyanazine and, therefore, has a smaller probability of generating a response from non-cyanazine compounds that may be present. However, in comparison, the ELISA data show the same relative concentration trends and plume boundaries as the GC/MS data. The triazine and, for the most part, the cyanazine ELISA data show promise for application as screening methods; the development of a validation procedure for the false positive cyanazine samples by GC/MS is currently in progress.

A wide range of contaminant concentrations at the RMA was identified using ELISA. The occurrence of pesticide concentrations in areas downgradient of the manufacturing and disposal sites on RMA most likely can be explained by transport in ground water. The installation of the barrier systems was done to intercept the downgradient transport of these contaminants offsite.

Conclusions

The results of this research have shown that despite the highly complex chemical matrix present at RMA, ELISA performed extremely well as a screening technique for large-scale ground-water contamination. The triazine and cyanazine assays show no significant matrix interferences. The development of validation procedures for the cyclodiene assay will allow a more quantitative evaluation of potential matrix interferences, but relative concentration trends have been shown to agree well with historical data. The only sample-preparation step required was dilution, which represented a considerable simplification of the many extraction, concentration, and analytical steps commonly required for the instrumental analysis of these pesticides. ELISA was shown to work well on samples with a limited volume when other methods may not have shown adequate sensitivity. Although this research was performed using wells that usually are not sample limited, the time and effort involved in sampling the wells at RMA made efficient use of the collected volumes crucial.

The data obtained from the cyclodiene ELISA analyses exhibited a high degree of correlation among wells in close geographical proximity. Although direct GC/MS validation was not available for all samples analyzed by ELISA, the ELISA data obtained for the cyclodiene insecticides compares favorably with the historical GC data available for the sampled wells. The relative concentration trends are similar for both sample sets, and the extent of the contaminant plumes outlined is essentially the same for both methods of analysis. Future research in this area will seek to validate this work and identify any potential interferences.

The triazine ELISA was shown to be useful as a class-specific assay. The cross reactivity exhibited in this case resulted in the identification and subsequent evaluation of cyanazine as a previously unknown contaminant at RMA. This ability to respond immediately to the question of contaminant distribution and concentration is a significant benefit of using ELISA as a screening technique. Cyanazine was shown by both ELISA and GC/MS to be only a minor contaminant at RMA and, despite the complex chemical matrix present, the initial ELISA screen performed quickly, efficiently, and at reasonable cost. In direct comparison, the ELISA data for the triazine compounds and cyanazine clearly revealed the same relative concentration trends and plumes as the GC/MS data. Although the triazine and, for the most part, the cyanazine ELISA data show very promising trends for application at RMA, the development of a validation procedure for both the cyclodiene insecticides and the false-positive cyanazine samples by GC/MS will be crucial to this research.

If applied initially, ELISA could streamline a major sampling and analysis program as it would allow a large geographic area to be rapidly evaluated for target concentrations exceeding the designated action levels. The areas identified by ELISA as "clean" then could be sampled less rigorously while more time and analytical effort could be applied to areas of greater concern. Overall, in the areas of time, sample preparation, analytical difficulty, and expense, the ELISA methods evaluated have demonstrated excellent potential as a screening technique for large-scale sites with suspected ground-water contamination.

Acknowledgment

The authors acknowledge the advice and experience ofJaye Lunsford and Cecil Slaughter (U.S. Geological Survey office, Lakewood, Colorado) without which this research would not have been as successful.
[1]The use of brand names in this paper is for identification purposes only and does not constitute endorsement by the U.S. Geological Survey.

Literature Cited

(1) Thurman, E.M.; Goolsby, D.A.; Meyer, M.T.; Kolpin, D.W.; Environ.Sci.Technol., 1991, 25, 1794-1796.

(2) Thurman, E.M.; Goolsby, D.A.; Meyer, M.T.; Mills, M.S.; Pomes, M.L.; Kolpin, D.W.; Environ.Sci. Technol., 1992, 26, 2440-2447.

(3) Kolpin, D.W.; Thurman, E.M.; Goolsby, D.A.; Environ.Sci.Technol. 30, 1996, 335-340.

(4) Aga, D.A.; *Analytical Applications of Immunoassays in Environmental and Agricultural Chemistry: Study of the Fate and Transport of Herbicides;* PhD Dissertation, University of Kansas: Lawrence, KS, 1995.

(5) Laws, E.A.; *Aquatic Pollution*; Wiley and Sons: New York, NY, 1980, pp 561-567.

(6) Garlock, E.T., Ed.; In *Eagle Watch*; Denver, CO, **1992**, Vol. 4, Issue 8.

(7) Konikow, L.F.; Thompson, D.W.; In *Groundwater Contamination*, National Research Council Geophysics Study Committee, National Academy Press, Washington, D.C. **1984**, pp 93-103.

(8) Van Emon, J.M.; Lopez-Avila, V.; *Anal. Chem.* **1992**, *64*, (2), 79A-88A.

(9) Thurman, E.M.; Meyer, M.T.; Pomes, M.L.; Perry, C.A.; Schwab, P.; *Anal. Chem.* **1990**, *62*, 2043-2048.

(10) Aga, D.S.; Thurman, E.M.; *Anal. Chem.* **1993**; *64*, (20), 2894-2898.

(11) Mills, M.S.; Thurman, E.M.; *Anal. Chem.*, **1992**, *64*, (17), 1985-1990.

(12) Meyer, M.T., Mills, M.S.; Thurman, E.M.; *J. Chromatog.* **1993**, *629*, 55-59.

(13) Aga, D.S.; Thurman, E.M.; In *New Frontiers in Agricultural Immunoassay*; Kurtz, D.A.; Skerrit, J.H.; Stanker, L.; Eds., AOAC International: Arlington, VA; **1995**, Chapt. 9, pp 123-136.

(14)   *Drinking Water Regulations and Health Advisories*; U.S. Environmental Protection Agency, Office of Water, Feb. 1996, EPA 822-R-96-001.

(15) Meyer, M.T.; *Geochemistry of Cyanazine and Its Metabolites: Indicators of Contaminant Transport in Surface Water of the Midwestern United States,* PhD Dissertation, University of Kansas: Lawrence, KS, 1995.

Chapter 19

Comparison of Immunoassay to High-Pressure Liquid Chromatography and Gas Chromatography—Mass Spectrometry Analysis of Pesticides in Surface Water

M. I. Selim[1], C. Achutan[1], J. M. Starr[1], T. Jiang[2], and B. S. Young[2]

[1]Institute for Rural and Environmental Health, Department of Preventive Medicine and Environmental Health, College of Medicine, 137 IREH, 100 Oakdale Campus, University of Iowa, Iowa City, IA 52242-5000
[2]Millipore Corporation, 80 Ashby Road, P.O. Box 9125, Bedford, MA 01730-9125

This study was designed to evaluate the performance of EnviroGard immunoassay kits for the field analysis of pesticides in surface waters in Malaysia. The primary objective was to determine possible effects of the surface water matrix or environmental conditions on the sensitivity or reliability of the EnviroGard immunoassay kits. The kits used in this study were: alachlor, triazines, diazinon, cyclodienes, paraquat, and 2,4-diclorophenoxyacetic acid. Field work consisted of enzyme immunoassay (EIA) of surface water samples using the EnviroGard plate kits in conjunction with a Millipore Microwell strip reader. Split samples of surface waters, blanks, and spikes were processed in the field using solid phase extraction (SPE) disks (C_8 or C_{18}) and transported to University of Iowa, where extracted pesticides were eluted and analyzed using high pressure liquid chromatography (HPLC) and gas chromatography / mass spectrometry (GC/MS). HPLC and GC/MS data were corrected for pesticide recovery and stability on the C_8 or C_{18} during transit. Good agreement was obtained between the field EIA and laboratory HPLC or GC/MS analysis of field spiked samples for all kits with the exception of paraquat. The results of this study indicate that the EnviroGard enzyme immunoassay is an efficient and reliable field testing tool for pesticides in water.

The rural populations in developing countries, are at an increased health risk from exposure to agricultural chemicals that are known or suspected to be carcinogenic (*1*). Estimates by the United Nations indicate that there could be more than 2 million cases of pesticide poisoning worldwide with an excess of 40,000 deaths (*2*). Acute neurological and dermal diseases from occupational exposures are well described among farmers in both the United States and in developing countries (*3-7*). It has become critical for many government agencies and private organizations to detect pesticide levels in food, water, and soil. For this reason, rapid, cost-effective and easy-to-use screening and analytical techniques are needed, particularly in developing countries where expensive instrumentation may be a barrier. The immunoassay field kits seem to meet these criteria. Numerous studies have been done to determine pesticide residues in water, food, and soil by enzyme immunoassay (EIA) compared with established GC and HPLC analysis (*8-15*). Bushway et. al. (*8*) studied atrazine content in a broad range of food products and concluded that enzyme immunoassay was an excellent screening method for atrazine. They also found that EIA provided reproducible and inexpensive analysis compared with other methodologies such as HPLC. In studies done comparing EIA determination with HPLC analysis of atrazine residues in water and soil, it was found that the recovery of atrazine in water was better than in soil (9). It was also found that the immunoassay has cross-reactivity with several triazines, thereby providing an advantage for metabolite detection. Confirmation by GC or HPLC/MS was necessary as the kit was non-selective because of cross-reactivity with structurally similar metabolites of atrazine (*11*).

Solid phase extraction (SPE) followed by GC/MS analysis has also been used for the evaluation of the EIA analysis of triazines in surface waters also (*12,16*) In these studies, good agreement was found between the EIA results and the SPE-GC/MS analysis. The EIA was found to be an economical technique (about $15 per sample) and effective for the screening of a class of pesticides in water such as triazines (*12*).

The primary objective of this study was to investigate the sensitivity and reliability of the EnviroGard Enzyme Immunoassay (EIA) method as a field screening test for the analysis of alachlor, triazines, diazinon, cyclodienes, paraquat, and 2,4-diclorophenoxyacetic acid in raw surface waters in Malaysia. The field evaluation involved EIA analysis and solid phase extraction (SPE) of split samples of raw and treated surface water. Field SPE disks were transported to the U.S. for laboratory analysis using HPLC and GC/MS.

Materials And Methods

Materials and Chemicals. The EnviroGard enzyme immunoassay kits were provided by Millipore Corporation (Bedford, MA). Empore C_{18} (47 mm) disks were manufactured by 3M (St. Louis, MO). ENVI-8 (C_8, 47 mm) disks were obtained from Supelco (Bellefonte, PA). Neat pesticide standards; 2,4-D (99%), paraquat (98%), alachlor (99%), diazinon (99%), atrazine (98%), and chlordane (isomer mixture) were obtained from Chem Service (West Chester, PA). A standard solution of γ-chlordane, 1000 μg/mL methanol (Absolute Standards, Hamden, CT), was used for quantitative analysis of chlordane isomers. Water, methanol, acetonitrile, methylene chloride, and acetone were all pesticide residue or HPLC grade and were obtained from Fisher Scientific (Fair Lawn, NJ). Cetyl trimethyl ammonium bromide, diethylamine, and 1-hexanesulfonic acid sodium salt were obtained in the highest purity avialable from Aldrich (Milwaukee, WI). Hydrochloric acid, orthophosphoric acid, sodium hydroxide, and ammonium hydroxide were all of analytical grade and obtained from Fisher Scientific.

Study Design. The experimental work in this project consisted of field and laboratory components. The purpose of the laboratory component was to develop and validate the SPE and the instrumental methods, as well as to evaluate the performance of the EIA kits under controlled laboratory conditions. The field work consisted of EIA analysis and SPE of split samples using reversed phase extraction disks (C_8 or C_{18}). The SPE disks were transported for HPLC and GC/MS analysis at the Analytical Toxicology Laboratory (ATL), University of Iowa. In order to account for any loss of analytes during transit, a stability study was conducted in the laboratory to determine the effect of storage time and temperature on the recovery of various pesticides concentrated on the SPE disks. All field and laboratory protocols were developed, validated, and documented prior to the field work.

Selection of Pesticides for EIA Screening. A survey of available literature on pesticides used in Malaysia was conducted in order to identify the most commonly used pesticides and the areas of potential monitoring needs. The compiled list of pesticides consisted of major pesticides, including DDT, triazines, chlordane, heptachlor, dieldrin, paraquat, malathion, aldrin, phenoxy acid herbicides, lindane, endrin, and endosulfan. These pesticides are used mainly in the rubber and palm oil plantations, or on growing vegetables, fruits and rice. Thus, the primary source of surface water contamination was expected to be through surface water run-off from these agricultural areas.

Solid Phase Extraction (SPE). Different SPE extraction methods were used; for phenoxy acid herbicides (2,4-D, MCPA, 2,4,5-T, Dichloroprop, 2,4,5-TP), paraquat, and other pesticides listed in Table 1.

SPE of Phenoxy Acid Herbicides. Water samples were prepared for SPE by adjusting the pH to 2 using phosphoric acid. High turbidity water samples were filtered through Millipore AP25 glass fiber prefilters (Millipore, Bedford, MA). Approximately 500 mL of the filtered samples were collected into a stock bottle of defined tare. The C_{18} disks were conditioned by passing 10 mL methanol followed by 20 mL residue free water at a flow rate of 5-10 mL/min. The water samples were then passed through the conditioned sorbent at a flow rate of 20-25 mL/min. The exact volume of water throughput was determined by reweighing the stock bottle. The C_{18} disks were washed with 10 mL residue-free water at a flow rate of 5-10 mL/min. to remove salt residues from the samples. Vacuum suction was continued for 20 minutes to dry the disks with air flow. Field SPE disks were placed in clean 49 mm Petri dishes, sealed and stored over ice packets during transit, or refrigerated, until ready for analysis. Elution was carried out by placing the dry filter on the vacuum manifold, adding 10 mL acetonitrile to the disk reservoir, allowing 5 minutes for complete wetting of the dried sorbent, and allowing the eluent to filter under atmospheric pressure, or slight vacuum, at approximately 2 mL/min. The eluent was collected in a clean test tube, evaporated to dryness, and redissolved into 500 µL acetonitrile-water (2:8). The eluate was filtered through a 0.2 µm PTFE filter (Gelman Sciences, Ann Harbor, MI) and analyzed by HPLC.

SPE of Paraquat. The SPE method for paraquat was based on the US Environmental Protection Agency Method 549.1 and the ENVI-8 application Note No. 60 from Supelco. The C_8 disks were conditioned by sequentially passing 10 mL methanol, 2 x 10 mL HPLC grade water, 10 mL conditioning solution A (0.5 g of cetyl trimethyl ammonium bromide + 5 mL concentrated ammonium hydroxide in 1 liter of HPLC grade water), 2 x 10 mL HPLC grade water, and 10 mL conditioning solution B (1-hexanesulfonic acid sodium salt + 10 mL concentrated ammonium hydroxide in 500 mL HPLC grade water). Water samples were prepared for analysis by adjusting the pH to 10.5 ± 0.2, using 10% (w/v) sodium hydroxide solution, or 10% (v/v) hydrochloric acid solution. The samples were then added to the disk reservoir and allowed to filter under vacuum at a rate of 30-40 mL/min. Approximately 1 mL methanol, (just enough to cover the filter), was added followed by 4 mL disk eluting solution (13.5 mL of orthophosphoric acid + 10.5 mL diethylamine per liter of HPLC grade water). Vacuum was adjusted to allow slow (2 mL/min.) solution filtration through the disks. Another 4 mL of the disk eluting solution was drawn through the disk into a previously silyated collection tube and 100 µL of the ion pair solution (3.75 g 1-hexanesulfonic acid sodium salt in 50 mL disk eluting solution) was added to the eluate. The final volume was adjusted using the elution solvent. Approximately 1 mL of the eluate was filtered through a 0.2 µm PTFE filter (Gelman Sciences, Ann Harbor, MI) and analyzed by HPLC.

SPE of Miscellaneous Pesticides and Metabolites. Surface water samples were prepared for SPE by adjusting the pH to 6-8, using 1M NaOH or 1M HCl. High turbidity water samples were filtered through Millipore AP25 glass fiber prefilters (Millipore, Bedford, MA). The C_{18} disks were mounted on the extraction manifold and conditioned by sequentially passing 10 mL acetonitrile, 10 mL methanol, and 10 mL HPLC grade water without allowing the disk to dry. Field samples (or controls) were added to the extraction disk from a preweighed bottle. The vacuum was adjusted to allow sample flow at approximately 30-50 mL/min. The exact volume of water extracted was determined by reweighing the sample bottle. The extraction disks were washed by passing 10 mL of HPLC grade water. Vacuum suction was continued for 20 min. until the disks were completely dry. For laboratory SPE, the disks were eluted by passing 10 mL acetonitrile at a flow rate of 2 mL/min. The extract was evaporated under nitrogen and reconstituted to 500 µL iso-octane for GC/MS analysis. Field SPE disks were placed into 49 mm Petri dishes, following the drying step, and stored refrigerated or transported over ice packet until ready for analysis.

HPLC and GC/MS Confirmation Methods. HPLC and GC/MS methods were developed and used for qualitative and quantitative analysis of the pesticides in this study.

HPLC Analysis of Phenoxy Acid Herbicides. HPLC analysis of phenoxy acid herbicides was performed on an HP1090 HPLC (Hewlett-Packard, Palo Alto, CA) equipped with a diode array detector and HPLC ChemStation operating software. The analytical column was a Phenomenex (Torrance, CA) 200 x 2.1 mm,

packed with 3 μm Hypersil ODS stationary phase. The mobile phase consisted of:
(A) 0.005 M KH_2PO_4 in .001% CH_3COOH and (B) Acetonitrile-methanol (1:1). The
column was initially equilibrated at 15% B for 20 minutes. With the run start, the
15% B was programmed to change to 50% B at 16 min., to 55% at 20 min., and
back to 15 % B at 21 min. The mobile phase flow was 0.35 mL/min. and the column
temperature was 45 °C. The diode array detector was set at λ =280 nm, bandwidth
12 nm, reference at 460 nm.   The retention times for 2,4-D, MCPA, 2,4,5-T,
Dichloroprop, and 2,4,5-TP were 18.09, 19.36, 20.76, 21.67, and 25.12 min.,
respectively.

HPLC Analysis of Paraquat.   The HPLC analysis of paraquat was
performed on the same HPLC and analytical column described above. The mobile
phase consisted of 0.1% aqueous solution of hexane sulfonic acid, sodium salt,
containing 3.5 mL triethylamine. The pH was adjusted to 2.5 using orthophosphoric
acid. Mobile phase flow was 0.3 mL/min. and column temperature was 40 °C. The
diode array detector was set at λ = 257 nm.

GC/MS Analysis of Other Pesticides. The GC/MS system used consisted of
an HP 5890 Series II gas chromatograpgh and an HP 5989A MS Engine. The GC
was equipped with a fused silica capillary column 30 m x 0.25 mm, 0.25 μm DB-Wax
column (J&W Scientific, Folsom, CA). The column temperature was held at 40 °C
for 4 min. and programmed to 220 °C at 15 °C /min. and held for 30 min.   The
injection temperature was kept at 180 °C. The samples were injected in the splitless
mode using an HP 7673A automatic injector. The MS conditions were as follows:
transfer line 250 °C; source 250 °C; manifold temperature 100 °C; ionization voltage
70 eV and electron multiplier voltage 200 V above the daily tune up value. Selected
ion monitoring (SIM) was based on the retention times and a minimum of three
characteristic ions and their relative abundance. The dwell time ranged from 15 to
100 msec. The shortest dwell was used with the very narrow, early-eluting, peaks.

EIA Procedures. The EnviroGard plate kits were selected for this study since they
were available for all the compounds of interest. These kits were used in conjunction
with a Millipore Microwell Strip Reader (Millipore, Bedford, MA). Data collection
and analysis were carried out using an IBM compatible laptop computer and Bio-Tek
KinetiCalc EIA analysis software (Bio-Tek Instruments, Highland Park, VT). Field
EIA kits were shipped on ice packs from the Millipore Headquarters in Bedford, MA,
to the Millipore office in Kuala Lumpur, Malaysia. The kits were refrigerated upon
arrival in Kuala Lumpur then transported to the field on ice packs.
    The antibodies used in these kits are polyclonal or monoclonal.   These
antibodies bind the target and closely related compounds including the enzyme
conjugate.   The competition between the analyte and its enzyme conjugate for a
limited number of antibody binding sites determines the inverse relationship between
the signal and the concentration of analytes, upon which quantitation is based.
    The procedure and reagents used for the EIA analysis were as described in the
kit inserts.   The exact procedure varied slightly from assay to assay in terms of
reagent amount and incubation time. In general, 100 μL water samples or controls

were incubated along with 100 µL of enzyme conjugate solution in the 96-well plate, which is precoated with the antibody. After one hour, the plates were washed with tap water and shaken to empty. One hundred µL of substrate and chromogen solutions were added to each well and the plates were incubated for a an additional 30 min. Finally, 100 µL of stop solution was added to each well to terminate the reaction and the absorbance measured at 450 nm using Millipore Microwell Strip Reader.

Field Sampling and Analysis in Malaysia. The primary goal of the field sampling was to obtain a set of representative surface water samples with potential pesticide contamination, to evaluate the performance of the EIA kits under the prevailing matrix and field conditions in Malaysia. The heavily agricultural area of Cameron Highlands was most suited to meet this goal, particularly due to its proximity to some major rivers and the presence of several water treatment facilities.

The primary field sampling and analysis activities were carried out between June 12 and 22, 1995. Thirteen sampling stations were selected throughout the Cameron Highlands area. These included five water treatment facilities, four rivers, three streams, and one from farm run off. In addition, two sampling stations were also selected at the intake of a major treatment facility near the city of Kuala Lumpur. A total of 27 field samples and controls were processed for SPE and EIA analysis during the field work. The control samples included eight matrix spikes consisting of three river, one farm run off, and four water treatment intake samples.

Except for the water treatment facilities, 5-liter samples were collected at each sampling station. Three 5-liter samples were collected for each water treatment facility; at the raw water intake, after treatment at the plant, and at a central distribution point. All samples were processed for SPE and EIA analysis simultaneously following collection.

Quality Control and Quality Assurance. All field and laboratory standard operating procedures for sampling and analysis were developed, evaluated, and documented prior to field work in Malaysia. Multilevel calibration was routinely used for all instrumental analysis after each group of ten samples. A check standard (an intermediate calibration standard) was used to verify instrument calibration after each five samples. Recalibartion was conducted if the calculated concentration was $> \pm$ 5% of the expected concentration of the check standard. Method performance was routinely checked using reagent blanks and matrix spikes. Deionized water blanks, spiked deionized water, and matrix spikes were used to validate the performance of the combined SPE - HPLC, GC/MS analyses. Negative controls and calibrators were routinely used with the EnviroGard kits to verify their performance to specifications, as described in the kit insert. A minimum of two replicates of each sample were run with the EnviroGard test and three replicates were run with the SPE method.

Results and Discussion

In order to achieve the primary objective of this study it was important to compare the field results of EnviroGard EIA test with reliable state-of-the-art laboratory methods such as the HPLC and GC/MS. The SPE technique was found to be ideally suited for extracting split samples in the field for later laboratory analysis. Reverse phase (C_8 or C_{18}) disks were selected over cartridges due to the susceptibility of the latter for clogging, particularly with high turbidity samples such as those found in Malaysia's surface waters. The C_{18} disks were previously evaluated and used in our laboratory and showed that they, too, are susceptible to clogging with surface water, limiting their capacity to less than 500 mL. Therefore in the field work, high turbidity raw water samples were prefiltered using Millipore AP25 glass fiber filters to facilitate flow through the C_8 or C_{18} disks.

The effect of storage time on the recovery of selected pesticides from C_8 or C_{18} disks at 25 °C and 4 °C are listed in Table I. All pesticides concentrated on the C_8 or C_{18} disks showed gradual decrease in recovery with storage time. The recovery was slightly lower for disks stored at room temperature (25 °C), compared to those kept refrigerated (4 °C). The recovery of paraquat from C_8 disks was reduced with storage time and temperature as shown in Table I. However, the decrease in recovery was more pronounced with paraquat stored at room temperature, where the recovery dropped to less than 10% after three weeks of storage. Similarly, the recovery of phenoxy acid herbicides stored on C_{18} disks at 25 °C and 4 °C showed approximately 50% decrease in recovery after one week of storage. Again, the recovery was slightly lower for disks stored at room temperature. Accordingly, the SPE disks (C_8 or C_{18}) processed in the field were refrigerated immediately, following the concentration step, and were transported from Malaysia to the US on ice packs and refrigerated upon arrival at the laboratory. The stability data was then used to apply a correction factor to account for the loss in recovery during transit time This time was approximately one week for most samples.

Consistent with the field EIA no pesticides were detected in the field samples using HPLC and GC/MS in total ion scan. Selected ion monitoring GC/MS analysis of field samples showed the presence of a variety of pesticides (e.g. prometryn, metolachlor, diazinon, alachlor, propachlor, and the triazine group). However, the concentrations of all pesticides identified were below the limits of reliable GC/MS quantification (20-400 ppt) and were close to the lower limit of detection by EIA. Thus, the use of field spikes was very useful for evaluating the performance of the EnviroGard EIA kits using actual field samples.

The field spiking solution contained all compounds listed in Table I, as well as terbutylazine, and five phenoxy acid herbicides (2,4-D, MCPA, 2,4,5-T, Dichloroprop, and 2,4,5-TP). The EIA equivalence for each kit was calculated by adding the products of the concentration of each cross-reacting compound and its known cross-reactivity data for the EnviroGard kit. For example, in the presence of atrazine, propazine, and terbutylazine in the spiking solution, the EIA equivalence for atrazine was calculated using the following formula:

EIA Equiv. = Atrazine Concentration + Propazine Concentration x (Cross-reactivity of Atrazine / Cross-reactivity of Propazine) + Terbutylazine Concentration x (Cross-reactivity of atrazine / Cross-reactivity of Terbutylazine)

For the EnviroGard triazine kit, the cross-reactivities for atrazine, propazine, and terbutylazine at 50% B_0 were 0.25, 0.20, and 2.50, respectively. Similarly, EIA equivalence was calculted for the 2,4-D, alachlor, and the cyclodiene kits.

Table I. Effect of Storage Time on the Percent Recovery of
Pesticides from C_{18} or C_8 Disks Stored at 25° and 4°C

| Pesticide | *Percent Recovery* | | | | | | |
| | *Day* | *Day 5* | | *Day 15* | | *Day 25* | |
	25 °C	*4 °C*	*25 °C*	*4 °C*	*25 °C*	*4 °C*	*25 °C*
PCNB[a]	65.6	72.9	41.3	75.6	54.3	49.6	55.9
Heptachlor	48.8	40.4	24.1	35.1	30.0	27.6	32.7
Aldrin	38.8	29.7	16.7	20.7	19.8	19.5	24.6
Gamma BHC	77.7	75.3	50.7	81.6	60.1	49.1	54.6
Alachlor	80.5	76.3	60.5	78.7	64.9	46.2	50.9
Propazine	65.9	46.2	44.1	40.2	55.3	34.2	34.8
Metolachlor	79.5	76.3	60.5	78.7	64.9	46.2	50.9
Atrazine	61.0	47.2	40.6	46.4	43.2	45.6	51.8
Chlordane	77.7	63.7	46.3	48.0	46.9	45.4	53.6
Dieldrin	73.2	59.0	42.9	51.2	45.0	35.1	39.9
Endrin	76.4	69.9	49.2	60.2	52.4	43.7	48.2
Triphenylphosphate[a]	73.5	94.5	74.5	97.0	82.9	54.9	61.3
Paraquat[b]	87.2	63.2	62.5	46.6	36.9	31.0	6.5

[a] PCNB & Triphenylphosphate: internal standards
[b] Extracted using C_8

The results of field EIA and laboratory SPE-HPLC analysis of phenoxy acid herbicides in spiked field samples are listed in Table II. The differences in spiked concentrations are due to the different volumes of water samples used. Both HPLC and EIA results for phenoxyacid herbicids are in good agreement with the expected spiked concentrations. The recoveries based on SPE and HPLC analysis ranged from 86 to 107% while the recoveries based on field EIA analysis ranged from 89 to 119%. Neither 2,4-D nor MCPA were detected in any of the field samples before spiking.

Table II. Comparison of the Field EIA and the Laboratory HPLC Analysis
for Phenoxy Acids Herbicides Spiked in Raw Surface Water

Field sample ID	*EIA Equiv. of Spiked Conc. (µg/L)*	*EIA Equiv. of Lab HPLC Conc. (µg/L)*	*%Reocv. Based on HPLC*	*Field EIA Conc. (µg/L)*	*% Recov. Based on EIA*
BR1R	38.13	40.37	106	34	89
BV1R	33.54	35.13	105	30	89
LT1R	33.87	35.21	104	31	92
RTFM	41.09	40.24	98	41	100
SGUL	38.13	39.22	103	42	110
SGTK	38.13	32.73	86	40	105
SGTR	35.45	35.82	101	39	110
DI00	31.93	34.23	107	38	119

The percent recovery based on field EIA and laboratory SPE-GC/MS analysis for triazines, alachlor, diazinon, and cyclodienes in the spiked field samples are listed in Table III. For triazines, the measured GC/MS concentrations were slightly higher than the expected spiked concentration, except for three spikes which were slightly lower. The recovery based on the measured GC/MS concentrations ranged from 56 to 125% with mean recovery of 106%. The triazines recovery based on field EIA results ranged from 74 to 111 % with a mean recovery of 87%. In general, there was good agreement between the GC/MS and the EIA results for triazines. The slight increase in GC/MS recovery is possibly due to overcorrecting for the stability of triazines during storage and transit time.

As shown in Table III the EIA results for alachlor agree well with the GC/MS results. The alachlor recovery obtained with SPE-GC/MS analysis ranged from 77.6 to 122%. For metolachlor, the SPE-GC/MS recovery ranged from 91 to 136%. The EIA recovery for alachlor equivalence ranged from 85 to 156%. The higher than expected EIA recoveries are within the assay variance.

The percent recoveries of diazinon based on the SPE-GC/MS analysis ranged from 93-192% with a mean recovery of 146%. The recoveries before correction for storage stability were 86-119%, indicating the possibility that the higher recoveries obtained after correction are due to overcorrection. The field EIA for diazinon were consistently low (recovery 45-63%), except for four samples that showed highly variable recovery, ranging from 108 to 286%. Field EIA analysis showed the presence of interfering compounds in the four samples before spiking. However, subtracting the levels of background signals did not significantly affect the values obtained with the spiked matrices. These four samples consisted of three river samples and one farm run-off, which were highly turbid and of apparent poor quality even after prefiltration.

Table III. Comparison of the EIA Field Results and Laboratory GC/MS Analysis of Triazines, Alachlor, Diazinon, and Cyclodienes in Spiked Field Samples

Sample ID	Triazines		Alachlor		Diazinon		Cyclodienes	
	GC/MS	*EIA*	*GC/MS*	*EIA*	*GC/M*	*EIA*	*GC/MS*	*EIA*
SGUL	114	85	116	102	139	286	78	152
LT1R	122	108	117	93	157	45	90	78
BV1R	92	74	100	89	121	52	61	80
SGTK	117	111	113	156	151	108	86	120
SGTR	125	94	123	153	175	233	90	128
RTFM	115	78	120	112	135	188	96	239
BR1R	56	77	78	112	93	53	54	87
RT1R	121	76	124	100	157	63	101	63
DI00	93	79	106	85	192	50	103	68

Cyclodienes (chordane, dieldrin, heptachlor, aldrin, and endrin) were the least frequently detected pesticides in the field samples. The cyclodiene recoveries based on the SPE-GC/MS analysis ranged from 54 to 103 %. The recoveries based on field EIA ranged from 63 to 152%, except for a farm run off sample that gave a recovery of 239%.   Since the recoveries based on the SPE-GC/MS analysis are within reasonable range (54 - 103%), and no significant level of other known organic compound(s) were detected in any of these samples, it is possible that the increased EIA recovery is caused by the high level of fine particulates, as well as dissolved organic and inorganic matter in the these field samples.

The paraquat study was problematic because of the extremely low recoveries by both HPLC and EIA analysis. We were not able to detect paraquat in the any of the field samples and spikes. The EIA results showed very low concentration, even in spiked deionized water samples. This may be atributed to rapid degradation of paraquat under field conditions.

Conclusion

The results of this study demonstrated that field EIA results are in good agreement with the laboratory measured SPE - HPLC and GC/MS data. Although there is some variability in the EIA measurements, there was no discrepancy or false positive or negative measurements within the EIA detection limits. The GC/MS analyses were more specific for compound identification. However, the EIA screening provided a fast, easy, and economical field screening test. Except for the paraquat test, all the EnviroGard kits tested in this study appear to meet their expected performance under the environmental field conditions used in this study.

Literature Cited

1.   WHO. *Public Health Impact of Pesticides Used in Agriculture*; Geneva, 1990.

2.   International Organization of Consumers Unions and Pesticide Action Network. *Problem Pesticides, Pesticide Problems*; IOCU, Penang, Malaysia, 1988.

3.   Ernst, W.; Maclkad, F.; El-Sebae, A.; Halim, Y.   *Monitoring of Organochlorine Compounds in Some Marine Organisms from Alexandria Region*; Proceedings of the International Conference on Environmental Hazards of Agrochemicals in Developing Countries; 95-108.

4.   EL-Sebae, A.H. In *Ecotoxicology and Climate*; Bourdeau, P.; Haines, J.A.; Klein, W.; Krishna Murti, C.R., Eds.; John Wiley and Sons Ltd.: New York, NY, 1989.

5.   Popendorf, W.; Donham, K.J. In *Patty's Industrial Hygiene and Toxicology*; Clayton, G.D.; Clayton, F.E., Eds; John Wiley and Sons Ltd.: New York, NY, 1990.

6.   Blair, A.; Malker, H.; Cantor, K.P.; Burmeister, L.; Wiklund, K. *Scand J Work Environ Health*. **1985**, *11*, pp 397-407.

7.   Schenker, M.; McCurdy, S. In *Cancer Prevention: Strategies in the Workplace*; Becker, C.E.; Coye, M.J., Eds.; Hemisphere Publishing Corp.: Washington, DC, 1986, pp 29-37.

8. Bushway, R.J.; Perkins, B.; Savage, S.A.; Lekousi, S.L.; Ferguson, B.S. *Bull Environ Contam Toxicol*; **1989**, *42*, pp 899-904.

9. Bushway, R.J; Perkins, B; Savage, S.A; Lekousi, S.L; Ferguson, B.S. *Bull Environ Contam Toxicol;* **1988**, *40*, pp 647-654.

10. Goh, K.S; Hernandez, J; Powell, S.J; Greene, C.D. *Bull Environ Contam Toxicol;* **1990**, *45*, pp 208-214.

11. Goh, K.S; Hernandez, J; Powell, S.J; Garretson, C; Troiano, J; Ray, M; Greene, C.D. *Bull Environ Contam Toxicol;* **1991**, *46*, pp 30-36.

12. Thurman, E.M; Meyer, M; Pomes, M; Perry, C.A; Schwab, A.P. *Analytical Chemistry*, **1990**, *62(18)*, pp 2043-2048.

13. Harrison, R.O; Ferguson , B.S. *Quantitative Enzyme Immunoassay of Pesticides in Water at Part Per Billion Levels.* Proceedings of the FOCUS Conference on Eastern Regional Ground Water Issues; October 17-19, 1990.

14. Goolsby, D.A; Thurman, E.M; Clark, M.L; Pomes, M.L. *Immunoassay as a Screening Tool for Triazine Herbicides in Streams.* ACS Symposium Series 451, American Chemical Society: Washington, DC, 1990.

15. Leavitt, R.A; Kells, J.J; Bunkelmann, J.R; Hollingworth, R.M. *Bull Environ Contam Toxicol*, **1991**, *46*, pp 22-29.

16. Gruessner, B; Shambaugh, N,C; Watzin, M.C. *Environ Sci Technol*, **1995**, *29*, pp 251-254.

Chapter 20

Monitoring of Atrazine in River and Coastal Sea Waters by Immunoassay, Solid-Phase Disk Extraction, and Gas Chromatography– Mass Spectrometry

Jordi Gascón, Jaume S. Salau, Anna Oubiña, and Damià Barceló[1]

Department of Environmental Chemistry, Centro de Investigación y Desarrollo, Consejo Superior de Investigaciónes Cientificas, c/Jordi Girona 18–26, 08034 Barcelona, Spain

A magnetic-particle based immunoassay (IA) for atrazine was compared with gas chromatography-mass spectrometry (GC-MS) for the monitoring of atrazine in coastal sea waters of varying salinity. Water samples were measured directly by IA and also after solid-phase disk extraction (SPE) by GC-MS. Correlations were established between the data obtained by IA directly from water samples, at various salinity levels (from 0 to 35 g/L) and from GC-MS following SPE. In the Ebre river water the atrazine levels determined varied from 20 to 225 ng/L whereas their common cross-reactants, deethylatrazine and simazine, were detected at concentrations varying from 0 to 110 ng/L and from 0 to 150 ng/L, respectively. At salinity values between 0.5 and 8 g/L, the atrazine levels varied between 20 and 120 ng/L.

The s-triazine herbicides are among the most commonly detected pesticides in water. Their widespread use combined with overapplications, accidental spills, runoff from mixing-loading areas, and faulty waste disposal creates environmental concerns. The stability of these chemicals (atrazine half life in soil is about 50 days) together with their solubility (for atrazine, 33 mg/L) and their mobility in surface and ground water prompts s-triazine herbicides to reach estuarine areas carrying contamination to the sea. S-triazine herbicides have been detected in coastal sea water samples from the Mediterranean, through different rivers such as Ebre (Spain), Po (Italy), Rhône (France), and Axios (Greece) (1, 2), and in the Mississippi river (USA) (3-4).

Implementation of new analytical techniques for monitoring trace quantities of contaminants in sea and river water samples is now an important issue in environmental research. Because of their specificity, high sensitivity, adaptability for field use and ability to recognize a wide range of substances, immunochemical techniques can be particularly suited to this type of measurements. Numerous immunoassays and related techniques have been developed during the last years, for a broad range of pesticides and contaminants of industrial origin (for review see 5 and 6). Although today some of them are commercially available, the acceptance of immunoassay as routine screening method depends upon validation compared to chromatographic methods. Quantitative analysis of atrazine in water matrices has been carried out with little matrix interferences (7). Recently the good correlation existing between ELISA (enzyme-linked immunosorbent assay) techniques and GC-

[1]Corresponding author

MS to determine triazines in water has been demonstrated (8). Similar agreement was obtained on a validation study comparing ELISA and liquid chromatography followed by postcolumn derivatization fluorescence detection to analyze carbaryl in ground and surface waters (9).

Interferences in immunoassay methods can be categorized into two major classes: i) those that affect binding of antigen by competing for the specific binding site on the antibody (often referred to as cross-reactants) and ii) those that affect the binding event between the antibody and an antigen in a general way. Recent review articles (6, 10) have emphasized the need to study the performance of ELISA techniques in real environmental samples and the difficulties encountered with some complex matrices. In general, for competitive immunoassay, matrix effects are manifested as a reduction of the color development. These effects may result from nonspecific binding of the analyte to the matrix, nonspecific binding of the matrix to the antibody or enzyme, or from denaturation of the antibody and/or enzyme. Sea water presents some features that may interfere with the immunological reaction, however, in a previous paper (11) we have demonstrated that ELISA techniques can be used for the determination of atrazine in artificial sea water samples.

The first objective of this work was to evaluate the influence of the sea water samples and selected physico-chemical properties, such as ionic strength and pH, on the RaPID-ELISA for the determination of atrazine in samples of varying salinity and pH. The second objective was to demonstrate the usefulness of this technology to perform studies with real estuarine and coastal water. Atrazine is currently measured by solid-phase extraction (SPE) using Empore C_{18} disks or liquid-liquid extraction (LLE) followed by gas chromatography (GC) either with nitrogen-phosphorous detection (NPD) or mass spectrometry (MS). Our group has previously used the RaPID-ELISA for detection of atrazine in various types of water matrices (ground, distilled, and artificial sea water samples) (8, 12). To our knowledge, no publications have been reported that applied the RaPID-ELISA or other ELISA techniques to real coastal water samples of varying salinity and performed an intercomparison of the data with another independent method, involving SPE with C_{18} Empore disks followed by GC-MS.

Experimental Section

Apparatus. All spectrophotometric measurements were determined using the RPA-I Analyzer (Ohmicron Corp., Newtown, PA, USA). The RaPID-magnetic particle-based ELISA from Ohmicron was purchased through J.T. Baker (Deventer, NL). A two-piece magnetic separation rack consisting of a test tube holder which fits over a magnetic base containing permanently positioned rare earth magnets is required. This two-piece design allows for a 60-tube immunoassay batch to be set up, incubate, and magnetically separated without removing the tubes from the holders (13). Adjustable pipettes, Gilson P-1000 and P-200 (Rainin, Woburn, MA, USA), and Eppendorf repeating pipettes (Eppendorf, Hamburg, Germany) were used to dispense liquids.

Sample Preparation. Water samples used throughout this study were distilled water, estuarine, and Ebre river water (Amposta, Tarragona, Spain). Samples with different salinity values were prepared by dissolving the corresponding amounts of sea salts (Sigma Chemical Co., St. Louis, MO, USA) in Milli-Q water (Millipore, Bedford, MA, USA). Measurements of the sea salt content of the different real environmental samples were performed by using a portable conductimeter from Crison (Alella, Barcelona, Spain). Previously a calibration graph was constructed measuring the conductivity of solutions containing known amounts of sea salts.

For estuarine and Ebre river water samples from Amposta (pH = 7.9 - 8.2 and salinity 0.50 g/L), about 2.5 L of water was collected, filtered through 0.45 µm filters (Millipore, Bedford, MA, USA) to remove suspended particles, and stored at 4°C until analyzed. Estuarine-river and coastal water samples were collected at several points of the estuarine area showing different salinity values. These samples were first filtered through fiber-glass filters (Millipore Corp.) of 0.70 μm and subsequently through 0.45 μm filters to eliminate particulate matters as described (*11*). The SPE off-line method used a Millipore 47-mm filtration apparatus. The membrane extraction disks were manufactured by 3M (St. Paul, MN, USA) under the Empore trademark and are distributed by J.T. Baker and Analytichem International. The disks used in these experiments were 47 mm in diameter and 0.5 mm thick. Each disk contains about 500 mg of C_{18} bonded silica material. The disk, placed in a conventional Millipore apparatus, was conditioned with 2 x 10 mL of methanol under vacuum avoiding the solid phase becoming dry and 1 liter of water was extracted with the vacuum adjusted to yield a 1 hour extraction time just to dryness. The disks containing the pesticides were used for transportation and storage. Immediately before the analysis, the pesticides trapped in the disk were eluted with methanol, evaporated just to dryness with nitrogen and re-dissolved in ethyl acetate for analysis by GC-MS.

To study the effect of the salinity on the immunoassay determinations, solutions containing different sea salt concentrations (0, 2.5, 5, 10, 15, 20, 25, 30 and 35 g/L) were prepared by dissolving sea salts in Milli-Q water. Another batch of solutions were prepared with varying pH values (3, 4, 4.5, 5, 5.5, 6, 6.5, 7, 7.5, 8, 8.5 and 9) by adding small amounts of 1N HCl or 1N NaOH. Each of these solutions was used to prepare standard curves and to analyze their parallelism by ELISA. For the RaPID High-Sensitivity atrazine immunoassay we only checked 0 and 35 g/L sea salt concentrations while pH effect was studied on solutions at 2.5, 5.7, 7.0 and 9.1. Water from the Ebre estuarine-river was also tested to confirm the absence of matrix effects.

The selected area for the pilot monitoring survey of trace-levels of herbicides was the Ebre delta, Tarragona, Spain, which is shown in Figure 1. Samples of water were collected at stations 1, 2, 3 and 4 with the following names: Ebre river (station 1), Encañizada lagoon (station 2), Encañizada lagoon for fishery (station 3), and Tancada lagoon (station 4). Water samples were sampled monthly from April to June 1991. The complete analytical protocol has been described elsewhere (*8*). In addition to the pilot survey, a continuos monitoring was carried out during 1995 and 1996 at station 1 (see Fig. 1). Also, sea water samples of varying salinity were collected from the estuarine down to the coastal sea water.

RaPID Assay for Atrazine. Two different kits were used: High-Sensitivity Atrazine RaPID Assay and the Atrazine RaPID Assay. All samples were assayed according to the RaPID Assay package insert. The RaPID magnetic particle-based solid-phase Enzyme-linked Immunosorbent Assay (RaPID-ELISA) has polyclonal antibodies coated on paramagnetic beads. A total of 250 μL (or 200 μL) of the sample water to be analyzed is added to a disposable test tube, along with 250 μL of pesticide (atrazine) hapten-horseradish peroxidase (HRP) enzyme conjugate, and 500 μL of rabbit anti-pesticide magnetic particles (anti-atrazine) attached covalently. Both pesticides of the sample and enzyme labeled pesticide compete for antibody sites on the magnetic particles. Tubes were vortexed and incubated for 30 min (or 15 min) at room temperature. The reaction mixture was magnetically separated using a specially designed magnetic rack. After separation, the magnetic particles were washed twice with 1.0 mL of distilled water to remove unbound conjugate and eliminate any potential interfering substances. Pesticide and enzyme labeled pesticide remained bound to the magnetic particles in concentrations proportional to their original concentration. The presence of labeled pesticide was detected by adding a total of 500 μL of a 1:1 mixture of a solution containing substrate and chromogen (hydrogen peroxide and 3,3',5,5'-tetramethylbenzidine (TMB)). The tubes were vortexed to resuspend particles and incubated for another 20 min at room temperature to allow color development. The color reaction was stopped by the addition of 500 μL of 2M of sulfuric acid solution. The final concentrations of pesticide for each sample were determined using the RPA-I RaPID Photometric Analyzer by determining the

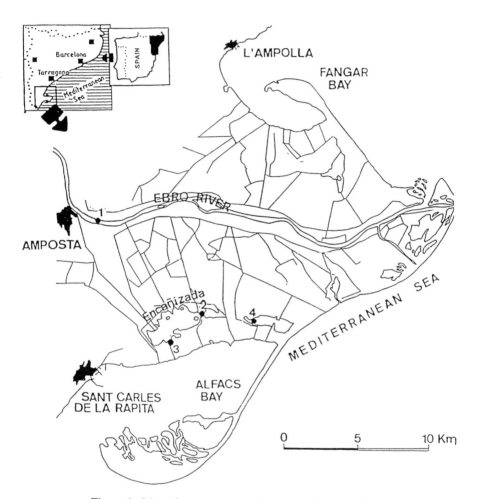

Figure 1. Map of the area where the monitoring took place.

absorbance at a wavelength of 450 nm. The observed results were compared to a linear regression line using a log of the pesticide concentration (x) versus logit $\%B/B_0$ (y) standard curve (B/B_0 is defined as the absorbance at 450 nm observed for a sample or standard divided by the absorbance at the zero standard). The standard curve was prepared from calibrators containing known levels of atrazine at 0, 35, 250 and 1000 ng/L (for the High Sensitivity Atrazine RaPID Assay) and 0, 0.1, 1.0 and 5.0 μg/L (for Atrazine RaPID Assay). All standards were analyzed in duplicate. Test kits were stored at 4°C. Temperature and time are important assay parameters that must be controlled for an ELISA to work properly. In all cases, immunoassay reagents and samples were allowed to equilibrate to room temperature before use, and reaction times were consistent throughout the experiment. A negative control had to be tested concurrently with each set of water samples. The negative control was used to standardize the water sample's absorbance measurements.

GC-MS. A Model 8065 GC system and a model MD 800 GC-MS system from Fisons Instruments (Manchester, England) interfaced with a LAB-BASE data system were used for GC-MS with electron impact (EI). The DB-5 column used contained 5% phenyl-95% methyl polysiloxane (J & W Scientific, Folsom, CA, USA). The column was programmed from 70 to 300°C (10 min) at 6°C/min. The extracts were injected in SIM mode. Calibration under GC-MS is shown in Table I.

Table I. Calibration equations under GC-MS after SPE using Empore C_{18} disks

Compound (m/z Ions)	Upper limit	Lower limit	Equation	r^2
Atrazine (200,215)	1.7 ng	0.7 ng	y = −122454+1777377x	0.997
Simazine (201,186)	1.4 ng	0.1 ng	y = −90989+670068x	0.996
Deethylatrazine (172, 187)	2.1 ng	0.2 ng	y = −46003+2721759x	0.995

GC-NPD. The extracts were injected onto the column of a gas chromatograph (GC 5300 Mega series, Carlo Erba, Milan, Italy) equipped with a nitrogen-phosphorus detector (NPD). A 15 m x 0.15 mm i.d. fused silica capillary column coated with chemically bonded cyanopropylphenyl DB-225 (J & W Scientific) was used. Hydrogen was the carrier gas at 60 kPa and helium the make up gas at 110 kPa. The temperatures of injector and detector were held at 270°C. The column was programmed from 60 to 90°C at 10°C/min and from 90 to 220°C at 6°C/min.

Results and Discussion

General Remarks. The use of C_{18} Empore disks, such as the work described in an earlier paper from one of our groups for the determination of atrazine in artificial (14) and natural sea water samples, (15) needs few considerations. Due to the smaller pore size (7 μm), problems of blocking the sites on the disk, are common. Thus, two prefiltration steps, using 0.45 and 0.7 μm are needed. In addition, the high flow rate generally used with the disks is decreased dramatically when sea water samples are analyzed. Empore disks contain 500 mg of C_{18} phase thus, when preconcentrating 1 liter of water the recoveries of the low-capacity analytes such as deethylatrazine, are above 70% (16). However, when 2-4 liters of water are preconcentrated so that triazine compounds may be measured at low levels, the recovery of deethylatrazine drops to values < 20% (15). However, the use of Empore disks is still advantageous in terms of easy handling and transport from the field to the laboratory to perform the final analysis. Also, the storage of preconcentrated samples up to three months at − 20°C without any loss of atrazine and related compounds (17) is helpful in saving space. By comparing this method with ELISA, we are evaluating the performance of two procedures for analyzing atrazine in sea water samples and give recommendations about their future application in these water matrices.

Acceptance of immunoassay technology requires validation and confirmation of the results obtained by some other robust technique. Competitive immunoassays for pesticides rely on the ability of the analyte to inhibit binding of the antibody to a certain competitor, usually a hapten-derivatized enzyme tracer. Because of this inhibition, the measured enzymatic activity, linked to the immunocomplex, is decreased when compared to the negative controls where analyte is absent (maximum binding of the antibody to the enzyme-tracer). The enzymatic activity of the immunocomplex is measured by employing a convenient substrate that, generates a chromophore after enzymatic reaction. The amount of analyte in a sample is calculated by interpolating the resulting absorbance on a standard curve usually prepared in buffer. In the first case, analytical chemists may have difficulties accepting that a decrease in color is attributed to the presence of the target analyte. Although antibody binding to the analyte is well recognized to be a very specific reaction, it is also true that factors (other than cross-reacting compounds) may inhibit the binding of the antibody to the enzyme tracer resulting in a decrease of the color development and consequently leading to false positive results. Factors such as anions, cations, pH, and organic content are responsible of what is known as matrix effect and may interfere non-specifically with the immunochemical reaction. In fact, a tendency to overestimate the amount of pollutants in environmental samples has often been observed (8, 9, 18, 19) by ELISA when compared to chromatographic methods. Although antibodies are made to work under physiological conditions immunochemical reactions have occurred satisfactorily in many other aqueous environments, or even in gas phase media (20). However, under these conditions, the affinity of the antibody versus the analyte or the enzyme tracer may vary, thus changing the kinetics of the immunoreaction. Therefore, before starting to perform immunoassays in a new matrix it is advisable to determine if the kinetics of the antibody reaction in such media remains identical to that of the standards on the assay buffer.

Cross-reactivities Studies. In a previous work, the cross-reactivity (CR) of the RaPID-ELISA for Atrazine on different matrices was studied (Table II) (8). Antibody specificity was determined by measuring the cross-reactivity of compounds structurally similar to atrazine, i.e., simazine and the transformations products (TPs) of atrazine. The cross-reactivities were based upon atrazine (=100%) and were calculated from the calibration according to the formula:

$$\% \text{ CR} = (\frac{\text{atrazine IC}_{50}}{\text{cross - reacting triazine IC}_{50}}) \times 100$$

Table II summarizes the specificity data with the major atrazine cross-reactants: deethylatrazine, deisopropylatrazine, simazine.

It was found that the response of atrazine was greatest with the lowest IC_{50} concentration of 0.302 µg/L. The most remarkable data corresponds to the high CR found for deethylatrazine in all the studied water matrices. Simazine and deisopropylatrazine gave much lower values of CR. Note that the binding was strongest for compounds that have structures most closely resembling atrazine, that is, 4-ethyl-amino and 6-isopropyl-amino group. For example, a substitution from an isopropyl to two ethyl groups (simazine) increased the IC_{50} to 1.749 µg/L in distilled water. Substitution of hydrogen for isopropyl (deisopropylatrazine) increases the IC_{50} dramatically to 2.059 µg/L in distilled water. Moreover, substitution of hydrogen for ethyl (deethylatrazine) only slightly changed the IC_{50} (0.371 µg/L).

Table II. Specificity of the Atrazine and RaPID-ELISA in various aqueous matrices

Compounds	Distilled water		Ground water		Estuarine water	
	IC$_{50}$[a]	CR (%)[b]	IC$_{50}$	CR (%)	IC$_{50}$	CR (%)
ATR	0.302 (6)[c]	100.0	0.302 (5)	99.67	0.405 (8)	74.57
DEA	0.371 (5)	80.79	0.458 (6)	65.50	0.486 (7)	61.72
DIA	2.059 (7)	14.57	2.265 (8)	13.25	2.190 (5)	13.69
SIM	1.749 (7)	17.15	2.088 (7)	14.37	2.100 (7)	14.40

[a] IC$_{50}$: 50% inhibition concentration (50% B/B_0) (in μg/L). [b] CR (%): Percentage cross-reactivity is determined by estimating the amount of cross-reacting compound required to displace 50% of the enzyme conjugate compared to the amount of atrazine. [c] CV (%): Coefficient of variation (n=6). ATR: atrazine, DEA: deethylatrazine, DIA: deisopropylatrazine, SIM: simazine

Effect of pH and Salinity. As we point out in the introduction section, chlorotriazine herbicides and their corresponding metabolites are usually found in surface river and estuarine waters at levels varying from 0.05 to 1.0 μg/L. In a previous study we have shown using SPE followed by GC-MS that the concentration of atrazine in estuarine-river and coastal water samples decreases when salinity increases (*15*) and therefore immunoassay technology appears to be suitable alternative to routinely monitor low levels of atrazine in estuarine water. Main features of sea water samples are the ionic strength, usually about 0.7, and pH, generally varying from 7.5 to 8.5. As it is shown in Figure 2, the evaluation of the effect of these parameters demonstrated that there was no effect on the RaPID-Atrazine assay performance when salinity content in water was raised to 35‰ (g/L). Standard curves run at different salinity values (0 to 35‰) were parallel in all immunoassays employed in this study (see Figure 2a) suggesting that reliable analyses could be made on the river-estuarine waters without the need of changing immunoassay protocol or applying clean-up procedures. Analogously, these immunoassays were not affected by extreme pH conditions since the standard curves did not show variations on a pH range between 3 and 9 (see Figure 2b). Only a slight decrease in the absorbance was observed when analyzing atrazine in pH 2.0 water samples with the High-Sensitivity RaPID-Atrazine kit (see Figure 2c), although the slope of the assay did not show any significant difference when compared to the standard curve run at pH 7.0. On the other hand, the salinity effect on the High-Sensitivity RaPID-Atrazine assay was not significant as shown in Figure 2d.

Ebre river and coastal water samples were collected at different locations within the estuarine area showing salinity levels ranging from 0.5 to 7.9 g/L. Since no strong effect of the salinity on the immunoassay performance was observed (Figure 2d), we analyzed the real samples by immunoassay, comparing the results obtained with those of the chromatographic method. Samples were processed by GC-MS after the clean-up procedure described in the experimental section. Two different steps of the purification process was performed with the High-Sensitivity Atrazine RaPID assay: i) after passing the samples through 0.7 and 0.45 μm filters and ii) after extracting atrazine with C$_{18}$ extraction disks. Higher correlation between ELISA and GC-MS was observed with the samples processed after SPE extraction (r = 0.983) (Figure 3). Although a good correlation was observed, the ELISA tends to overestimate the results obtained by GC-MS when triazine concentration is low (Figure 4). Possibly, the RaPID-ELISA, presents a higher deviation at this level concentration (20-50 ng/L).

Table III shows the concentrations of the various herbicides analyzed using SPE and GC-MS. First, we can notice that the concentration of deethylatrazine is relatively high in all samples although it is always lower than that of atrazine. Considering that the recovery of deethylatrazine is slightly lower than the other analytes (76%) (*16*)

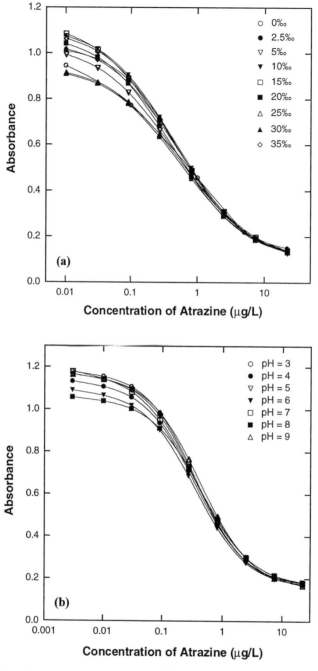

Figure 2. Standard curves run under different salinity (a) or pH values (b) with the RaPID-Atrazine immunoassay. Figure 2 (c) presents the influence of different pH values or different salinity (d) for the RaPID High-Sensitivity Atrazine immunoassay.

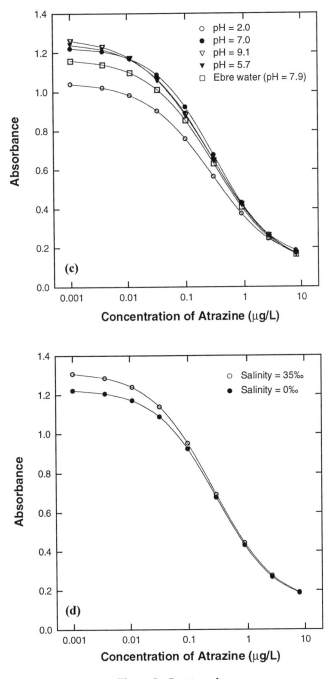

Figure 2. *Continued*

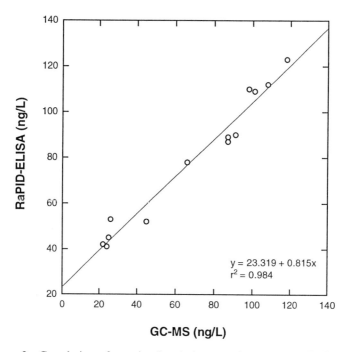

Figure 3. Correlation of atrazine levels in coastal sea waters obtained for
GC-MS and ELISA after SPE C$_{18}$ Empore disks (r = 0.983).

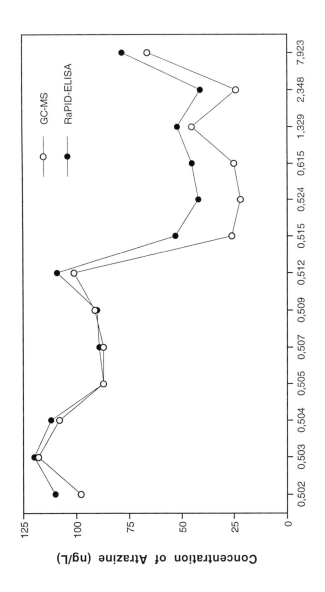

Figure 4. Comparative results of atrazine levels in coastal sea waters obtained by GC-MS and ELISA in the salinity profile.

and that the results of Table III are not corrected for recovery, the real concentration of deethylatrazine will be slightly higher than in Table III, but in any case the values will be much lower than those of atrazine.

Table III. Concentration of atrazine (ATR), deethylatrazine (DEA), simazine (SIM), in ng/L in Ebre river and coastal sea waters determined after solid-phase extraction followed by GC-MS

Salinity (g/L)	ATR	DEA	SIM
7.923	66	6	60
2.348	24	0	21
1.329	45	3	37
0.615	25	1	22
0.524	22	0	18
0.515	26	1	22
0.512	101	4	51
0.509	91	3	41
0.507	87	4	41
0.505	87	4	43
0.504	108	4	48
0.503	118	4	50
0.502	98	2	33

It can be noticed that the concentration of deethylatrazine is quite stable during the salinity profile whereas both atrazine and simazine diminish when the salinity value increases (except a 7.9 g/L of salinity). This is a remarkable fact and indicates that deethylatrazine can be much more easily transported to coastal sea water samples probably due to its higher solubility, 3200 mg/L, versus 33 and 6 mg/L for atrazine and simazine, respectively.

Monitoring of River Waters. The results of a first pilot monitoring survey at the Ebre river using the different methods are shown in Table IV (April-June, 1991).

Table IV. Concentration values (μg/L) and CV (%) of Atrazine and Total Chlorotriazines[a] (n = 6)

Month-Station	GC-NPD		RaPID-ELISA		% Recovery	
	Triazines	Atrazine	Triazines	Atrazine	Triazines	Atrazine
April-1	0.064 (11)	0.036 (9)	0.041 (4)	0.036 (3)	64.1	100.0
April-3	0.032 (10)	0.020 (7)	0.022 (7)	0.019 (5)	67.9	92.2
May-2	0.086 (8)	0.008 (9)	0.045 (4)	0.010 (4)	52.3	117.6
May-3	0.120 (5)	0.012 (6)	0.074 (8)	0.011 (7)	61.4	86.6
June-1	0.326 (12)	0.188 (10)	0.195 (6)	0.168 (6)	59.5	89.4
June-3	0.120 (9)	0.057 (10)	0.067 (10)	0.063 (9)	55.6	110.5
June-4	0.067 (9)	0.016 (8)	0.054 (8)	0.013 (7)	80.5	80.5

[a] Total chlorotriazines are atrazine and simazine, with the exception of sample May-3, where the concentrations of atrazine, simazine and deethylatrazine were 0.013, 0.040 and 0.067 μg/L, respectively. The values for atrazine by the RaPID-ELISA have been calculated after applying the percentage cross-reactivity indicated in Table II.

These data have been obtained in the following way. First, the RaPID assay values for total chlorotriazines were obtained and afterwards, those showing a positive number where analyzed by GC-NPD. The exact concentration values for atrazine, simazine and in one sample for deethylatrazine were obtained by GC-NPD. GC techniques were used to compare and evaluate the immunoassay results. Then, based on the percentages of cross-reactivity values obtained in Table II, we were able to determine the real concentration of atrazine by ELISA using the corresponding correction factors. When all these values were obtained, we have indicated the percentage of agreement or recoveries in the same Table for chlorotriazines, namely simazine and atrazine. There is more agreement for atrazine which is expected since this magnetic-ELISA test is specific for atrazine and not for all chlorotriazines. Second, the recoveries for atrazine are relatively better and can be used for inter-comparison with GC-NPD. Coefficients of variation below 10% were observed. For total chlorotriazines the coefficient of correlation between RaPID assay and GC-NPD measurements was 0.9936 whereas for atrazine a coefficient of correlation of 0.9968 was obtained. The correlation obtained for the total chlorotriazines is acceptable and allows us to use the total data obtained in this ELISA test for total chlorotriazines (generally for atrazine and simazine, which are the most ubiquitous chlorotriazine contaminants detected in Ebre river)

From the data in Table IV we should indicate that we could not obtain correlation with all the samples. In total, 12 samples were positive for atrazine and simazine using GC-NPD analysis. However station 2 had too low values in April and May, no atrazine was detected in station 4 in April, and atrazine was too low in May and June in station 1 and 2, respectively. These values considered too low were of 0.010 μg/L of atrazine or lower and a similar values or lower for simazine. Except for one case where the atrazine concentration by GC-NPD is 0.008 μg/L, the rest of the samples in Table IV have concentrations for atrazine and chlorotriazines higher than 0.02-0.03 μg/L. Most probably this is one of the limitations of this magnetic particle-based ELISA method when estimating such low concentration of atrazine in natural estuarine waters.

Table IV indicates that the immunoassay method detects lower amount of total chlorotriazines compared to the chromatographic method. This may be due to the fact that this magnetic-ELISA method detects only a certain percentage of simazine and deethylatrazine. For estuarine waters from the Ebre delta area values of 14.40% and 61.72% of cross-reactivity are obtained for simazine and deethylatrazine, respectively. However, the concentration of atrazine in the estuarine waters from the Ebre delta area is comparable using GC-NPD and immunoassys.

Another monitoring survey was carried out between March 1995 and June 1996, but only in Amposta (Tarragona, Spain) (station 1, Figure 1). Until April of 1995 the RaPID-Atrazine assay was applied, but the concentration of atrazine during these months was too low for the detection limit of this immunoassay kit. Then, the High-Sensitivity Atrazine RaPID assay, with a detection limit of 0.015 μg/L, was applied. Acceptable correlation was obtained in most of the months ($r^2 = 0.94$). Exceptions were March and April 1995. During the first two months we did not use the High-Sensitivity Atrazine RaPID assay, so the values of atrazine are overestimated due to the use of the conventional RaPID assay for atrazine. Only after using the High-Sensitivity RaPID for atrazine did the values between GC-MS and ELISA correlate well. Figure 5a shows the values obtained by GC-MS for all the analytes monitored and Figure 5b shows the data obtained by RaPID-ELISA.

Conclusions

In summary we can conclude that SPE followed by either RaPID-ELISA and GC-MS are complementary methods. First, we can indicate that ELISA can be used as a screening procedure to have the first values for atrazine in estuarine and coastal sea water samples of varying salinity. When the concentrations of the common cross-reactants are high, then further calculations are needed to obtain a good correlation between ELISA and gas chromatographic techniques. The second aspect is that when the samples are filtered and preconcentrated by SPE Empore disks, the results obtained by RaPID-ELISA are comparable to those obtained by GC-MS indicating

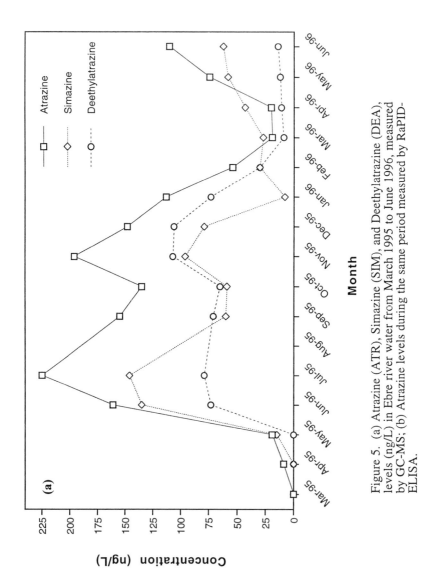

Figure 5. (a) Atrazine (ATR), Simazine (SIM), and Deethylatrazine (DEA), levels (ng/L) in Ebre river water from March 1995 to June 1996, measured by GC-MS; (b) Atrazine levels during the same period measured by RaPID-ELISA.

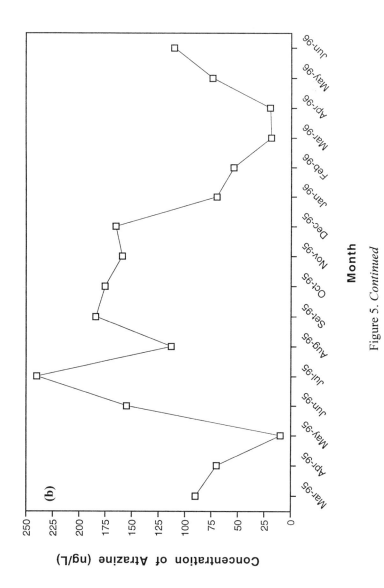

Figure 5. *Continued*

that the ELISA method is as accurate as a chromatographic technique. Finally, it should be indicated that it will be of interest to develop other RaPID assays with higher sensitivity to those presently in use in order to determine common herbicides like simazine and deethylatrazine, which are ubiquitous contaminants through different rivers and coastal areas.

Acknowledgments

This work has been supported by the Commission of the European Communities (Contract EV5V-CT94-0524) and by the CICYT (AMB95-1230-CE). Also, we thank to Ohmicron Corp. and Merck.

Literature Cited

1. Readman, J.W.; Albanis, T.; Barceló, D.; Galassi, S.; Tronczynski, J.; Gabrielides, G.P. *Mar. Pollut. Bull.* **1993**, *26*, 613.
2. Durand, G.; Bouvot, V.; Barceló, D. *J. Chromatogr.* **1992**, *607*, 319.
3. Pereira, W.E.; Hostettler, F.D. *Environ. Sci. Technol.* **1994**, *27*, 1542.
4. Mills, M.S.; Thurman, M.E. *Environ. Sci. Technol.* **1994**, *28*, 600.
5. Sherry, J.P. *Crit. Rev. Anal. Chem.* **1992**, *23*, 217.
6. Meulenberg, E.P.; Mulder, W.H.; Stoks, P.G. *Environ. Sci. Technol.* **1995**, *29*, 553.
7. Rubio, F.M.; Itak, J.A.; Scutellaro, A.M.; Selisker, M.Y.; Herzog, D.P. *Food Agric. Immunol.* **1991**, *3*, 113.
8. Gascón, J.; Durand, G.;Barcelo, D. *Environ. Sci. Technol.* **1995**, *29*, 1551.
9. Marco, M.P.; Chiron, S.; Gascón, J.; Hammock, B.D.; Barceló, D. *Anal. Chim. Acta* **1995**, *311*, 319.
10. Lucas, A.D.; Gee, S.J.; Hammock, B.D.; Seibert, J.N. *J. AOAC Int.* **1995**, *78*, 585.
11. Gascón, J.; Oubiña, A.; Ferrer, I; Önnerfjord, P.; Marko-Varga, G.; Hammock, B.D.; Marco, M.P.; Barceló, D. *Anal. Chim. Acta,* **1996**, in press.
12. Gascón, J.; Martínez, E.; Barceló, D. *Anal. Chim. Acta* **1995**, *311*, 357.
13. Itak, J.; Selisker, M. Y.; Herzog, D.P. *Chemosphere* **1992**, *24*, 11.
14. Barceló, D.; Durand, G.; Bouvot, V.; Nielen, M. *Environ. Sci. Technol.* **1993**, *27*, 271.
15. Durand, G.; Barceló, D. *Talanta* **1993**, *40*, 1665.
16. Barceló, D.; Chiron, S.; Lacorte, S.; Martinez, E.; Salau, J.S.; Hennion, M.C. *Trends Anal. Chem.* **1994, ** *13*, 352.
17. Martinez, E.; Barceló, D. *Chromatographia* **1996**, *42*, 72.
18. Gruessner, B; Shambaugh, N.C.; Watzin, M.C. *Environ. Sci. Technol.* **1995**, *29*, 251.
19. Brady, J.F.; LeMasters, G.S.; Williams, R.K.; Pitman, J.H.; Dauber, J.P.; Cheung, M.W.; Skinner, D.H.; Turner, J.R.; Lange, J.; Sobek, S.M. *J. Agric. Food Chem.* **1995**, *43*, 268.
20. Ngeh-Ngwainbi, J.; Foley, P.H.; Kuan, S.S.; Guilbault, G.G. *J. Am. Chem. Soc.* **1986**, *108* 5444.
21. Aga, D.S.; Thurman, E.M.; Yockel, M.E.; Zimmerman, L.R.; Williams, T.D. *Environ.Sci.Technol.*, **1996**, *30*, 592.

Chapter 21

A Paramagnetic Particle-Based Enzyme-Linked Immunosorbent Assay for the Quantitative Determination of 3,5,6-Trichloro-2-pyridinol in Water

Jeanne A. Itak[1], William A. Day[1], Angel Montoya[2], Juan J. Manclús[2], Amy M. Phillips[3], Dwayne A. Lindsay[3], and David P. Herzog[1]

[1]Ohmicron Corporation, 375 Pheasant Run, Newtown, PA 18940
[2]Laboratorio Integrado de Bioingenieria, Universidad Politécnica de Valencia, Apdo. de Correos 22012, Camino de Vera s/n, 46071 Valencia, Spain
[3]DowElanco, 9330 North Zionsville Road, Indianapolis, IN 46268

A competitive enzyme-linked immunosorbent assay (ELISA) for the quantitation of 3,5,6-trichloro-2-pyridinol (TCP), the major biological and environmental degradation product of chlorpyrifos and triclopyr, was developed. Magnetic particles were used as the solid phase to attach monoclonal anti-TCP antibodies. The ELISA has an estimated least detectable dose of 0.25 parts per billion (ppb; ng/mL) in water. Specificity studies indicate that the assay is specific for TCP and can distinguish it from the parent compounds as well as many other related and unrelated agricultural compounds. Results compare favorably with GC/MSD measurements (r = 0.959).

Interest in monitoring pesticide residues has increased in recent years as a result of concerns over the potential for water, soil and food contamination. Some attention has also focused on the monitoring of certain pesticide metabolites as indicators of contamination or exposure, to study metabolic pathways under field use conditions or to monitor for degradation of parent pesticides during shipment or storage of environmental samples. An example is 3,5,6-trichloro-2-pyridinol (TCP), the major biological and environmental degradation product of the insecticide chlorpyrifos and the herbicide triclopyr in soil, water and plant surfaces (*1,2*). The toxicity of TCP itself has been defined for several species (*3-5*). At neutral pH, TCP is ionic and appears to be more leachable than either parent compound and hence may serve better as an indicator of water contamination.

Traditionally, TCP analysis requires specific and labor intensive sample preparation steps. This paper describes the development of a paramagnetic particle-based ELISA which can be used for the quantitation of TCP in environmental water samples at detection levels in the sub-parts per billion (ppb; ng/mL) without any sample preparation. Magnetic particle-based ELISAs for the quantitation of other

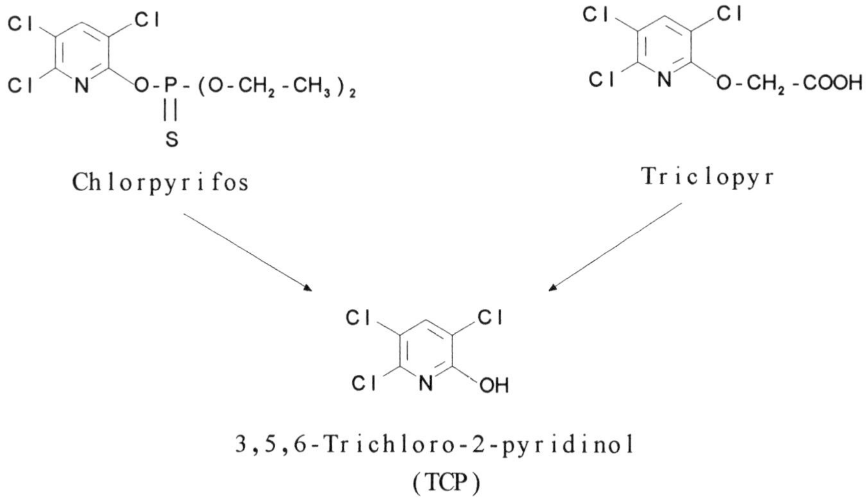

Chlorpyrifos

Triclopyr

3,5,6-Trichloro-2-pyridinol
(TCP)

pesticide residues have been described previously (6-9). In these systems, the antibody is covalently coupled to the paramagnetic-particle solid phase. Because the antibody is covalently bound, sensitivity and precision problems associated with passive adsorption to polystyrene surfaces are eliminated. These problems include desorption or leaching of proteins passively adsorbed to microtiter plates and coated tubes, and well-to-well variability within microtiter plates (10-13). With magnetic particle-based assays, the paramagnetic particles are uniformly dispersed throughout the reaction mixture, allowing for precise addition of antibody and rapid reaction kinetics. The TCP immunoassay presented takes less than one hour to perform. Photometric determination of the final colored product is done with a specially designed microprocessor-controlled photometer with extensive data reduction capability that directly reports parts per billion (ppb, ng/mL, μg/L) concentrations of TCP in the sample (Ø).

Materials and Methods

Amine terminated superparamagnetic particles were obtained from PerSeptive Diagnostics, Inc. (Cambridge, MA). Glutaraldehyde and horseradish peroxidase (HRP) were obtained from Sigma Chemical (St. Louis, MO) and the TMB Microwell Peroxidase Substrate was obtained from Kirkegaard & Perry Labs (Gaithersburg, MD). 3,5,6-Trichloro-2-pyridinol, chlorpyrifos, chlorpyrifos-methyl, clopyralid, fluroxypyr, fluroxypyr-2-pyridinol, 2-methoxy-3,5,6-trichloropyridine, picloram and triclopyr were obtained from DowElanco (Indianapolis, IN). Other pesticides and metabolites were purchased from ChemService (West Chester, PA) and Crescent Chemical Co., Inc. (Hauppauge, NY). All other reagents were reagent grade or chemically pure as supplied by commercial sources.

The following apparatus were utilized: Magnetic Separation Unit, RPA-I Analyzer™ and vortex mixer obtained from Ohmicron Corporation, Newtown, PA;

adjustable precision pipette, Gilson P-1000 (Rainin; Woburn, MA); and repeating pipette (Eppendorf; Hamburg, Germany).

Water samples were characterized using Hach Test Kits for nitrate/nitrite (cat. #14081-00), salinity (cat. #24218-00) and chlorine, hardness, iron and pH (cat. #2230-02).

The TCP monoclonal antibody, LIB-MC2, was derived from a hapten prepared by hydrolysis of the thiophosphate ester of a chlorpyrifos hapten, as previously described (*14,15*).   The procedure for coupling anti-TCP antibody to magnetic particles was also detailed in prior literature (*9*).   The TCP-HRP conjugate was synthesized according to methods outlined previously for a chlorpyrifos hapten and hapten-protein conjugates (*14*).

The immunoassay procedure required adding 250 µL of standard or sample with 250 µL of horseradish peroxidase labeled TCP analog and 500 µL of anti-TCP antibody coupled paramagnetic particles to disposable test tubes.   The tubes were vortexed and incubated for 20 min at room temperature.   The reaction mixture was separated by placing the tube rack into the magnetic base and the supernatant was decanted.   Particles were washed twice with 1 mL washing solution (preserved deionized water).   The peroxidase substrate solution, peroxide/TMB (3,3',5,5'-tetra-methylbenzidine) was added, 500 µL per tube, and allowed to develop for 20 min. The color reaction was stopped with 500 µL of 0.5% sulfuric acid.   To ensure accurate results, a calibration curve (0.0, 0.5, 2.5 and 6 ppb TCP), in duplicate, was included with every run.   The concentrations of TCP for each sample were determined using the RPA-I Analyzer set at 450 nm.   The RPA-I compares the observed sample absorbances to a regression line using a log-logit standard curve derived from the calibrator absorbances and reports parts per billion TCP in the sample.

Residue analysis for TCP was performed by capillary gas chromatography with mass selective detection.   The method required that 25 mL water samples be extracted twice with 1.0 mL of 2.0 N hydrochloric acid, 10 g sodium chloride and 5.0 mL of 1-chlorobutane and shook for 30 minutes on a reciprocating shaker.   Samples were centrifuged for 5 minutes at 2500 rpm and the 1-chlorobutane layers were combined and concentrated to less than 1 mL (not to dryness) on an evaporator set at 35°C with a flow rate of 200 mL/min.   Sample volume was adjusted to 1 mL with 1-chlorobutane.   Then, 100 µL of MTBSTFA (*N-tert*-butyldimethylsilyl)-*N*-methyl-trifluoro-acetamide) derivatizing reagent was added and the samples were vortexed or sonicated for 10-15 seconds.   Samples were allowed to react in an oven at 60°C for 1 hour.   After   cooling,   the   samples   were   analyzed   by   capillary   gas chromatography/mass spectrometry using a J&W DB 1701 capillary column and an HP 5971 mass selective detector.

Results and Discussion

The TCP immunoassay described uses a competitive assay format.   Since the enzyme-labeled TCP competes with the unlabeled (sample) TCP for the antibody sites, the color developed is inversely proportional to the concentration of TCP in the sample. To describe color inhibition, it is common to report displacement in terms of a *B/Bo* measurement, defined as the absorbance observed for a sample or standard (B) divided by the absorbance at a zero analyte concentration (Bo).

Figure 1 illustrates the mean standard curve for the TCP calibrators, collected over 50 assay runs, linearly transformed using a log/logit curve fit with error bars representing 2 standard deviations. The range of the immunoassay calibration curve is 0.5 to 6 ppb TCP with an estimated least detectable dose (LDD) of 0.25 ppb. The assay LDD, defined as the lowest concentration that can be distinguished from zero, was based on an average 90% B/Bo estimation from 50 assay runs (*16*). The 0.25 ppb LDD estimate is greater than 4 standard deviations from the "true" zero, determined by calculating the standard deviation and mean absorbance value for three sets of twenty replicates of the zero standard. Quantitation with the immunoassay should be limited to within the range of the standard curve. To analyze samples with higher TCP concentrations, water samples were diluted in the zero standard for analysis and samples concentrations calculated by multiplying results by the appropriate dilution factor.

Table I summarizes a precision study that was conducted with four concentrations of TCP in four environmental water samples: two wells, a spring and a creek. The water samples are described in Table II. TCP was added at 1, 2, 4 and 5 ppb. Five replicates of each level were assayed within a single run for each of 5 days. The within and between day and total variation was determined by the method of Bookbinder and Panosian (*17*) using Statistical Analysis Software (*18*). Coefficients of variation (CVs) were less than 10% for all levels tested.

The accuracy of the TCP assay was assessed by evaluating the same four environmental water samples each spiked with known amounts of TCP at four levels.

Table I. Precision of TCP Measurement by Immunoassay

	Concentration Level (ppb)			
Sample	1.0	2.0	4.0	5.0
N	25	25	25	25
Mean ppb	1.00	2.24	4.34	5.18
% CV within assay	9.1	6.6	6.6	5.7
% CV between assay	1.1	3.4	1.1	1.1
%CV total	9.2	7.3	6.6	5.6

Table II. Characterization of Water Samples Used in Precision and Accuracy Experiments

Sample	iron, mg/L	nitrate, mg/L	nitrite, mg/L	pH	Hardness $CaCO_3$, g/L	Resistivity μOhms
1	<0.1	26.4	0.50	8.0	221	363
2	<0.1	3	<0.03	7.5	68.5	137
3	<0.1	1	<0.03	7.5	68.5	127
4	<0.1	26.4	0.13	7.0	51.0	50

Added amounts of TCP were recovered quantitatively in all cases. The average recoveries of three (3) assays with the 4 water samples is summarized in Table III.

Table III. Accuracy of TCP Measurement by Immunoassay (n = 12)

TCP added (ppb)	Mean (ppb)	SD (ppb)	% Recovery
1.0	1.02	0.06	102
2.0	2.09	0.14	105
4.0	4.23	0.13	106
5.0	5.17	0.21	103
Average			104

Substances that could potentially be found in water were evaluated for interferences in the TCP assay. The assay was said to tolerate a given concentration of a specific compound if a blank (no TCP) sample gave an assay result below the LDD (i.e. less than 0.25 ppb), and recovery of the 2 ppb TCP spike was ± 20% of the fortification value. With the exception of sulfate, the maximum concentration tested for each compound was 500 ppm. The concentrations of interferences chosen for these studies would most likely exceed levels found in groundwater samples (*19, 20*). Table IV summarizes the results for the maximum tolerated concentrations of each compound tested. When an ocean water sample (0.5 M sodium chloride) was spiked with TCP at 0.5 ppb increments along the range of the curve (0.5 to 5.5 ppb) recovery averaged 101% (SD = 7); similarly, a water sample containing 6.5 ppm humic acid was spiked at 0.5 ppb increments along the range of the curve with an average recovery of 93% (SD = 4). Data is given in Table V. Sample pH was also shown to have no adverse effect on blank results or spike recoveries from pH 3 to 11.

Results from the previous experiments suggest that the assay is robust and free from interferences from components commonly found in groundwater. To determine recovery and possible interferences, 142 water samples from around the United States were evaluated as blanks and spiked with 2.5 ppb TCP. A mean recovery of 102% (SD = 7) with a range of 84 to 118% was observed.

Two studies were conducted to compare results of the ELISA method with a traditional analysis procedure (capillary gas chromatography/mass spectrometry; GC/MSD). In the first study, 14 water samples were fortified with TCP at 1, 2.5 and 5 ppb. The GC/MSD method averaged 93% recovery; the immunoassay averaged 99% recovery for the 14 samples. A correlation coefficient (r) of 0.994 was obtained and the equation of the best fit line was y = 0.88x + 0.34. In a second experiment, 20 field samples collected from a rice paddy during a triclopyr dissipation study were analyzed by GC/MSD and TCP immunoassay. Most of the samples contained both TCP and triclopyr, with triclopyr concentrations ranging from <10 to 310 ppb. A correlation coefficient (r) of 0.943 was obtained for these samples with a best fit line equation of y = 1.1x + 0.03. A comparison of all 34 samples by the ELISA method

Table IV. Results of Maximum Tolerated Concentrations of Probable Water Components

Compound	Maximum Concentration (ppm)	Neat Result (ppb)	2 ppb Spike Recovery (%)[a]
Calcium	500	nd[b]	110
Copper	100	nd	111
Iron	50	nd	109
Magnesium	500	nd	92
Manganese	500	nd	112
Mercury	500	nd	99
Nickel	100	nd	89
Nitrate	500	nd	106
Peroxide	500	nd	109
Phosphate	500	nd	105
Silicates	500	nd	110
Sulfate	10,000	nd	109
Sulfite	250	nd	103
Thiosulfate	500	nd	99
Zinc	100	nd	99

[a] Value is average of two results: the diluent/zero fortified with the compound of interest and spiked with TCP, and distilled water fortified with the compound of interest and spiked with TCP.
[b] nd = not detected (<0.25 ppb)

Table V. Recovery of TCP from Ocean Water and a High Humic Acid Water Sample

TCP added (ppb)	Ocean		Humic Acid	
	TCP, ppb	% Recovery	TCP, ppb	% Recovery
0.5	0.56	112	0.44	88
1.0	1.02	102	1.01	101
1.5	1.57	105	1.45	97
2.0	2.03	102	1.71	86
2.5	2.73	109	2.38	95
3.0	2.91	97	2.71	90
3.5	3.56	102	3.18	91
4.0	4.12	103	3.73	93
4.5	4.68	104	4.03	90
5.0	4.58	92	4.69	94
5.5	4.80	87	5.16	94
Average		101		93

(y) and the GC/MSD method (x), illustrated in Figure 2, shows excellent agreement between the two methods; $r = 0.959$, $y = 0.97x + 0.17$.

Table VI summarizes the cross-reactivity data of TCP, chlorpyrifos, triclopyr and a variety of other pesticides in the TCP assay. The LDD for each compound was approximated at 90% B/Bo; the IC_{50} was determined at 50% B/Bo. The percent cross reactivity was determined by estimating the amount of compound required to displace 50% of the enzyme conjugate compared to the amount of TCP required for 50% displacement. Accurate cross-reactivity estimates for the parent compounds, chlorpyrifos and triclopyr, proved to be a difficult issue. An ultra-pure chlorpyrifos standard (99.9%) was obtained that showed virtually no reactivity in the assay. In contrast, chlorpyrifos-methyl (99.8% pure) showed significantly more reactivity. Two different triclopyr standards, one analyzed to be 99.6% pure, the other 99.8% pure, indicated dramatically different cross-reactivity estimates in the assay, as shown in Table VII. These results, along with the correlation study in which TCP was accurately determined in samples containing up to 310 ppb triclopyr, strongly suggest the presence of very low levels of TCP as a manufacturing degradate in the parent compound standards. Because the analytical standards were tested at concentrations up to 10,000 ppb in the assay, TCP present at as low as 0.005% in any of the standard materials would be detected, i.e. a 10,000 ppb solution prepared with an

Table VI. Cross-reactivity of Pesticides and Related Agricultural Compounds in the TCP Immunoassay

Compound	LDD[a], ppb (90% B/Bo)	IC_{50}[b] ppb	% Cross-Reactivity
3,5,6-Trichloro-2-pyridinol	0.25	2.31	100
Fluroxypyr-2-pyridinol	62.6	531	0.44
Chlorpyrifos-methyl	97.7	1094	0.21
6-Chloro-2-pyridinol	662	7343	0.03
Profenfos	1000	>10,000	<0.01
5-Chloro-2-pyridinol	1188	>10,000	<0.011
Triclopyr	1336	>10,000	<0.01
Metolachlor	2126	>10,000	<0.01
Chlorpyrifos	10,000	>10,000	<0.01

[a] LDD is the least detectable dose calculated at 90% B/Bo.
[b] % Cross reactivity is determined by estimating the amount of compound required to displace 50% of the enzyme conjugate compared to the amount of TCP required for 50% displacement.

No reactivity was seen with the following compounds up to 10,000 ppb: alachlor, aldicarb, ametryn, atrazine, azinphos-methyl, benomyl, carbaryl, carbendazim, carbofuran, clopyralid, 2,4-D, diazinon, dinoseb, fenitrothion, fluroxypyr, glyphosate, lindane, malathion, MCPA, methamidophos, methomyl, 2-methoxy-3,5,6-trichloropyridine, oxamyl, parathion, parathion-methyl, phosmet, picloram, pirimcarb, pirimphos-ethyl, pirimphos-methyl, propachlor, terbufos and thiophanate-methyl.

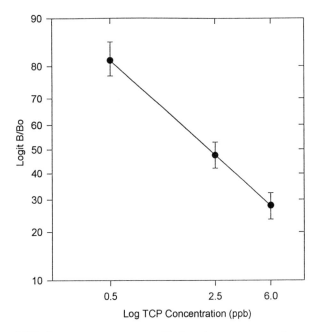

Figure 1.   TCP dose response curve.   Each point represents the mean of 50 determinations.   Vertical bars indicate $\pm$ 2 SD about the mean.

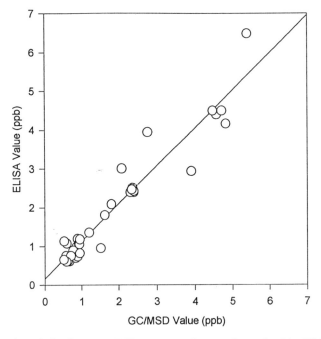

Figure 2. Correlation between TCP concentrations as determined by ELISA and GC/MSD methods. n = 34, r = 0.959, y = 0.97x + 0.17.

Table VII. Apparent Cross-Reactivity of Triclopyr in the TCP Immunoassay

Purity	LDD (ppb)	IC_{50}	% Reactivity
99.6%	65.8	580	0.4
99.8%	1336	>10,000	<0.01

analytical standard of triclopyr, chlorpyrifos or chlorpyrifos-methyl containing 0.005% TCP would contain 0.5 ppb TCP. Therefore, the cross-reactivity values shown for triclopyr, chlorpyrifos and chlorpyrifos-methyl can be considered to be only conservative estimates of their reactivity in the immunoassay until actual concentrations of TCP which may be in the analytical standards can be determined.

Conclusion

The performance characteristics of this ELISA have shown it to be a powerful tool to selectively analyze for TCP in water samples even in the presence of high concentrations of the parent compounds. The assay described was free from interferences with numerous compounds commonly found in water and over a wide pH range. It has been shown to be accurate and precise and correlates well with a traditional analytical method. Future applications of this assay could extend to the analysis of soil and food samples for TCP, or in even human exposure monitoring programs for triclopyr or chlorpyrifos contamination.

References

1.   Espinosa-Mansilla, A.; Salinas; Zamoro, F. *J Agric Food Chem.* **1995**, *43*, 146-150.
2.   Woodburn, K.B.; Green, W.R.; Westerdahl, H.E. *J Agric Food Chem.* **1993**, *41*, 2172-2177.
3.   Barron, M.G.; Woodburn, K.B. *Rev Environ Contamin Toxicol.* **1995**, *144*, 1-93.
4.   Marshall, W.K.; Roberts, J.R.   *Ecotoxicology of Chlorpyrifos;* National Research Council, NRCC No. 16079, Ottowa, Ontario, Canada, 1978, 314.
5.   Allender, W.J.; Keegan, J. *Bull Environ Contamin Toxicol.* **1991**, *46*, 313-319.
6.   Herzog, D.P.; Mihaliak, C.A.; Jourdan, S.W.; Lawruk, T.S. *J Clin Ligand Assay,* **1995**, *18*, 150-155.
7.   Itak, J.A.; Selisker, M.Y.; Jourdan, S.W.; Fleeker, J.R.; Herzog, D.P. *J Agric Food Chem.* **1993**, *41*, 2329-2332.
8.   Lawruk, T.S.; Lachman, C.E.; Jourdan, S.W.; Fleeker, J.R.; Herzog, D.P.; Rubio, F.M. *J Agric Food Chem.* **1993**, *41*, 747-752.
9.   Rubio, F.M.; Itak, J.A.; Scutellaro, A.M.; Selisker, M.Y.; Herzog, D.P. *Food and Agric Immunol.* **1991**, *3*, 113-125.
10. Howell, E.H.; Nasser, J.; Schray, K.J. *J Immunoassay.* **1981**, *2*, 205-225.
11. Engvall, B. In: *Methods in Enzymology;* Van Vunakis, H. and Langone, J.J., Ed.; Academic Press USA, Inc: New York, 1980, Vol. 70; pp. 419-439.

12. Lehtonen, O.P.; Viljanen, M.K. *J Immunol Meth.* **1980**, *34*, 61-70.
13. Harrison, R.O.; Braun, A.L.; Gee, S.J.; O'Brien, D.J.; Hammock, B.D. *Food Agr Immunol.* **1989**, *1*, 37-51.
14. Manclús, J.J.; Primo, J.; Montoya, A. *J Agric Food Chem*, **1994**, *41*, 1257-1260.
15. Manclús, J.J.; Montoya, A. *Anal Chim Acta*, **1995**, *311*, 341-348.
16. Midgely, A.R.; Niswender, G.D.; Rebar, R.W. *Acta Endocrinol*, **1969**, *63*, 163-179.
17. Bookbinder, M.J.; Panosian, K.J. *Clin Chem*, **1986**, *32*, 1734-1737.
18. SAS Institute, Inc. *SAS User's Guide: Statistics;* SAS Institute, Inc., NC, 1988; Version 6.03 Edition.
19. American Public Health Association. *Standard Methods for Examination of Water and Wastewater;* American Public Health Association: Washington, DC, 1989.
20. Wolfe, N.L.; Mingelgrin, U.; Miller, G.C. In *Pesticides in the Soil Environment: Processes, Impact and Modeling;* Cheng, H.H., Ed; Soil Society of America, Inc.: Madison, WI, 1990, pp 106.

Chapter 22

Rapid Determination of Dioxins in Drinking Water by Enzyme Immunoassay

H. Wang[1], L. Wang[1], J. E. George III[1], G. K. Ward[1], J. J. Thoma[1], R. O. Harrison[2], and B. S. Young[2]

[1]Environmental Health Laboratories, 110 South Hill Street, South Bend, IN 46617–2702
[2]Millipore Corporation, 80 Ashby Road, P.O. Box 9125, Bedford, MA 01730–9125

An enzyme immunoassay (EIA) was developed to rapidly analyze trace levels of 2,3,7,8-tetrachlorodibenzo-p-dioxin (2,3,7,8-TCDD) in water samples. Water samples were extracted by the solid phase extraction method (SPE) using 47 mm, C_{18} Empore extraction disks (3M). Dioxins were eluted from the disks with dichloromethane. The extracts were dissolved in methanol through a solvent exchange step. EnviroGard reagents and a microwell strip reader (Millipore) were used to perform the dioxin enzyme immunoassay. The working range of the dioxin enzyme immunoassay was found to be 15 pg/L to 100 pg/L in water. The precision and accuracy of EIA were determined by performing five replicates of reagent water spiked at a concentration of 25 pg/L. The recovery of the dioxin assay ranged from 74% to 122%, and %CV for five replicates was less than 15%. The accuracy of EIA results was also confirmed by ion trap GC/MS/MS (Varian). In general, EIA provides a relatively easy and cost effective means for measuring trace levels of dioxins in drinking water samples.

Dioxins, particularly 2,3,7,8-TCDD (Figure 1), have achieved great notoriety and evoked the greatest fears as probable human carcinogens. EPA has regulated 2,3,7,8-TCDD in drinking water with a maximum contamination level (MCL) of 30 pg/L.

Dioxins in water and other sample matrices are usually determined by the EPA Method 1613 developed by the EPA Office of Science and Technology (1). This method, however, requires expensive instrumentation (HRGC/HRMS) and a highly trained analyst. A GC/EI/MS/MS technique has been developed for the ultra-trace detection of dioxins and furans (2-4). The sensitivity of this method is 1 pg of native analyte. High specificity for polychlorinated dibenzo-p-dioxins and furans (PCDD/F's) has been demonstrated by using an ion trap detector with MS/MS capability.

When considering costs to monitor the dioxin contaminants in drinking water and assess the health risk, the analytical technology should be reasonably affordable by regional and large metropolitan public water systems and also available for widespread utility. Enzyme Immunoassays (EIAs) have been successfully used in the field and laboratory for the rapid determination of pesticides in environmental water and soil samples (5-9). Because enzyme immunoassay is based on an antibody coated plate and colorimeter to measure 2 2 enzymatic color development, it is a simple and quick method which requires very little training to run. In addition, the sample preparation is simpler and shorter due to the specificity and sensitivity of the dioxin antibody. Therefore, the dioxin enzyme immunoassay will significantly reduce the cost compared to classical EPA methods. A competitive inhibition enzyme immunoassay based on a mouse monoclonal antibody which is specific for 2,3,7,8-TCDD and related dioxin and furan congeners was developed (10). Specificity of this assay roughly parallels the toxicity of the dioxin and furan congeners which have been tested. The feasibility of the enzyme immunoassay as an analytical system for dioxins has been shown by repetition of standard curves having detection limits below 100 pg/tube.

The goal of our research was to develop a method for rapid determination of low level 2,3,7,8-TCDD and 2,3,7,8-substituted dioxins in water using solid phase extraction (SPE) followed by a dioxin enzyme immunoassay. The specific objectives were the following: 1) determine the detection range and stability of the dioxin enzyme immunoassay, 2) demonstrate the extraction efficiency of SPE disks for extracting dioxins from water, 3) demonstrate the precision, accuracy and matrix interferences of the dioxin enzyme immunoassay, and 4) develop the quality control procedures. These studies would provide a very interesting bench mark for comparison to conventional methods and allow the EPA to further demonstrate the potential capability of enzyme immunoassay for the monitoring of ground water and surface water.

Materials and Methods

The stock solution of 2,3,7,8-TCDD was obtained from AccuStandard, Inc. An independent quality control sample was obtained from ULTRA Scientific. The EnviroGard Dioxin plate kit was supplied by Millipore Corporation. A Varian Saturn III GC/MS/MS equipped with a Varian large volume injector (LVI) was used to confirm the results of EIA.

Safety. 2,3,7,8-TCDD has been found to be acnegenic, carcinogenic and teratogenic in laboratory animal studies. Disposable plastic gloves, lab coat, safety glasses should be used. Workers must be trained in the proper method of removing contaminated gloves and clothing without contacting the exterior surfaces. Primary solutions should be prepared in a hood.

Extraction. Two liters of water sample were extracted using a 47 mm, C_{18} Empore extraction disk (3M). Dioxins were eluted with dichloromethane and the eluant was dried with sodium sulfate. The extract was evaporated to dryness under a nitrogen stream. The residue was redissolved by the addition of 100 μL of methanol with 5-15 seconds of vigorous mixing. The dioxin standards were also treated through the solvent exchange step.

Procedure of EIA. EnviroGard reagents and microwell strip reader (Millipore Corporation) were used to perform the dioxin enzyme immunoassay. Mouse antibodies which are specific for 2,3,7,8-TCDD and related congeners were immobilized on the walls of plastic microwells. The strip format was first planned to allow for the placement of the negative control, four calibration standards, samples, and quality control samples. One hundred fifty microliters of reagent water was added to a microwell, and then 50 μL of standards or prepared sample extracts in methanol was mixed with reagent water in the microwell. The microplate was covered and incubated for 30 minutes at room temperature. Following a washing step, the well was then incubated with a 200 μL aliquot of competitor-HRP conjugate. After 30 minute incubation, the unbound conjugate was washed away with reagent water and 100 μL of substrate was added to result in the formation of blue color. After 30 minutes, 100 μL stop solution was added to each well to stop the color development. The optical density in each well was measured by the microwell strip reader at a wavelength of 450 nm. The concentration of 2,3,7,8-TCDD was determined by relating the absorbance response to that of standards.

GC/MS/MS Confirmation. The extracted analytes were separated by injecting a aliquot of the concentrated extract to a high resolution fused silica capillary column and identified and quantitated by Ion Trap MS/MS. Reference product ion spectrum and retention time for dioxins are obtained by the measurement of calibration standards under the same GC/MS/MS conditions used for sample analysis.

Results and Discussion

Working range and reproducibility. We first evaluated the linear working range of the standard curve and then demonstrated the reproducibility of standard curves by generating five standard curves and calculating the coefficient of variation of these five observations. The standards were run in duplicate, and the absorbance values were converted to %Bo by dividing the absorbance of a non-zero standard by the absorbance value of the negative control. The working range of the assay was from 15 pg/L to 100 pg/L (Figure 2). The stability of the dioxin standard curve was evaluated by comparing the standard curves generated over six months. The reproducibility of the dioxin standard curves is presented in Table I. The percent coefficient of variation (%CV) was less than 15% and indicates good reproducibility.

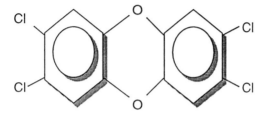

Figure 1. Structure of 2,3,7,8-TCDD.

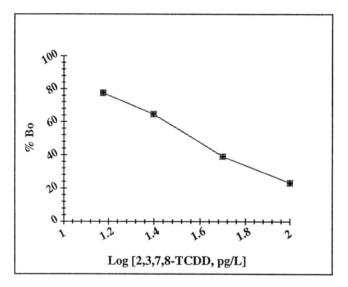

Figure 2. Standard curve of 2,3,7,8-TCDD. Each point represents the mean of five replicate determinations.

Extraction efficiency. The solid phase extraction method was established to achieve concentration factors required to meet the detection limit of 15 pg/L in water. The dioxin extraction efficiency was tested by extracting seven replicates of water samples at the spiked concentration of 30 pg/L which is the MCL for 2,3,7,8-TCDD set by USEPA. The concentrations of 2,3,7,8-TCDD in these extracts were determined by GC/MS/MS. The recovery and %CV of the extraction are presented in Table II. This study demonstrates that recoveries between 70% and 85% with a %CV of less than 7.5% can be achieved using our extraction procedure.

Table I. The Stability of Dioxin Calibration Curve

		$\%B_O$			
Obs. #	Month	(15pg/L)	(25 pg/L)	(50pg/L)	(100 pg/L)
1	1	77.4	55.9	34.2	20.1
2	2	77.2	65.3	39.8	21.2
3	3	74.2	65.1	41.8	27.0
4	4	75.4	59.1	31.9	18.8
5	6	83.4	68.7	42.5	25.8
Mean		77.5	64.6	39.0	23.2
SD.		3.56	3.47	4.22	3.33
%CV		4.59	5.37	10.8	14.3

**Table II. Recovery of 2,3,7,8 - TCDD Extraction as Determined by GC/MS/MS
(Spiked Concentration: 30 pg/L in Reagent Water)**

Obs. #	Cal. Conc. pg/L	% Recovery
1	25.7	85.7
2	21.4	71.3
3	24.6	81.9
4	20.9	69.6
5	22.9	76.2
6	22.2	74.0
7	21.0	69.9
Mean(n=7)	22.3	75.5
SD. (n=7)	1.73	5.75
% CV (n=7)	7.74	7.74

Precision and Accuracy. Five replicates of reagent water spiked with certified standards within the working range were tested to evaluate the precision and accuracy of this assay. Table III shows the validation results of the EIA by performing five replicates of reagent water spiked at a concentration of 25 pg/L. The recovery of the dioxin assay ranged from 74% to 122%, and %CV for five replicates was less than 15%. The results presented in Table III illustrate that the precision and accuracy of this assay were very good. However, the recoveries of some spiked samples were higher than 100%. These false positive results might be caused by residual methylene chloride in the extracts because methylene chloride may kill the antibody and reduce absorbance vaule.

Table III. Precision and Accuracy of Dioxin Enzyme Immunoassay (Spiked Concentration : 25 pg/L in Reagent Water)

	Cal. Conc. pg/L	% Recovery
Rep-1	18.6	74.3
Rep-2	23.8	95.1
Rep-3	30.4	121.7
Rep-4	23.7	95.0
Rep-5	26.4	105.5
Mean (n=5)	24.6	98.3
SD. (n=5)	3.53	14.1
% CV (n=5)	14.4	14.4

Drinking water samples from different water sources and matrix spikes were also tested to demonstrate the accuracy and the matrix effects of this method. The drinking water samples were obtained from municipal water sources including ground water, surface water and bottled water. The results of dioxin enzyme immunoassay were compared with the GC/MS/MS method on all the samples in order to evaluate accuracy, matrix effects, and method performance. Before spiking with dioxin, the seven water samples were analyzed by GC/MS/MS in order to demonstrate that no dioxin residue was present. The analysis of EIA for these samples showed that the response of the samples were similar to that of the negative control, thus the water matrix did not affect this assay. The accuracy of EIA was also tested by analyzing these samples spiked at the final concentration of 30 pg/L of 2,3,7,8-TCDD. The spiked samples were tested by EIA and GC/MS/MS. The results were summarized in Table IV. Compared with the GC/MS/MS results, EIA was consistent with the results obtained by GC/MS/MS indicating that the enzyme immunoassay has good correlation with GC/MS/MS for dioxin analysis. The slightly higher recovery of EIA than GC/MS/MS might be caused by some solvent residues and sample matrix interferences.

The specificity of the dioxin antibody has been shown to be primarily directed toward selected tetra- and pentachlorodibenzodioxins, with reduced recognition of the corresponding furans (Table V). This antibody is specific for 2,3,7,8-TCDD and 2,3,7,8-substituted dioxins and furans but detection levels are different for different congeners. Further studies are underway to evaluate the interferences in the dioxin enzyme immunoassay.

Table IV. Comparison of EIA with GC/MS/MS
(Spiked Concentration: 30 pg/L)

Sample #	Matrix	Location (States)	EIA % Recovery	GC/MS/MS % Recover	Relative % Difference
1	GW	IN	82.7	73.3	11.8
2	SW	LA	76.7	65.3	16.0
3	BW	IL	56.7	71.3	22.9
4	BW	IL	88.6	75.3	16.3
5	BW	WI	87.8	80.3	8.7
6	BW	IN	63.2	60.7	4.3
7	BW	IL	69.2	61.3	12.2

GW: Ground Water; SW: Surface Water; BW: Bottled Water.

Table V. Crossreactivity of Dioxin, Furan, and PCB Congeners
in the Enzyme Immunoassay
(Reactivity Relative to 2,3,7,8-TCDD)

Compound Name	%Crossreactivity
2,3,7,8-Tetrachlorodibenzo-p-dioxin	100
1,2,3,4,7,8-Hexachlorodibenzo-p-dioxin	124
1,2,3,7,8-Pentachlorodibenzo-p-dioxin	95
2,3,7,8-Tetrachlorodibenzofuran	27
1,2,7,8-Tetrachlorodibenzo-p-dioxin	14
2,3,7-Trichlorodibenzo-p-dioxin	14
1,2,3,4,7,8-Hexachlorodibenzofuran	1.7
1,7,8-Trichlorodibenzo-p-dioxin	<2
3,4,5-3',4' Pentachlorobiphenyl	0.7
2,7-Dichlorodibenzo-p-dioxin	0.4
3,4,-3',4' -Tetrachlorobiphenyl	0.2
1,2,3,4,6,7,8,9-Octachlorodibenzo-p-dioxin	<0.1
1,2,3,4,6,7,8,9-Octachlorodibenzofuran	<0.1

Quality Control. The quality control procedure was designed based on the validation data. For each batch, the negative control and calibration standards need to be tested, and then the samples are run. If the concentrations of samples are out of working range, the samples require dilution in order to fall within the working range, and the analysis of these samples must be repeated. A laboratory method blank (LMB) is analyzed in each batch to determine that all glassware and reagent interference are under control. LMB values must be less than the detection limit of the dioxin assay (15 pg/L). A laboratory fortified blank (LFB) is also analyzed in each batch at a concentration of 30 pg/L. The recovery of the LFB analysis must be within 70-130%. Matrix spike samples (spiked at 30 pg/L) are analyzed for every 20 samples processed in the same batch to determine if the matrix affects method performance. The percent recoveries of matrix spike samples must be within 60-140%. In addition, an external QC sample prepared at a concentration of 30 pg/L is analyzed at the end of every batch. The recovery of QC sample must be within 70-130%.

Conclusions

Enzyme Immunoassay is a very powerful screening tool for determining trace levels of dioxins in drinking water. This method is capable of analyzing for trace levels of 2,3,7,8-TCDD and other dioxins using very little equipment. A detection limit of 15 pg/L in water samples was achieved. This method offers good precision and accuracy and significant improvements in speed and cost compared to the expensive conventional HRGC/MS method. Since EIA is cost effective, simple, quick and sensitive, it can be used effectively to analyze dioxins in the laboratory and used for large-scale screening of water samples in the field.

References

(1) USEPA Office of Water Regulations and Standard, Method 1613: Tetra-through Octa Chlorinated Dioxins and Furans by Isotope Dilution HRGC/HRMS, Reversion B, October, 1994.

(2) Hamelin, G.; Brochu, C. and Moore, S. *Organohalogen Compounds*, **1995,** *Vol. 23*, pp 125-129.

(3) Hayward, D. G. *Organohalogen Compounds*, **1995,** *Vol. 23*, pp 119-124.

(4) Plomley, J. B.; Mercer, R. S. and March, R. E. *Organohalogen Compounds*, **1995,** *Vol. 23*, pp 7-12.

(5) Aga, D. S. and Thurman, E. M. *Anal. Chem.*, **1993,** *Vol. 65*, pp 2894-2898.

(6) Brady, J. F.; LeMasters. G. S.; Williams, R. K.; Pittman, J. H.; Daubert, J. P.; Cheung, M. W.; Skinner, D. H.; Turner, J.; Rowland, M. A.; Lange, J. and Sobek, S. M. *J. Agric. Food Chem.*, **1995,** pp 268-274.

(7) Bushway, R. J.; Perkins, B.; Savage, S. A.; Lekousi, S. J. and Ferguson, B. S. *Bull. Environ. Contam. Toxicol.*, **1988,** *Vol. 40*, pp 647-654.

(8) Wang, H.; George, J. E.; Thoma, J. J.; Wang, L. and Young, B. S. WEF Specialty Conference Series Proceedings, Environmental Laboratories: Testing the Waters, Cincinnati, OH, 1995, pp 6-21 - 6-29.

(9) Wang, H.; George, J. E.; Thoma, J. J. and Young, B. S. Proceedings of Water Quality Technology Conference, New Orleans, LA, 1995, pp 205-212.

(10) Harrison, R. O.; Carlson, R. E; and Shirkhan, H. 15th International Symposium on Chlorinated Dioxin and Related Compounds, Edmonton Alberta, Aug 21-25, 1995.

Chapter 23

Studies on the Applications of an Immunoassay for Mercury

Larry C. Waters[1], Richard W. Counts[2], Rob R. Smith[1], and Roger A. Jenkins[1]

Divisions of [1]Chemical and Analytical Sciences and [2]Computer Sciences and Mathematics, Oak Ridge National Laboratory, P.O. Box 2008, Oak Ridge, TN 37831-6120

Immunoassays (IAs) are rapidly becoming a significant component of the arsenal of field analytical methods. Validation of such alternative methods is necessary for their acceptance by the regulatory agencies and potential users. As part of a program to evaluate the performance of such methods, the capacity of the BiMelyze (BioNebraska, Inc.) immunoassay-based method to measure mercury in soil and sludge was examined. Results showed that the immunoassay method performed as well as either X-ray fluorescence or neutron activation methods for the analysis of mercury in contaminated soil samples. The method was also shown to be capable of detecting as little as 2.5 ppm mercury in sludge from a waste water treatment plant.

Sample analysis is a major component of environmental restoration and waste management activities. Standard laboratory methods are both expensive and time-consuming. A major portion of the field samples taken to the laboratory for analysis either are negative for the analyte being tested, or are contaminated below the regulated level. Effective field analytical methods could eliminate much of this effort and expense. An objective of this program is to define commercial screening methods that are capable of measuring environmental contaminants of concern to the Department of Energy (DOE) (*1,2*). The approach involves selecting potentially useful technologies, experimentally evaluating the methods that utilize the technology, and transferring the validated methodology to appropriate users. Of specific concern to the DOE at its facilities at Oak Ridge, TN, is the mercury that is contaminating the flood plain of the East Fork Poplar Creek. Arising from releases in the 1950s by an upstream nuclear weapons plant, it has been estimated that as much as 80,000 kg of mercury still remain in the flood plain soils (*3*). An account of our experiences with an immunoassay-based method for the analysis of mercury in those soils, and in sludge generated at the city of Oak Ridge waste water treatment facility, will be presented.

Description of the Mercury Immunoassay Method

The kits used in our evaluation studies are the BiMelyze 96-Well Plate Mercury Assay Kit and the BiMelyze 16-Tube Mercury Assay Kit, with associated extraction kits. They were obtained from BioNebraska, Inc., Lincoln, NE. The basis for these kits is a monoclonal antibody to Hg^{2+}, produced and described by Wylie, et al (*4,5*). Materials required but not supplied in the 96-well plate kit include the pipets required for making sample dilutions and adding samples and reagents to the plates, and a plate reader. Because the tube kit utilizes dropper-topped bottles to deliver samples and reagents, there is no requirement for pipettors. Tubes are read with a differential photometer. Both kits require a user supplied balance to weigh the samples, and the nitric and hydrochloric acids used to extract the soil. Easy to follow instructions are provided with the kits. The plate format is well suited for the analysis of multiple samples in the laboratory while the tube format is more appropriate for onsite field use.

Preparation of Samples for Analysis. A 1-g sample of soil is shaken intermittently with 3 mL of a 2:1 mixture of concentrated HCl and concentrated HNO_3 for 10 min. This procedure converts poorly soluble salts of mercury, e.g., HgS, elemental mercury, and to some extent methyl mercury, to soluble forms of Hg^{2+}. The extract is partially neutralized by the addition of 7 mL of buffer, effecting a 1 to 10 dilution of the sample. (Caution: protective apparel should be worn and care taken during these extraction steps.) A further 1 to 1000 dilution is made (total dilution equals 1 to 10,000) before the extract is analyzed. Consequently, a 5-ppm sample is equivalent to 0.5 ppb in the assay. Water, in the absence of interferences, can be analyzed directly with a sensitivity of $\leq$ 0.25 ppb (see section on sludge analysis). Although it was not used in these studies, the extraction procedure has been modified to use 5 grams of soil with a concomitant increase in test sensitivity of 5 fold, i.e., to about 1 ppm (Carlson, L., BioNebraska, Inc., personal communication, 1993).

Immunoassay. The assays were performed as follows: 1) Aliquots of the diluted soil sample extracts, or of aqueous samples, are added to the tubes, or plate wells, which have been coated with -SH rich proteins. Hg^{2+} binds to these proteins in proportion to the concentration of mercury in the sample. Unbound mercury is removed by rinsing the tubes or plate wells. 2) A mercury-specific antibody is added that binds to the mercury on the tube or well surfaces. Unbound antibody is removed by rinsing the tubes or wells. 3) A conjugate, consisting of a peroxidase enzyme coupled to a second antibody that is specific for the mercury-specific antibody, is added. This conjugate forms a complex with the mercury-specific antibody. The more mercury bound in step 1 the more mercury-specific antibody is bound in step 2, and the more conjugate is bound in step 3. Unbound conjugate is removed by rinsing the tubes or wells. 4) A colorless substrate for the peroxidase component of the conjugate is added. Oxidation of this substrate causes it to become colored; the *more* mercury in the sample the *more* color is produced. The intensity of the color is actually proportional to the log of the mercury concentration. The mercury immunoassay is different from most of the other IAs for environmental contaminants in that the latter are formatted to allow the

Table *I.* Comparison of Immunoassay (96-well plate) to Neutron Activation and X-ray Fluorescence for Analysis of Mercury in Contaminated Soil Samples

Sample	NAA (ppm)	XRF (ppm)	IA Summary Average	IA Summary %RSD	IA (2 replicates per assay, ppm) Assay 1	Assay 2	Assay 3
1	<7.6	-	2.5	-	<5 , <5	<5 , <5	<5 , <5
2	8.6	13 , 15	16.4	27	18 , -	15 , 23	11 , 15
3	19	-	46.5	31	33 , 38	57 , 66	31 , 54
4	<5	7 , 30	9	13	9.7 , 11	8.3 , 8	9.2 , 7.9
5	<2.1	-	2.5	-	<5 , <5	<5 , <5	<5 , <5
6	8.3	23 , 33	33.8	10	- , -	38 , 32	30 , 35
7	56	63 , 81	77.8	37	75 , 112	46 , 113	67 , 54
8	<5.3	22 , 52	39	39	65 , 39	25 , 43	23 , 39
9	12	19 , 24	36.5	50	32 , 24	39 , 71	20 , 33
10	57	31 , 33	63.2	34	61 , 95	61 , 59	29 , 74
11	46	56 , 69	74.7	30	102 , 76	33 , 76	77 , 84
12	73	56 , 59	73.2	16	86 , 79	53 , 69	74 , 78
13	12	12 , 20	20	52	30 , -	24 , 28	9.1 , 8.9
14	<1.5	-	2.5	-	<5 , <5	<5 , <5	<5 , <5
15	18	24 , 24	29	21	20 , 23	34 , 33	29 , 35
16	<1.2	-	3.5	-	5.3 , 5.8	<5 , <5	<5 , <5
17	<1.3	-	2.5	-	<5 , <5	<5 , <5	<5 , <5
18	<3.3	-	2.5	-	<5 , <5	<5 , <5	<5 , <5
19	17	15 , 31	38.5	40	66 , 37	45 , 30	31 , 22
20	<7.7	-	2.5	-	<5 , <5	<5 , <5	<5 , <5
21	11	-	42.2	40	45 , 46	72 , 34	34 , 22
22	121	118 , 122	125.3	23	120 , 170	132 , 130	118 , 82
23-S	166	202 , 233	174	42	132 , -	129 , 174	300 , 135
24-S	589	601 , 684	740	8	795 , 705	690 , 780	660 , 810
25-S	116	90 , 116	102.5	29	105 , 160	92 , 78	88 , 92
26-S	87	150 , 159	139.7	21	156 , 150	178 , 142	100 , 112
27-S	206	235 , 259	262.7	31	344 , 264	240 , 372	168 , 188
28-S	94	80 , 106	109.8	48	92 , 124	49 , 108	204 , 82
29-S	121	131 , 131	107.7	46	92 , 174	74 , 164	92 , 50

analyte(s) to compete with the conjugate for binding, in which case *more* color indicates *less* analyte.

In their current formats, the incubation times for the four steps of the immunoassay are 10 minutes for the 96-well plate assay and 5 minutes for the 16-tube assay. Experienced users can run the 96-well plate assay, with dilutions and replicates of standards and test samples totaling 96, in about 4 hours. The 16-tube assay, with dilutions and replicates of standards and test samples totaling 16, can be completed in about 1.5 hours.

In our evaluations, 1-g samples of soil were used. With this amount of soil, a sensitivity of about 5 ppm is expected with a working range of up to about 80 to 100 ppm. Extracts of samples containing higher levels can be further diluted to within the working range and reanalyzed. As indicated in the previous section the use of 5 grams of soil increases the sensitivity to about 1 ppm.

Applications of the Immunoassay for Mercury

Both the plate and tube formats were used in these studies. Most of the soil samples used were retained samples taken from the East Fork Poplar Creek Flood plain in the city of Oak Ridge, TN. These samples had been previously analyzed by X-ray fluorescence (XRF) and/or neutron activation analysis (NAA). The predominant mercury species in these soil samples is HgS, with lesser amounts of elemental mercury and methyl mercury. Sludge samples were obtained from the city of Oak Ridge waste water treatment plant. Other samples were prepared by spiking blank soils with $HgCl_2$.

96-Well Plate Assay of Mercury in Soils. The 96-well plate format was used in a quantitative mode by using a plate reader capable of converting the results obtained from simultaneously run standards to a standard curve. From this curve, a plot of absorbance at 410 nm versus log of the concentrations of the standards, the mercury concentrations of the test samples were determined. A group of 29 field soil samples were analyzed and the results were compared with those previously obtained by XRF and/or NAA (Table *I*). The samples were independently assayed 3 times with 2 replicates/assay. Values obtained with the immunoassay (IA) that were below 5 ppm, i.e., below the linear range of the absorbance versus log concentration plot, are reported as < 5 ppm. Values above the linear range, i.e., > 80 ppm, were determined by using appropriate dilutions of those samples (indicated by the suffix S).

Statistical comparisons of the methods were performed with the data given in Table *I*. For this analysis, IA and NAA "less than" values were assigned one-half the given value and the XRF and IA values used were the averages of the multiple determinations. Because of their high values, the data points for sample 24-S were not plotted in figure 1 but were used in calculating best-fit lines. Comparison of the data obtained by IA with those obtained with XRF gave an R^2 value of 0.980 and a slope of 1.10 (Figure 1). When compared with NAA, the results obtained by IA gave a best-fit straight line with an R^2 value of 0.980 and a slope of 1.21 (Figure 1). If the results obtained by XRF are compared with those obtained by NAA, a straight line with an R^2 value of 0.978 and a slope of 1.08 is obtained (Figure 1). These results are interpreted to indicate that soil mercury contents determined by the immunoassay method correlate well with those determined by NAA and XRF and that the correlation of the immunoassay with each of the chemical/physical methods is as good as the correlation of those two methods with each other.

Tube Assay for Mercury in Soils. The tube format was used in a semiquantitative mode; the absorbance values obtained for the test samples were compared to those obtained for the reference soil samples included with the kit. These reference samples contained 0, 5 or 15 ppm mercury. By comparison, the test samples could be categorized as containing 0 to 5, 5 to 15 or greater than 15 ppm mercury. It should be noted that data that are relevant to any particular remedial action level can be obtained by simply changing the concentrations of the reference standards used.

Two independent experiments were performed to evaluate this method. Experiment 1 included 10 field samples and 3 spiked samples, in addition to the 3 standard samples supplied in the kit. Experiment 2 consisted of 8 field samples, 5 spiked samples and the 3 kit samples. Four of the field samples were common to both experiments. Instead of measuring absorbances directly in the assay tubes using a differential photometer, aliquots were transferred to 96-well plates and read with an available microplate reader. Aliquots of 200 μL were read in Experiment 1 and 100 μL aliquots were read in Experiment 2. Test samples were referenced against the kit standards. Wherever possible, analytical data obtained previously by NAA, XRF and IA are given for comparison. Results are given in Table *II*.

The tube assay gave the expected results with all of the samples tested, except for one sample, #14 in Experiment 1. It appears, on the basis of previous data obtained using the plate assay (Table *I*) and that shown in Table *II*, that the mercury content of sample #6 (both experiments) was significantly underestimated by NAA. Similarly, the mercury content of sample #8 (Experiment 1) also appears to have been underestimated by NAA, relative to IA and XRF. In these experiments, the correlation between the IA and XRF data is very good. Like the plate assay, the tube assay appears to be as accurate as either NAA or XRF for measuring mercury in soil. Results obtained for the four field samples that were analyzed in both experiments showed good reproducibility of the method.

Analysis for Mercury in Sludge. Historically, the city of Oak Ridge waste water treatment facility has disposed of post-digestion sludge by spreading it on nearby University of Tennessee pasture land. In some instances, subsequent monitoring of the land has indicated an above-background level of mercury. Thus, the facility operators were interested in identifying a rapid, low-cost method to analyze the sludge *before* it was spread on UT land.

The dark, odorless sludge had a pH of about 8 and was about 30% by volume particulate in nature. For the initial experiment, the sludge was separated into soluble and particulate fractions by low speed centrifugation. The soluble fraction was spiked with mercury in the range of 0 to 200 ppb and analyzed as if it were water or a diluted acid extract. As shown in Table *III*, mercury was detected only in the sample that was spiked at 200 ppb and the amount found was equivalent to only about 2 ppb. On the other hand quantitative recovery of mercury was observed with the spiked and acid extracted particulate fractions (remember that the particulate's extracts were diluted 1

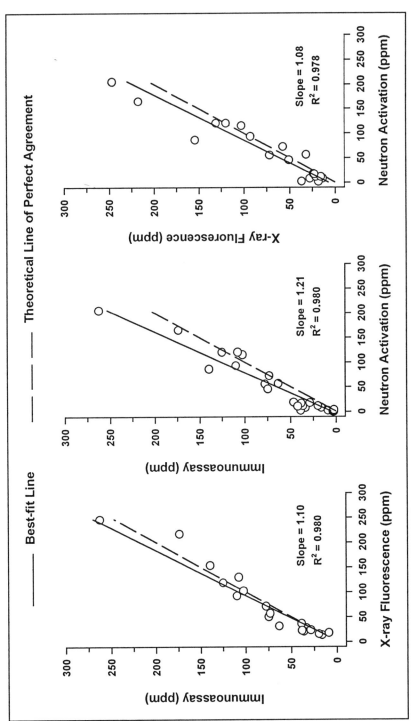

Figure 1. 96-well plate immunoassay for mercury in soil. A comparison with analyses by X-ray fluorescence and neutron activation.

Table *II*. Immunoanalysis of Mercury in Soil Using the Tube Assay

Sample No.	Spiked (ppm)	Concentration (ppm) by: NAA	XRF	IA[a]	Absorbance @ 410 nm	Concentration by Tube Assay (ppm)
Experiment 1[b]						
1	0[c]	-	-	-	0.077	0
2	5[c]	-	-	-	0.131	5
3	15[c]	-	-	-	0.216	15
4	-	116	90 , 116	102	0.357	> 15
5	-	< 3.3	-	< 5	0.104	0-5
6	-	11	-	42	0.293	> 15
7	-	< 2.1	-	< 5	0.096	0-5
8	-	< 5.3	22 , 52	39	0.292	> 15
9	-	< 1.5	-	< 5	0.127	0-5
10	-	< 7.7	-	< 5	0.088	0-5
11	-	87	150 , 159	140	0.337	> 15
12	-	19	-	46.5	0.259	> 15
13	-	121	118 , 122	125	0.264	> 15
14	7.5[d]	-	-	-	0.122	0-5[FN]
15	12.5[d]	-	-	-	0.164	5-15
16	100[d]	-	-	-	0.264	> 15
Experiment 2[e]						
1	0[c]	-	-	-	0.046	0
2	5[c]	-	-	-	0.068	5
3	15[c]	-	-	-	0.097	15
4	-	116	90 , 116	102	0.159	> 15
5	-	< 3.3	-	< 5	0.055	0-5
6	-	11	-	42	0.134	> 15
7	-	< 2.1	-	< 5	0.056	0-5
8	-	< 1.5	-	< 5	0.065	≈ 5
9	-	-[f]	-	-	0.059	0-5
10	-	-[f]	-	-	0.054	0-5
11	-	-[f]	-	-	0.050	0-5
12	0[d]	-	-	-	0.054	0-5
13	2.5[d]	-	-	-	0.054	0-5
14	5[d]	-	-	-	0.066	≈ 5
15	7.5[d]	-	-	-	0.071	5-15
16	25[d]	-	-	-	0.118	>15

(a)- results obtained by 96-well plate assay; (b)- a 200 µL aliquot was read; (c)- standard soils provided with kit; (d)- soils spiked in the laboratory; (e)- a 100 µL aliquot was read; (f)- stated to be "below detection limits of NAA method" = false negative result

Table *III*. Immunoanalysis of Mercury in Sludge

Samples[a]	Mercury Content		Absorbance @ 410 nm
Experiment 1			
Standards:			
Standard 1	0 ppb		0.000
Standard 2	0.5 ppb		0.026
Standard 3	2.0 ppb		0.240
Standard 4	8.0 ppb		0.346
Sludge:[b]			
Soluble 1	0 ppb		0.000
Soluble 2	0.5 ppb		0.000
Soluble 3	2.0 ppb		0.000
Soluble 4	8.0 ppb		0.000
Soluble 5	200 ppb		0.235
	(Added)	(Expected)	
Particulate 1	0 ppm	0 ppb	0.000
Particulate 2	5 ppm	0.5 ppb	0.062
Particulate 3	20 ppm	2 ppb	0.298
Particulate 4	80 ppm	8 ppb	0.344
Experiment 2			
Standards:			
Standard 1	0 ppb		0.000
Standard 2	0.25 ppb		0.021
Standard 3	0.5 ppb		0.125
Standard 4	2.0 ppb		0.318
Standard 5	8.0 ppb		0.381
Sludge:[b]	(Added)	(Expected)	
Whole 1	0 ppm	0 ppb	0.000
Whole 2	2.5 ppm	0.25 ppb	0.057
Whole 3	5 ppm	0.5 ppb	0.184
Whole 4	20 ppm	2 ppb	0.339
Whole 5	80 ppm	8 ppb	0.367

(a)- Samples were diluted (standards) or spiked (sludge) to the indicated concentrations.

(b)- The soluble fractions were analyzed directly without dilution; the particulate fractions and whole sludge were extracted with acid and diluted (1 : 10,000) prior to analysis.

to 10,000, i.e., to ppb, prior to assay). At least two possibilities can explain the poor recovery of mercury from the spiked soluble fraction of sludge. First, this fraction contains high levels of proteolytic enzymes that might degrade the -SH rich proteins coating the tubes. Second, the soluble fraction contains proteins that might, themselves, bind the added mercury. In either case, interference with binding of the mercury to the tubes would occur. The results obtained with the particulate fraction show that acid extraction will prevent or reverse these actions. This is further illustrated by a second experiment in which spiked, whole sludge was directly extracted with acid and as little as 2.5 ppm mercury was detected (Table *III*). These data show, furthermore, that the inspect sludge tested did not contain significant levels of mercury.

Summary of the Applications of an Immunoassay for Mercury

The present studies have shown that the immunoassay-based test kits produced by BioNebraska, Inc., are as effective as the two physical/chemical methods, XRF and NAA, for the detection of mercury in soil. Mercury can also be measured in sludge at a level of ≤ 2.5 ppm using the immunoassay-based method, provided the samples are acid extracted to prevent interferences.

Additional applications for the immunoassay have been developed by BioNebraska personnel and collaborators. These include the measurement of mercury in animal tissues, principally seafood (*6*), and a wipe test for measuring mercury on building and equipment surfaces (Carlson, L., BioNebraska, Inc., and Holcombe, L., Radian, Corp., personal communication, 1996). With all of these applications, the results obtained by IA correlated well with other, more traditional methods for mercury analysis.

Acknowledgments

The authors gratefully acknowledge the contributions of L. Carlson, M. Riddell and C. Schweitzer, BioNebraska, Inc., to these studies. Research was performed at Oak Ridge National Laboratory, managed by Lockheed Martin Energy Research Corp., for the Office of Technology Development, U.S. Department of Energy, under contract DE-AC05-96OR22464.

Literature Cited

1.   U.S. Congress, Office of Technology Assessment. *Complex cleanup: the environmental legacy of nuclear weapons production*; OTA-O-484, U.S. Government Printing Office, Washington, DC, February 1991.

2.   Riley, R.G.; Zachara, J.M.; Wobber, F.J. *Chemical contaminants on DOE lands and selection of contaminant mixtures for subsurface science research*; U.S. Department of Energy Report - DOE/ER-0574T, DOE Office of Scientific and Technical Information, Oak Ridge, TN, April 1992.

3.   Lindberg, S.E.; Kim, K-H.; Meyers, T.P.; Owens, J.G. *Environmental Sci. and Technol.* *1994*, 29, 126-135.

4.   Wylie, D.E.; Carlson, L.D.; Carlson, R.; Wagner, F.W.; Schuster, S.M. *Anal. Biochem.* **1991**, 194, 381-387.
5.   Wylie, D.E.; Lu, D.; Carlson, L.D.; Carlson, R.; Babacan, K.F.; Schuster, S.M.; Wagner, F.W. *Proc. Natl. Acad. Sci. USA.* **1991**, 89. 4104-4108.
6.   Carlson, L.; Holmquist, B.; Ladd, R.; Riddell, M.; Wagner, F.; Wylie, D. In *Immunoassays for residue analysis*; Beier, R.C.; Stanker, L.H., Eds.; ACS Symposium Series, 621; American Chemical Society: Washington, DC, 1996, pp 388-394.

Chapter 24

Nonextractable Pesticide Residues in Humic Substances: Immunochemical Analysis

A. Dankwardt[1], K. Kramer[1], R. Simon[3], D. Freitag[3], A. Kettrup[2,3], and B. Hock[1]

Departments of [1]Botany and [2]Ecological Chemistry, Technical University of München at Weihenstephan, D–85350 Freising, Germany
[3]Institute for Ecological Chemistry, GSF Research Center for Environment and Health, Schulstrasse 6, D–85356 Freising, Germany

Immunochemical methods are suited for determining non-extractable pesticide residues in humic substances. Soil samples may be assayed by a non-competitive, direct enzyme immunoassay (EIA). In this case, a signal directly proportional to the amount of non-extractable residues is obtained. 2-Chloro-4-arylamino-6-alkyl-1,3,5-triazines were applied as model compounds for non-extractable triazine residues in dissolved humic material and assayed by a competitive EIA. Cross-reactivities were determined as a measure of the affinity of antibodies toward the model compounds. 2-Chloro-6-isopropyl derivatives were recognized almost equally well as free atrazine. Investigations with a recombinant antibody showed comparable recognition of atrazine and arylamino-s-triazines, although higher concentrations of the analytes were necessary. Samples from photolytic degradation experiments with arylamino-s-triazines were assayed in parallel by EIA and HPLC and yielded comparable results.

The fate of pesticides in the environment is characterized by different pathways. In addition to uptake by plants and degradation, pesticides may be distributed either in the air as a gas or an aerosol, in surface and ground water by run-off and leaching from the soil, or in the soil itself where it is more or less strongly bound (for review cf. *1*). In the case of atrazine, it was shown by Capriel et al. (*2*) that nine years after the application of ^{14}C-labelled atrazine under field conditions, approximately 50% of the initially applied radioactivity was still present in the soil in the non-extractable form. Furthermore, this fraction contained the parent herbicide in addition to its metabolites.

The strong persistence of some pesticides has encouraged recent development of screening methods for the evaluation of non-extractable pesticides. Non-extractable pesticide residues are usually investigated by combustion, hydrolysis, supercritical fluid extraction or high temperature distillation, followed by HPLC, GC/MS, or heteronuclear NMR (*3-5*). However, these methods are expensive and time consuming. Immunochemical methods as screening tools for the determination of pesticides have become increasingly popular because they are fast, less expensive, and easy to carry out (e.g. *6-8*). It is shown in this paper that this technology can be adapted to the analysis of non-extractable residues.

Antibodies as Ligands for Non-extractable Residues

Immunochemical analysis takes advantage of the fact that antibodies (Ab) only bind to a restricted part of an antigen, the antigenic determinant. This applies directly to non-extractable residues which may be presented to antibodies as antigenic determinants on the surface of humic substances and other compounds. A model is shown in Figure 1. As antibody binding *per se* does not generate a signal that can be detected by simple means (Figure 1b), a suitable marker such as a fluorescent dye or an enzyme is required (Figure 1c). This label is either coupled directly to the antibody (direct immunolabelling) or to a second antibody, which recognizes an entire group of antibodies, e.g. rabbit antibodies. Antibody-residue binding is then detected by means of a fluorescent signal or an enzymatic reaction.

Atrazine Residues in Soil. Hahn et al. (*9*) developed an immunolabelling procedure to detect non-extractable atrazine residues in soil. Antigen-binding fragments (Fab) (Figure 2) were prepared from polyclonal antibodies (pAb) directed against atrazine and coupled to the fluorescent dye Rose Bengal B. This Fab-dye conjugate lacked the constant part (Fc) of the antibody which was assumed to be responsible for a significant portion of the unspecific binding to soil particles (9). The fluorescence signal of the labelled Fab was related to the amount of bound atrazine in native soil samples determined by GC after supercritical methanol extraction. However, it proved to be essential to block unspecific binding of antibodies which varied greatly between different soil samples. This was accomplished by pre-incubation with non-specific, unlabelled immunoglobulins from pre-immune sera.

These qualitative studies were then extended by Dankwardt and Hock (*10*) to obtain a quantitative assay. A non-competitive, direct enzyme immunoassay (EIA) with peroxidase-labelled antibodies or Fab fragments, respectively, was used. The enzyme-labelled antibodies and Fab fragments directed against atrazine are assumed to recognize the pesticide residues bound to the soil matrix and to yield a signal directly proportional to the amount of non-extractable residues. The use of labelled Fab is thought to improve the detection of residues in microcavities, as they can enter small pores due to their smaller molecular size. However, only those non-extractable triazine residues can be determined that still expose groups such as the ethyl or isopropyl side chains of the atrazine molecule and are therefore available to the binding of the antibodies. Triazine residues which are completly integrated into the structure of the humic polymer are not recognized.

Different antibodies against triazines were tested (*10*), including the ones which have been employed earlier for the immunolabelling of atrazine in soil (*9*). Fab fragments were produced from the antibodies by digestion with papain and separated from the Fc part of the antibody on a DEAE-Sephacel column (BioRad, Munich, Germany) using a NaCl gradient (0.025-0.25 mol/L) (*11*). Complete digestion and successful separation was checked by polyacrylamide gel electrophoresis (*11*). The coupling of the antibodies and the Fab fragments to peroxidase was carried out according to the periodate method by Nakane and Kawaoi (*12*). The carbohydrate groups of the peroxidase are oxidized with periodate to reactive aldehyd groups, which then react with the free amino groups of the antibody molecules. The resulting Schiff base is reduced by adding ascorbic acid (*13*).

To demonstrate the feasibility of the determination of non-extractable residues by EIA, model soil particles were investigated in the first step. XAD particles (polystyrene resin used for the isolation of humic acids from aquatic and terrestrial systems) were used as model soil particles. They were coated with humic acids

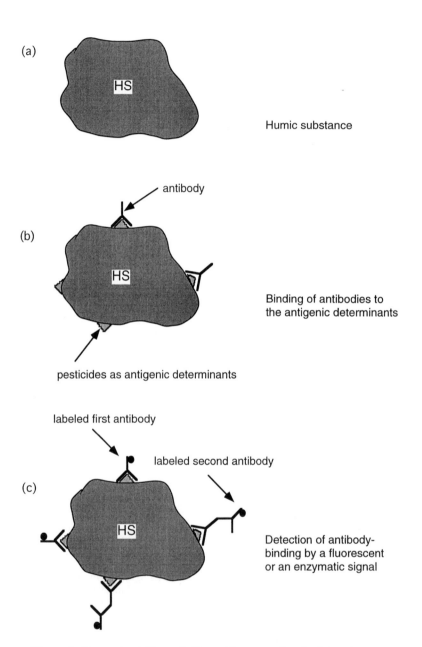

Figure 1. Non-extractable pesticide residues as antigenic determinants.

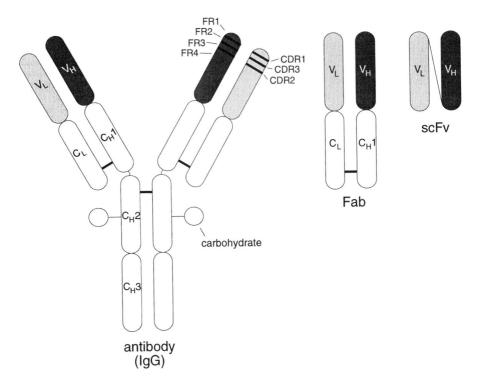

Figure 2. Schematic representation of a monoclonal antibody (IgG), a Fab fragment and the corresponding scFv-derivative. FR: Frame regions, CDR: Complementarity determining regions.

Table I. Investigation of Model Soil Particles with Non-extractable Atrazine Residues by Direct, Non-competitive EIA

Non-extractable atrazine (μg atrazine/g XAD)	EIA with anti-atrazine Ab (absorbancies)	EIA with control Ab (absorbancies)
0	0.33 ± 0.04	0.35 ± 0.03
0.1	0.49 ± 0.06	0.30 ± 0.04
1	0.57 ± 0.06	0.30 ± 0.02
10	0.75 ± 0.08	0.33 ± 0.03

Table II. Investigation of Soil Samples for Extractable and Non-extractable Atrazine Residues by EIA

Sample No.	Crop	Extractable atrazine (μg/kg) by competitive EIA and HPLC		Signal in the non-competitive EIA (absorbancies)	Non-extractable atrazine (μg/kg)
1	corn	15	20	0	0*
2	barley - corn	807	687	0	0*
3	several years of corn	1500	1380	0.5	400*
4 (reference)	no crop (barn)	n.d.	0	0	0+
5 (reference)	several years of corn	n.d.	50	0.25	200+

*Amount of non-extractable atrazine estimated by non-competitive EIA using GC data from samples 4 and 5 as a reference. +Amount of non-extractable atrazine determined after extraction with supercritical methanol by GC (reference samples).

(isolated from bog water (*14*), dissolved organic carbon = 500 mg/100 mL resin) and incubated with different amounts of s-triazine standard solutions (*15*). After centrifugation, the supernatant with the unbound s-triazine residues was removed and the concentration was determined by a competitive EIA (*16*). This concentration was compared to the concentration of the standard solutions before the experiment. The difference yielded the amount of atrazine bound to the humic acid-coated model particles. The model soil particles containing the non-extractable residues were washed several times and then assayed by the non-competitive, direct EIA. A signal increasing with the amount of triazines bound to the model soil particles was observed (Table I). The EIA was carried out in parallel with control antibodies to determine the unspecific binding. The binding sites of the control antibodies were saturated with atrazine by incubation with an excess concentration of the herbicide. Therefore, no specific binding sites were available. A background signal of about 0.3 absorbancies was observed, which did not increase with increasing amounts of atrazine bound to the model soil particles (Table I). The observed background signal is attributed to the unspecific binding of the antibodies to the humic acids.

In the next step, top soil samples from fields in Bavaria, Germany, were investigated. The samples were extracted with water for 12 hours on a shaker at 25 °C, and the extractable atrazine residues were determined by a competitive EIA and HPLC (*16*). The soil samples after extraction were then assayed by non-competitive EIA for non-extractable residues using peroxidase-labelled Fab fragments directed against atrazine. Three samples (No. 1, 2, 4) did not contain any non-extractable residues, the other two samples (No. 3 and 5) were positive (Table II). Sample 1 was a soil sample which did not contain any extractable atrazine. Sample 2 and 3 contained about 800 and 1500 µg/kg of extractable residues as determined by EIA. GC/MS data on the amount of non-extractable residues for samples 4 and 5 after extraction with supercritical methanol were available (No. 4: No residues, No. 5: 200 µg/kg, Haisch and Henkelmann, Bavarian State Institute of Soil Cultivation and Crop Production) and taken here as reference samples. Sample 4 was a control sample obtained from a plot beneath a barn, sample 5 was taken from a field after several years of cultivating corn. The non-competitive immunoassay indicated for sample 3 about twice as much non-extractable residues as for sample 5, leading to an estimated value of 400 µg/kg, while sample 2 did not contain any bound residues. The discrepancy between the ratio of extractable and non-extractable residues of sample 2 and 3 is explained by different pre-cultures. Corn was grown on plot 3 during previous years with yearly application of atrazine. On plot 2 other crops were grown before the cultivation of corn, which means that no atrazine was applied in the years before. Therefore, no formation of non-extractable atrazine residues occurred. The sampling date in June was probably too early to obtain soil samples from plot 2 with non-extractable residues, which could have formed already in the same year.

Atrazine Residues in Water-soluble Humic Substances. The occurrence of non-extractable pesticides is not limited to the solid fraction of the soil; pesticides may also be bound to water-soluble, mostly low-molecular weight, humic substances (*17*, *18*), also known as refractory organic substances (*14*). Therefore, the mobility and leaching of pesticides in soil are promoted *via* water-soluble humic substances, resulting in their final appearance in the ground and surface water.

Model Compounds. Studies of the behaviour and the formation of non-extractable residues under field conditions proved to be difficult because of the complexity of humic substances, which often prevents a direct characterization of the bound residue. As a convenient alternative, model compounds may be used, which are prepared in the laboratory and for which structural information is available, in order to study structural aspects or the fate of non-extractable residues. Bertin et al. (*19*) used a humic-like polymer produced by chemical oxidation of catechol in the

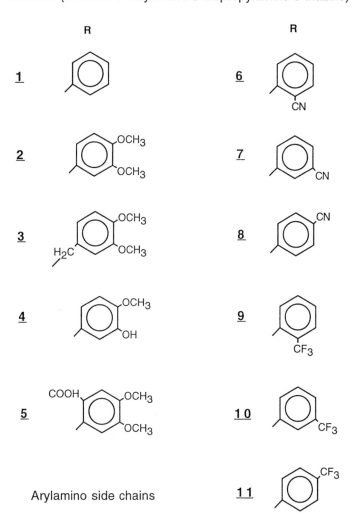

Atrazine (2-chloro-4-ethylamino-6-isopropylamino-s-triazine)

Arylamino side chains

Figure 3. Atrazine and the side chains R (**1-11**) of arylamino-s-triazines (derived from atrazine) as model compounds for non-extractable residues.

presence of atrazine. A portion of the atrazine was unextractable and formed bound residues. The model humic material was used to study the uptake of non-extractable residues by corn plants.

We have used arylamino-s-triazines as model precursors for non-extractable residues on the basis of an arylamino-s-triazine structure proposed by Andreux et al. (*20*) and Simon (*21*). These and closely related compounds can be assayed by a competitive EIA originally developed for the determination of atrazine in rain, surface water, and soil extracts (*16, 22*). The rationale behind this point is the potential applicability of the atrazine EIA for the determination of non-extractable residues in soluble humic substances.

Different arylamino-s-triazines derived from atrazine (Figure 3) were investigated by EIA. Calibration curves were obtained for these compounds using a polyclonal sheep atrazine antibody (S2) and a peroxidase tracer. The tracer was synthesized from the atrazine derivative 4-aminohexane carboxylic acid-6-isopropylamino-2-chloro-1,3,5-triazine. Most of the model compounds bound with similar affinities to the antibody (Figure 4) as indicated by 50% B/B_0 values (middle of the test) lying within a narrow range (0.15-0.27 µg/L). Compound **5** is a notable exception; it contains a carboxylic group at the aromatic ring that apparently interferes with the antibody binding. On the other hand, compound **10** exhibits a slightly higher affinity. However, when the chlorine at the 2′ position or the 6′ isopropyl group is replaced, the antibody binding is significantly weaker (cross-reactivity < 0.1-5%, not shown). Similar low cross-reactivities (0.4-15%) were also found with the metabolites of atrazine, deethylatrazine, deisopropylatrazine, and hydroxyatrazine (not shown). The metabolites also lack important functional groups for antibody recognition such as the chlorine or the isopropyl group.

Structures derived from the arylamino side chains but carrying an amino group instead of the triazine moiety did not exhibit cross-reactivity with the antibody. This proves that antibody binding is due to the specific recognition of the triazine part of the molecule and not to the binding of the antibody to the arylamino part.

Recombinant antibodies for the analysis of pesticide residues. It has been stressed before that the main limitation of the immunochemical approach is the availability of suitable antibodies (*23*). There is no doubt that access to recombinant antibodies (rAb) will significantly improve the situation. As soon as antibody libraries of sufficient size become available, they may be screened for rAb with improved binding properties. This is expected to significantly speed up antibody generation as compared to the conventional approach.

In order to show the feasibility of rAb binding, some of the arylamino-s-triazines were also investigated with the rAb K411B (*24*), which was derived from the mAb K4E7 (*25*). In brief, the antibody genes were isolated from the respective hybridoma cell line, cloned into a pCANTAB 5E vector (Pharmacia, Uppsala, Sweden) and expressed in *E. coli*. The rAb K411B is a single-chain fragment (scFv), an antibody molecule which is reduced to the variable regions of the heavy and light chain (Figure 2). The heterodimer is stabilized by covalent connection with a $(Gly_4Ser)_3$ peptide linker. After bacterial expression the rAb was affinity purified from culture supernatants and cell lysates using antibodies directed against a peptide coded by the E tag sequence which is placed downstream of the scFv insertion in the vector pCANTAB 5E. The anti-E tag antibodies were immobilized on a Protein G column (Pharmacia). The rAb were applied to the column and after a washing step eluted using low pH (0.1 mol/L Glycin, pH 2.8).

A similar affinity of the rAb towards atrazine and the two arylaminotriazines **1** and **6** was observed (Figure 5). However, the middle of the test (50% B/B_0) was moved from 0.2 µg/L for the EIA with the polyclonal sheep antibody to about 50 µg/L

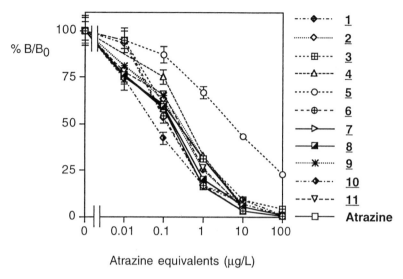

Figure 4. Binding curves of atrazine and the arylamino-s-triazines **1-11** with antiserum S2, directed against atrazine.

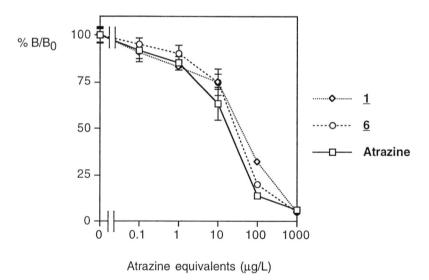

Figure 5. Binding curves of atrazine and the arylamino-s-triazines **1** and **6** with rAb K411B, directed against atrazine.

for the EIA with rAb K411B. The mAb K4E7, from which rAb K411B is derived, shows a 50% B/B_0 value for atrazine of 2 μg/L, if the same format with an immobilized conjugate (atrazine-protein conjugate) (*24*) is applied. In this case the antibodies are not bound to the microtiter plate but are in solution and bind either to the immobilized atrazine-protein conjugate or to the dissolved analyte (atrazine or aryltriazines) in the sample, depending on the concentration of analyte in the sample. The lower affinity of rAb K411B is attributed to PCR-based alterations of the rAb sequence, which is assumed to change the antibody affinity although the cross-reactivity pattern is not altered.

Photodegradation of arylamino-s-triazines. Finally, photolytic degradation experiments were carried out with arylamino-s-triazines **1** and **2** (*21*). Aqueous solutions (300 mL) were irradiated with a xenon lamp at 30 °C. Samples of 1 mL were taken during the experiments and the concentrations of the aminoaryl-s-triazines were determined in parallel by HPLC and EIA with pAb S2. Similar concentrations were found by HPLC and EIA (Table III) except for the first samples of **1**, which showed higher concentrations after analysis by HPLC. The observed concentrations of **1** and **2** decreased from 5 mg/L to 1.5 mg/L and from 400 μg/L to 150 μg/L atrazine equivalents, respectively.

Conclusion and Outlook

It has been shown in this paper that EIAs can be used to determine pesticide residues bound to humic substances. This applies equally to soil particles and dissolved humic material. In both cases, non-extractable residues are recognized as antigenic determinants. Arylamino-s-triazines were used as model compounds for non-extractable triazine residues in dissolved humic material. Concentrations as low as 0.02 μg/L atrazine equivalents could be detected in a sample of only 1 mL. If non-extractable residues of atrazine consist of similar arylamino-s-triazine structures, they should be detectable by EIA. The occurrence of such triazine-derived compounds in the environment is very likely. For example, the triazine metabolite deethylatrazine can be frequently found in soil, ground and surface water (*26-28*). Primary amines as in deethylatrazine may form covalent bonds with humic substances as proposed for dichloraniline and its metabolites (*29, 30*) and atrazine and its metabolites (*20*).

In order to screen water samples or soil leachates for the occurrence of residues bound to dissolved humic material or to investigate degradation pathways, EIA may be a helpful tool to assay humic substances samples for these compounds. They can also be useful in investigating the importance of humic-bound residues in pesticide degradation pathways.

Work is now in progress to examine more complex model compounds and pesticides bound to natural humic and fulvic acids by EIA. As a next step we will assay real samples for non-extractable residues in dissolved humic substances from areas where atrazine has been applied.

Significant improvements are expected from the availability of a broader spectrum of antibodies emerging in the recombinant field. Whereas changes of existing antibody properties are usually not feasible at the protein level, alterations at the DNA level are possible by genetic engineering followed by expression in simple hosts, such as *E. coli*. Antibodies with pre-determined specificities and affinities will be selected from antibody libraries to match the demands of environmental analysis. The next step is the generation of rAb libraries by randomizing CDR (complementarity determining regions).

Table III. Decrease of Arylamino-s-triazines During Photodegradation

Sample No.	Irradiation time (h)	Concentration (mg/L) by HPLC ($\Delta C/C = 0.1$)	Concentration (mg/L) by EIA
1-0	(0.0)	5.28	4.02 ± 0.24
1-1	(3.0)	5.02	3.88 ± 0.15
1-2	(6.3)	4.09	3.54 ± 0.14
1-3	(10.8)	3.56	3.25 ± 0.08
1-4	(22.5)	2.64	2.60 ± 0.13
1-5	(25.1)	2.38	2.29 ± 0.11
1-6	(28.1)	2.11	2.06 ± 0.13
1-7	(31.5)	1.58	1.84 ± 0.03
2-0	(0.0)	0.39	0.35 ± 0.03
2-1	(1.9)	0.22	0.22 ± 0.02
2-2	(3.6)	0.08	0.22 ± 0.02
2-3	(5.7)	0.19	0.20 ± 0.01
2-4	(6.7)	0.15	0.18 ± 0.01

Acknowledgements

We thank the Deutsche Forschungsgemeinschaft for financial support (Ho 383/30-3). We are grateful to Dr. G. Abbt-Braun and Prof. Dr. F. Frimmel (University of Karlsruhe) for the preparation of the model soil particles and their cooperation. We also thank Dr. A. Haisch, Mr. G. Henkelmann (Bavarian State Institute of Soil Cultivation and Crop Production) and Mr. R. Hofmann (Technical University of München) for the soil samples and the HPLC, GC data and Ms. K. Georgieva (Technological University of Sofia) for synthetic work.

Literature Cited

1. Hock, B.; Fedtke, C.; Schmid, R.R. *Herbizide. Entwicklung, Anwendung, Wirkungen, Nebenwirkungen*; Georg Thieme Verlag: Stuttgart, New York, 1995.
2. Capriel, P.; Haisch, A.; Khan, S.U. *J. Agric. Food Chem.* **1985**, *33*, 567.
3. Khan, S.U. *J. Agric. Food Chem.* **1982**, *30*, 175.
4. Capriel, P.; Haisch, A.; Khan, S.U. *J. Agric. Food Chem.* **1986**, *34*, 70.
5. Haider, K.; Spiteller, M.; Wais, A.; Fild, M. *Intern. J. Environ. Anal. Chem.* **1993**, *53*, 125.
6. Sherry, J.P. *Crit. Rev. Anal. Chem.* **1992**, *23*, 217.
7. Hock, B. *Acta Hydrochim. Hydrobiol.* **1993**, *21*, 71.
8. Meulenberg, E.P.; Mulder, W.H.; Stoks, P.G. *Environ. Sci. Technol.* **1995**, *29*, 553.
9. Hahn, A.; Frimmel, F.; Haisch, A.; Henkelmann, G.; Hock, B. *Z. Pflanzenernähr. Bodenk.* **1992**, *155*, 203.
10. Dankwardt, A.; Hock, B. *GIT Fachz. Lab.* **1995**, *8*, 721.
11. Peters, J.H.; Baumgarten, H. *Monoklonale Antikörper*, Springer Verlag: Berlin, 1990.
12. Nakane, P.K.; Kawaoi, A. *J. Histochem. Cytochem.* **1974**, *22*, 1084.
13. MacKenzie, D.J. In: *Serological methods for the detection and identification of viral and bacterial plant pathogens.* R. Hampton, E. Ball, S. de Boer, Eds.; APS Press: St. Paul, Minnesota, 1990, pp 87-92.
14. Abbt-Braun, G.; Frimmel, F.H.; Lipp, P. *Z. Wasser-Abwasser-Forsch.* **1991**, *24*, 285.
15. Dankwardt, A., *Entwicklung und Anwendung von Immunoassays in der Rückstandsanalytik von Atrazin - Untersuchung von Regenwasser, Oberflächenwasser und gebundenen Rückständen*; Ph.D. Thesis, Technical University of München, Department of Botany; 1994.
16. Dankwardt, A.; Pullen, S.; Rauchalles, S.; Kramer, K.; Just, F.; Hock, B.; Hofmann, R.; Schewes, R.; Maidl, F.X. *Anal. Lett.* **1995**, *28*, 621.
17. Caron, G.; Suffet, I.H.; Belton, T. *Chemosphere* **1985**, *14*, 993.
18. McCarthy, J.F.; Jimenez, B.D. *Environ. Sci. Technol.* **1985**, *19*, 1072.
19. Bertin, G.; Schiavon, M.; Andreux, F.; Portal, J.M. *Toxicol. Environ. Chem.* **1990**, *26*, 203.
20. Andreux, F.G.; Portal, J.M.; Schiavon, M.; Bertin, G. *Sci. Total Environ.* **1992**, *117/118*, 207.
21. Simon, R. *Charakterisierung der Bindung von Amino-s-triazinen an Huminstoffkonstituenten und photochemisches Verhalten der Modellverbindungen für gebundene Rückstände;* Ph.D. Thesis, Technical University of München, Department of Ecological Chemistry; 1996.
22. Dankwardt, A.; Wüst, S.; Elling, W.; Thurman, E.M.; Hock, B. *Environ. Sci. Poll. Res.* **1994**, *1*, 196.

23. Kramer, K; Hock, B. *Lebensmittel Biotechnologie* **1995**, *12*, 49.
24. Kramer, K.; Hock, B. *Food Agric. Immunol.* **1996**, *8*, 97.
25. Giersch, T. *J. Agric. Food Chem.* **1993**, *41*, 1006.
26. Thurman, E.M.; Goolsby, D.A.; Meyer, M.T.; Kolpin, D.W. *Environ. Sci. Technol.* **1991**, *25*, 1794.
27. Schneider, R.; Weil, L.; Niessner, R. *Intern. J. Anal. Chem.* **1992**, *46*, 129.
28. Skark, C.; Zullei-Seibert, N. *Vom Wasser* **1994**, *82*, 91.
29. Adrian, P.; Lahaniatis, E.S.; Andreux, F.; Mansour, M.; Scheunert, I.; Korte, F. *Chemosphere* **1989**, *18*, 1599.
30. Völkl, W.; Choné, T.; Portal, J.-M.; Gérard, B.; Mansour, M.; Andreux, F. *Fres. Environ. Bull.* **1993**, *2*, 262.

Chapter 25

Analysis of Hexazinone in Soil by Enzyme-Linked Immunosorbent Assay

Rodney J. Bushway[1], Lynn E. Katz[2], Lewis B. Perkins[1], Anthony W. Reed[2], Titan S. Fan[3], and B. S. Young[3]

[1]Department of Food Science and Human Nutrition, University of Maine, 5736 Holmes Hall, Orono, ME 04469–5736
[2]Department of Civil and Environmental Engineering, University of Maine, 5706 Aubert Hall, Orono, ME 04469–5736
[3]Millipore Corporation, 80 Ashby Road, P.O. Box 9125, Bedford, MA 01730–9125

A tube enzyme immunoassay (EIA) procedure was developed for the determination of the triazine herbicide hexazinone in soil. The antibody was polyclonal and was prepared by employing metabolite A of hexazinone conjugated to bovine serum albumin as the immunogen. Hexazinone was extracted from soil by shaking with methanol-water 80/20 for 10 min and allowed to set overnight before reshaking for 5 min. Aliquots for EIA analysis were diluted in such a way as to always contain 8% methanol. Reproducibility results for both standards and samples were good. A correlation coefficient of 0.9562 was obtained for 78 soil samples analyzed by EIA vs. HPLC. Of the eight known metabolites of hexazinone, 7 were tested for cross-reactivity and 5 were shown to be cross-reactive.

It is commonly known that pesticides applied on land surfaces can enter surface and groundwater either through surface runoff or infiltration processes. Indeed, non-point source contamination of ground and surface water due to these processes is well documented and the effects of migration of pesticides into surface and groundwater is both of regional and national interest (1-3). One pesticide of particular interest since it is a triazine is hexazinone. Many triazines have been found in groundwater and are considered environmental problems (1,3).

Hexazinone (trade name Velpar) is primarily a contact herbicide that can be used as a foliar spray or applied directly to soils for weed control in blueberries, pineapple, sorghum, sugarcane, and forests. Recent studies have identified the presence of hexazinone in surface and groundwaters within blueberry areas of eastern, southern and western, Maine (4). A variety of degradative and non-degradative processes influence the extent of Velpar contamination, including biodegradation, photolysis and sorption (5-6). The importance of quantifying the extent of soil sorption has lead to the development of an EIA method for the determination of hexazinone in soil.

Experimental

Hexazinone and its Metabolites. Hexazinone (3-cyclo-hexyl-6-(dimethylamino)-1-methyl-1.3.5-triazine-2,4-(1H,3H)-dione and the following 7 metabolites: metabolite A

[3-(4-hydroxycyclohexyl)-6-(dimethylamino)-1-methyl-1,3.5-triazine-2,4-(1H,3H)-dione]; metabolite B [3-cyclohexyl-6-(methylamino)-1-methyl-1,3.5-triazine-2,4-(1H,3H)-dione]; metabolite C [3-(4-hydroxycyclohexyl)-6-(methylamino)-1-methyl-1,3,5-triazine-2,4-(1H,3H)-dione]; metabolite D [3-cyclohexyl-1-methyl-1,3,5-triazine-2,4,6-(1H,3H,5H)-trione]; metabolite E [3-(4-hydroxy-cyclohexyl)-1-methyl-1,3,5-triazine-2,4,6-(1H,3H,5H)-trione]; metabolite A-1 [3-(trans-2-hydroxy-cyclohexyl-6-(dimethylamino)-1-methyl-1,3,5-triazine-2,4-(1H,3H) dione]; and metabolite 1 [3-(4-oxycyclohexyl)-6-(dimethylamino)-1-methyl-1,3,5-triazine-2,4-(1H,3H)-dione] were gifts from E.I. DuPont de Nemours & Company, Experimental Station, Wilmington, DE. Structures of hexazinone and its metabolites are shown in Figure 1.

Soil Samples. Soil samples were collected from blueberry fields from Florida and eastern, western and southern Maine. They were dried and sieved (20 mesh) before being extracted.

Preparation of Hexazinone Standard for Immunoassay. A stock solution of hexazinone was prepared by weighing 20 mg of hexazinone into a 50 mL volumetric flask and bringing to volume with HPLC grade methanol. An intermediate standard solution was made by pipetting 50 μL of stock solution into a 50 mL volumetric flask and bringing to volume with HPLC grade water. Working standards (0.11, 0.22, 0.44, 1.1, 2.2, 4.4, 8.8, and 17.6 ppb) were prepared by serially diluting the intermediate standard into HPLC grade water.

Preparation of Hexazinone Standards for HPLC. The same hexazinone stock solution that was employed above for immunoassay was used for HPLC. An intermediate standard was prepared by pipetting 50 μL of the stock standard into a 50 mL volumetric flask and bringing it to volume with methanol-acetonitrile-water (20:40:40, v/v/v). Working standards (12.5, 62.5, 312.5, 1562.5, and 7812.5 ηg/mL) of hexazinone were prepared by making serial dilutions of the intermediate standard with the methanol-acetonitrile-water solution.

Production of Antisera and Immunogen. These procedures are described in a previous paper (4). Briefly, rabbits were injected intradermal and subcutaneous with the active ester of hexazinone hemisuccinate. The hexazinone hemisuccinate was prepared by refluxing hexazinone metabolite A and succinic anhydride in pyridine.

Extraction of Hexazinone in Soil. One gram of soil was weighed into a 25 mL polypropylene bottle followed by the addition of 5 steel ball-bearings and 10 mL of 80:20 MeOH ACS grade/water. Samples were shaken by hand for 10 min and allowed to set overnight before shaking again for 5 min. A 100 uL aliquot was removed for ELISA and a 5 mL aliquot was taken for HPLC analysis.

EIA Analysis of Hexazinone in Soil. A tube kit from Millipore Corp. (Bedford, MA) was employed for the analysis. The 100 μL aliquot was added to 0.9 mL of HPLC grade water (This makes the sample 8% methanol). A 200 μL aliquot of the sample and/or standards were added to no more than 10 EIA tubes followed by 200 μL of enzyme conjugate. Each tube was mixed briefly by swirling. After a 20 min incubation at room temperature, the tubes were rinsed 4 times under tap water and blotted dry before the addition of 500 μL of K-blue (Elisa Technologies, Lexington, KY) substrate to each tube.

Tubes were incubated at room temperature for 10 min before 300 μL of 1 N HCl which stops the reaction and changes the color from blue to yellow. Absorbance of each tube was read at 450 nm using an EnviroGard tube reader. If samples needed to

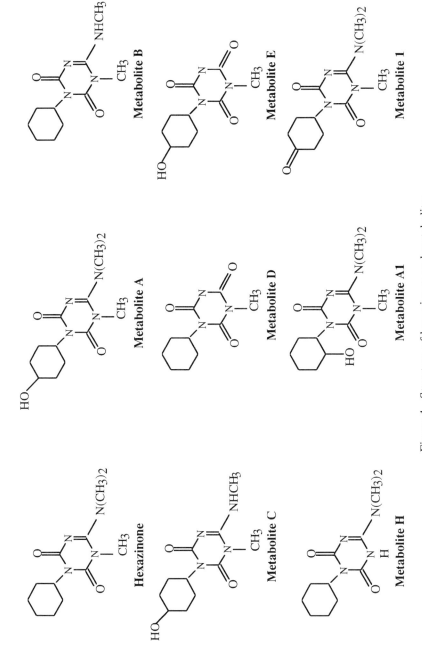

Figure 1. Structure of hexazinone and metabolites

be diluted to remain in the linearity range then they were diluted in an 8% methanol solution. In fact all standards and samples must be run in 8% methanol.

Quantitation of Hexazinone. Control tubes were run with each set of tubes to calculate %B values of standards and samples (absorbance at 450 nm of standard or sample/absorbance at 450 nm of negative control X 100). Standards were run at the beginning and end of each day with the average of both runs used to prepare the standard curve which was made by plotting %B versus the log of hexazinone concentration. Hexazinone levels in unknown water samples were interpolated from the standard curve.

Cross-Reactivity of Hexazinone Metabolites.   Cross-reactivity studies were performed in 8% methanol.

HPLC Operating Conditions. A Zorbax C18 column (stainless steel, 4.6 mm I.D. X 250 mm) (Phenomenex, Torrance, CA) was employed for the separation along with a mobile phase comprised of methanol-acetonitrile-water (20:40:40) at a flow rate of 1 mL/min using a Hewlett-Packard (HP) 1050 pump (Wilmington, DE). Injection (50 μL) were performed by a HP 1050 auto-injector. The samples were detected with a HP 1050 photodiode array detector set at 247 nm while a Vectra HP 486 Chem Station for windows was used to measure peak areas.

Analysis of Hexazinone in Soil by HPLC. The 5 ml aliquot from the soil extract was added to 100 ml of HPLC grade water. This mixture was passed through an activated C18 Sep-Pak (Waters Associates, Milford, MA) (Activation was done by passing 5 mL of HPLC methanol through a Sep-Pak followed by 5 mL of HPLC water) at 5 mL/min. After drying the Sep-Pak for 20 min under vacuum, it was eluted with 4 mL of a mixture of ethyl acetate-methyl tertiary butyl ether (20:80). The 4 mL were evaporated to dryness under nitrogen and the residue was dissolved in 1 mL of HPLC mobile phase. A 50 uL aliquot was injected into the HPLC.

Results and Discussion

A typical EIA standard curve is shown in Figure 2. The linear range was from 0.22 to 17.6 ppb with an IC_{50} (concentration of hexazinone at a %B value of 50) of 3.0 ppb. The lower limit of detection (LLD) for hexazinone in soil was determined to be 25 ppb and the lower limit of quantitation (LQD) to be 50 ppb (7). All standards and samples including controls were made up in 8% methanol. This was done to quicken the EIA analysis. Otherwise an evaporation step would have been needed because methanol at 8% does have an inhibitory effect on immunoassay even though it is small. Since the inhibitory effect of the methanol is slight, diluting the methanol to obtain even a smaller decrease in inhibition makes it impossible to obtain the LLD of 25 ppb. Thus the inhibitory effect of the 8% methanol is balanced out by performing all tests in that concentration of methanol.

The types of soils employed for spiking studies and aged soils analyzed come under 5 classes (loamy sand, loamy sand/sand, sand, sandy loam, and loam) based on texture (Table I). All but one soil type were acid which is normal for blueberry soils.

The cross-reactivity of this antibody is very broad reacting with most major metabolites. Exact cross-reactivity values are given in Table II. The values given in

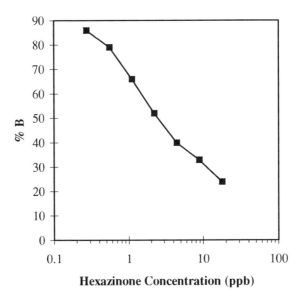

Figure 2. Hexazinone Standard Curve

Table II are based on compounds in 8% methanol and are less than what was given in a previous paper (4) where the compounds were dissolved in water. It appears that the cross-reactivity values for D and E did not change. This is not surprising since these two metabolites are not considered to be cross-reactive. Thus, metabolites A, A1, 1, B, and C demonstrate all the reactivity which is expected since they are the most structurally similar to hexazinone (Figure 1). This can be a problem if one wants to quantitate individual compounds and the others are present. However in a screening mode, in which more classical methods would be performed if the immunoassay was positive, total would be advantageous. Futhermore, some soils need extensive cleanup in order to quantify the metabolites of hexazinone, fractions could be collected and analyzed by EIA without performing the cleanup.

Recently, by employing a GC method (8) we have shown that the blueberry soil samples contain only 2 metabolites (B and D). Metabolite D has no cross-reactivity and; therefore, will not interfere with hexazinone analysis by EIA. Metabolite B demonstrates sufficient cross-reactivity to yield interference with hexazinone analysis by EIA but it must be at a concentration of 0.5 ppm or higher before it can interfere with hexazinone quantitation by EIA at a 1/10 dilution which is the smallest dilution ever made. Our analysis of metabolite B in 50 blueberry soils from Maine only showed 2 had sufficient B concentration to interfere with hexazinone analysis.

As with any analytical technique, precision within and between days is important. Reproducibility results for hexazinone standards and soil samples were good (Table III). For standards, the CVs ranged from 3.3 to 13% with all but one below 7% while actual field soils demonstrated CVs varying from 5.2 to 21% with an average of 13%.

The accuracy of the EIA method was tested using fortified soil samples spiked at 25, 50, 100, 250, 1000, and 5000 ppb (Table IV). Recoveries were excellent ranging from 76 to 96% with a mean of 86%. Coefficients of variations ranged from 12 to 25% with only 1 above 17% and that was the 25 ppb spike which was at the LLD of the method.

Table I. Characteristics of Soil Analyzed

Soil	pH	%OM	CEC	%Sand	%Silt	%Clay	Text.
1	6.0	0.5	7.3	84	<10	12	loamy sand
2	6.7	0.8	2.0	88	<10	<10	loamy sand/ sand
3	5.5	3.3	4.6	85	<10	<10	loamy sand
4	6.4	1.2	2.6	90	<10	<10	sand
5	6.3	2.7	3.4	88	<10	<10	loamy sand/ sand
6	6.2	2.4	4.1	89	<10	<10	loamy sand/ sand
7	5.7	0.6	1.1	97	<10	<10	sand
8	7.7	8.6	15.4	65	23	12	sandy loam
9	5.4	4.0	14.6	47	38	15	loam
10	4.2	8.6	11.6	58	35	7	sandy loam
11	5.2	13.2	16.4	61	28	11	sandy loam

Table II. Cross-reactants in the Hexazinone Tube EIA

Compound[a]	IC_{50}[b] (ppb)	LLD[c] (ppb)
Hexazinone	3.0	0.22
Metabolite A	6.3	0.27
Metabolite A1	15	0.55
Metabolite 1	4.0	0.55
Metabolite B	71	5.1
Metabolite C	140	11

[a]The following compounds showed no cross-reactivity at 1 ppm-- Metabolite D and E, simazine, cyromazine, prometyrn, procyazine, propazine, trietazine, terbutylazine, cyprazine, prometon, cyanazine, atrazine, analazine, amitraz, ametryn, amitrole, dichlone, bromacil, imidan, captan, iprodione, terbacil, lenacil, bentazon, capafol, and MH.
[b]Concentration that causes 50% inhibition.
[c]Lowest limit of detection at %B of less than 90.

Spiking studies are not always the best technique to use in ascertaining the extraction efficiency of a procedure. Thus, other extraction methods were tried on actual field samples. However, no other extraction method, including supercritical fluid extraction yielded as much hexazinone as did the 80:20 methanol/water with the ball-bearings.

A correlation study was performed between HPLC and EIA on soil samples collected from Maine and Florida. Results are shown in Table V and Figure 3. The correlation coefficient was 0.9526 for 78 soil samples while the equation was $y = 0.745x + 206$ which indicated a low bias for EIA, but there was no type of relationship between soil type and the values obtained by HPLC and EIA. However these results indicate good agreement between both techniques especially considering EIA is a screening method. Thus EIA for monitoring hexazinone in soil is a rapid, accurate, and cost-effective technique that can also detect some of the major metabolites if present in soil at sufficient levels and if these metabolites are not present then the EIA is a good method for hexazinone screening.

Table III. Reproducibility of the Hexazinone EIA

Samples (ppb)	CV[a] intra-assay	CV[b] interassay
Standard (0.22)	6.4	3.3
Standard (0.44)	4.3	3.7
Standard (1.10)	3.7	4.8
Standard (2.20)	4.4	4.8
Standard (4.40)	4.3	6.6
Standard (8.80)	6.5	4.7
Standard (17.6)	6.0	13
Soil 1 (343)	11	20
Soil 2 (813)	13	19
Soil 3 (538)	7.3	6.7
Soil 4 (7300)	9.6	5.2
Soil 5 (780)	21	18
Soil 6 (5580)	10	16

[a]Percent Coefficients of Variation based on 4 determinations in 1 day for standards and 5 determinations in 1 day for soils.
[b]Percent Coefficients of Variation based on 6 determinations in 6 different days for standards and 5 determinations for soils.

Table IV. Accuracy of Hexazinone EIA for Spiked Soil Samples

Hexazinone added (ppb)	Hexazinone found (ppb)	% Recovery	CV[a]
25	19	76	25
50	57	114	14
100	87	87	17
250	208	83	12
1000	860	86	13
5000	4817	96	13

[a]Percent Coefficients of Variation based on 6 determinations.

Table V. Comparison of EIA and HPLC Methods for Hexazinone in Soil

Sample	Hexazinone Concentration (ppb)		Sample	Hexazinone Concentration (ppb)		Sample	Hexazinone Concentration (ppb)	
	HPLC	EIA		HPLC	EIA		HPLC	EIA
Soil 1	143	54	Soil 27	847	1000	Soil 53	8521	4800
Soil 2	1036	1015	Soil 28	106	96	Soil 54	14706	9800
Soil 3	242	230	Soil 29	1207	980	Soil 55	4930	4900
Soil 4	967	900	Soil 30	15310	10000	Soil 56	5335	6300
Soil 5	197	64	Soil 31	642	540	Soil 57	909	680
Soil 6	127	54	Soil 32	9499	8000	Soil 58	268	275
Soil 7	178	74	Soil 33	14320	15000	Soil 59	101	95
Soil 8	184	120	Soil 34	5272	4300	Soil 60	979	920
Soil 9	1270	1600	Soil 35	1018	1410	Soil 61	3125	5000
Soil 10	1560	1900	Soil 36	181	230	Soil 62	4531	5400
Soil 11	660	1000	Soil 37	301	200	Soil 63	644	780
Soil 12	1450	1800	Soil 38	1104	920	Soil 64	910	900
Soil 13	253	110	Soil 39	2802	1740	Soil 65	687	540
Soil 14	1136	1600	Soil 40	3127	1860	Soil 66	1056	760
Soil 15	3370	4000	Soil 41	251	200	Soil 67	899	735
Soil 16	948	1200	Soil 42	249	245	Soil 68	5984	5600
Soil 17	178	190	Soil 43	180	68	Soil 69	8679	6000
Soil 18	216	325	Soil 44	435	465	Soil 70	163	65
Soil 19	181	94	Soil 45	7797	5000	Soil 71	929	1320
Soil 20	867	1450	Soil 46	8834	5250	Soil 72	727	460
Soil 21	850	1000	Soil 47	1911	1600	Soil 73	233	112
Soil 22	119	46	Soil 48	5503	4650	Soil 74	438	245
Soil 23	1270	1250	Soil 49	8293	6250	Soil 75	353	200
Soil 24	200	170	Soil 50	8293	4300	Soil 76	395	230
Soil 25	97	100	Soil 51	1264	1280	Soil 77	222	290
Soil 26	353	320	Soil 52	556	330	Soil 78	242	145

78 Soil Samples-- Equation $y = 0.745x + 206$; Correlation = 0.9526.

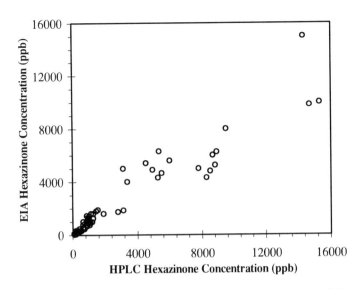

Figure 3. Comparison of HPLC and EIA for Hexazinone in Soil

Acknowledgements

This manuscript is number 2020 of the University of Maine Agricultural and Forestry Experiment Station. Partial support for this work was provided by the U.S. Geological Survey, through the University of Maine Water Resouces Institute, Project ME93-03-96-01. The contents do not necessarily reflect the views and policies of the Department of the Interior or the University of Maine, nor does mention of trade names or products constitute an endorsement by either organization.

References

1. Drinking Water Health Advisory: *Pesticides*; Cook, M. B., ed.; U. S. Environmental Protection Agency, Office of Drinking Water Health Advisories; Lewis Publishers: Chelsea, MI, 1989; pp 471-486.
2. Bedient, B. P.; Rifai, H. S.; Newell, C. J. *Ground Water Contamination Transport and Remediation;* Prentice-Hall Publishers: Englewood Cliffs, NJ, 1994; pp 20-90.
3. Funari, E.; Donati, L.; Sandroni, D.; Vighi, M. *Pesticide Risk in Groundwater*; Vighi, M.; Funari, E., eds., Lewis Publishers; New York, NY, 1995; pp 3-41..
4. Bushway, R. J.; Fan, T. S.; Katz, L. E.; Reed, A. W.; Ferguson, B. S.; Xu, C. Y.; Perkins, L. B.; Young, B. E. *Immunoassay for Residue Analysis: Food Safety*; Beier, R. C.; Stanker, L. H.; eds., American Chemical Society; Washington, D. C., 1996; pp 187-199.
5. Stone, D. M.; Harris, A. R.; Koskinen, W. C. *Environ. Toxicol. Chem.*; **1993**, 12, 399-404.
6. Sidhu, S. S.; Feng, J. C. *Weed Sci.* ; **1993**, 41, 281-287.
7. ACS Committee on Environmental Improvement, *Anal. Chem.*; **1980**, 52, 2242-2249.
8. Feng, J.C. *Can. J. Chem.* **1992**, 70, 1087-1092.

SAMPLE PREPARATION, CHEMOMETRICS, AND QUALITY CONTROL

Chapter 26

Development and Application of Molecular Imprinting Technology for Residue Analysis

Mark T. Muldoon and Larry H. Stanker

Food Animal Protection Research Laboratory, Agricultural Research Service, U.S. Department of Agriculture, 2881 F&B Road, College Station, TX 77845

Molecular imprinting technology has recently been applied to the analysis of environmentally-important compounds. Molecularly-imprinted polymers (MIPs) are formed by synthesizing polymers in the presence of a "print molecule", which is usually the analyte of interest or a closely-related structure. After polymerization, the print molecule is removed resulting in an "imprint" of the molecule. MIPs can be used as specific binding matrices for a variety of analytical applications. The most common application for MIPs have been as solid-phase adsorbents for HPLC. Recently, these MIPs have been employed as specific receptors in competitive ligand binding assays such as the Molecularly-Imprinted Sorbent Assay (MIA). MIPs have been incorporated into membranes for use in sensor applications. Finally, MIPs have been used as solid phase extraction materials in the preparation of complex biological samples for residue analysis. This paper reviews the process of MIP synthesis, the basis for analyte-MIP recognition, and current applications of the technology for residue analysis.

Molecular imprinting technology is a rapidly emerging field in which synthetic polymers, made in the presence of a "print molecule", are utilized as specific binding matrices. The print molecule is typically the analyte of interest or a closely-related analog. In contrast to the relatively non-specific mechanisms found in conventional solid-phase chromatography, molecularly-imprinted polymers (MIPs) are analyte-specific since the polymerization process results in the formation of a rigid binding pocket in which only the analyte of interest or closely related compounds can fit. Thus, MIPs are synthetic analyte-specific receptors and have been referred to as "plastic antibodies" since many of the properties that control binding may be similiar for both receptor systems (1, 2).

The development of molecular imprinting technology, as it applies to residue analysis, is in its infancy. In contrast, immunochemical technology has gained wide application for the analysis of low molecular weight compounds of environmental interest (*3*). However both immunoassays and MIP technology can compliment more traditional analytical methods. We are currently evaluating MIPs as part of our effort in developing rapid assays, including immunoassays, for the analysis of residues in environmental and biological matrices (*4*). This paper will review the processes involved in the synthesis of MIPs, the mechanisms of receptor-analyte recognition, and the applications of the technology for residue analysis.

Polymer Synthesis

The various steps involved in the synthesis of MIPs are illustrated in Figure 1 as they were applied for the generation of an anti-atrazine MIP (*5*). In the first step, the print molecule (e.g., atrazine) is mixed with the functional monomer. Functional monomers are polymerizable chemical units which possess specific chemical functionalities such as carboxyl- (shown in Figure 1), hydroxyl-, amino-, or aromatic groups. They can bind to the print molecule covalently and/or non-covalently. The particular functional monomer (or combination of monomers) used depends on the particular chemical properties of the analyte to which the MIP is being made. For the anti-atrazine MIP, we used the carboxylic acid, methacrylic acid as the functional monomer because of its ability to form strong hydrogen and electrostatic bonds with the ring and amino nitrogens of atrazine (*6*). Other functional monomers which have been commonly used are shown in Figure 2. In the second step, following the association and binding of the functional monomer(s) to the print molecule, the "print assembly" is then polymerized using excess co-monomer, or cross-linking agent. The co-monomer we used was ethylene glycol dimethacrylate (Figure 2). As a result, a highly cross-linked, rigid, glassy polymer is formed with functional groups that are "fixed" in a specific orientation around the print molecule. After polymerization, the print molecule is chemically cleaved (e.g., hydrolyzed) or extracted from the polymer resulting in an "imprint" of the print molecule. The process by which the print molecule is removed from the polymer depends upon whether it is bound covalently, non-covalently, or both, to the functional monomer. For most applications, the polymer is ground to a fine powder (< 50 μm) prior to use.

   Early studies conducted by Günter Wulff and coworkers at the Institut for Organische Chemie II der Universitat Düsseldorf, showed that it was possible to obtain synthetic polymers with binding cavities containing functional groups in a specific arrangement coinciding with that of the print molecule (*7, 8*). To demonstrate this, they synthesized a vinylphenylboronate (Figure 2) derivative of the print molecule. This was polymerized and the print molecules (sugar residues) were removed by chemical hydrolysis. The polymer was used as the solid-phase adsorbent for high performance liquid chromatography (HPLC) and shown to preferentially bind the specific sugar molecule used as the print molecule to the exclusion of enantiomers. This achievement led to the development of many other MIPs for a variety of print molecules and new methods to produce them.

   The wide application of MIP technology followed the development of MIPs using non-covalently bound print molecules. This technique was pioneered by Klaus

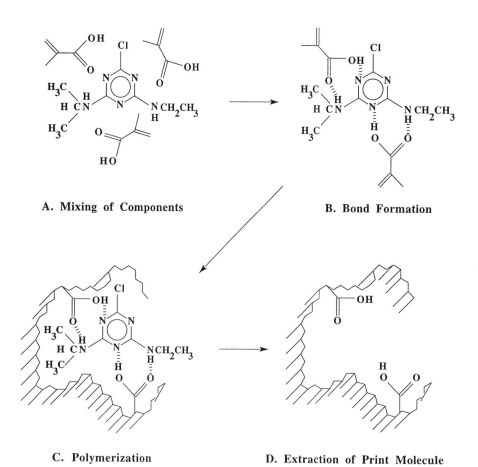

A. Mixing of Components

B. Bond Formation

C. Polymerization

D. Extraction of Print Molecule

Figure 1. Synthesis of a molecularly-imprinted polymer (MIP).

Mosbach and his coworkers at the University of Lund (*9*). Initial studies primarily used methacrylic acid as the functional monomer with ethylene glycol dimethacrylate (Figure 2) as the cross-linking agent. Much of this work used L-phenylalanine anilide, or closely-related amino acid derivatives, as print molecules (*9-13*). In most of these applications, the MIPs were used as solid-phase adsorbents for HPLC for the separation and analysis of racemic mixtures.

Many other synthetic approaches have recently been developed. In one approach, an *in situ* polymerization method was reported in which the print molecule (L-phenylalanine or diaminonapthalene) was non-covalently bound to the functional monomer and polymerization was carried out inside a stainless steel HPLC column (*14*). After polymerization, the column was connected to an HPLC system and the print molecule was extracted from the MIP column on-line. The column was subsequently used to separate the print molecule from related structures. Another innovative approach was used for the development of an MIP that recognized cholesterol (*15*). For this, cholesterol was covalently derivatized with polymerizable functionalities. The assembly was polymerized and the cholesterol residues were hydrolyzed from the polymer matrix. After removing the print molecule, rebinding of cholesterol involved non-covalent interactions. Other synthetic approaches have utilized biomolecules, as opposed to synthetic polymers, as the polymer matrix for molecularly-imprinting small molecules (*16*).

The optimization of the reaction conditions used for polymer synthesis has been the subject of extensive research (*15, 17-22*). The reaction variables studied include the types and relative amounts of print molecule, functional monomer, and cross-linking reagent used, and the type of chemical initiator and conditions (e.g., temperature and hv) used to start the polymerization process. The particular functional monomer used is particularly important since it is this component that is involved in forming an effective chemical bond with the print molecule. Equally important is the amount of cross-linking reagent used in polymerization. This component affects the structural rigidity of the polymer, thereby fixing the functional groups in a specific orientation corresponding to the print molecule. Other reaction variables that are important include, the type of solvent or "porogen" used (which affects polymer porosity), and the time and temperature at which polymerization is carried out.

Molecular Recognition in MIPs

Several studies on the mechanisms by which MIPs recognize the print molecule have been reported. Recognition is generally attributed to the formation of functional groups in a specific spatial arrangement within the matrix of the polymer which conforms to that of the print molecule (*11, 23*). This was studied using fourier transform-infrared (FT-IR) and nuclear magnetic resonance (NMR) spectroscopic methods (*24*). Furthermore, these studies described the processes by which small molecules orient themselves and subsequently bind to specific MIPs. In addition to specific analyte binding to functionalities in the polymer matrix, shape-selective cavities also contribute to binding and may be particularly important for inert print molecules such as aromatic hydrocarbons (*21, 24*). Depending on the functional monomer used in MIP synthesis, both covalent and non-covalent chemical bonds can be formed between an MIP and the print molecule and these are usually of the same

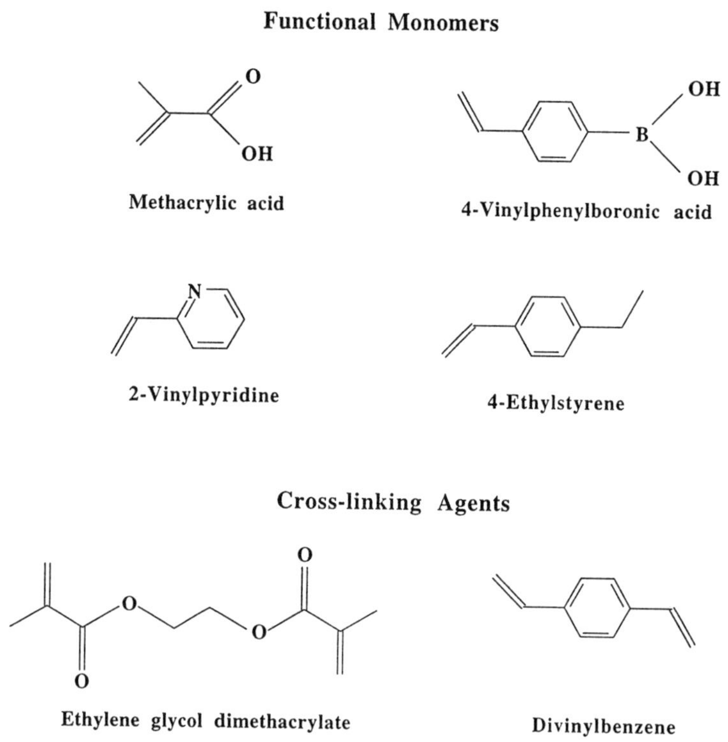
Functional Monomers
O
OH
Methacrylic acid
OH
B
OH
4-Vinylphenylboronic acid
N
2-Vinylpyridine
4-Ethylstyrene
Cross-linking Agents
O
O
O
O
Ethylene glycol dimethacrylate
Divinylbenzene

Figure 2. Commonly used reagents for molecularly-imprinted polymer (MIP) synthesis.

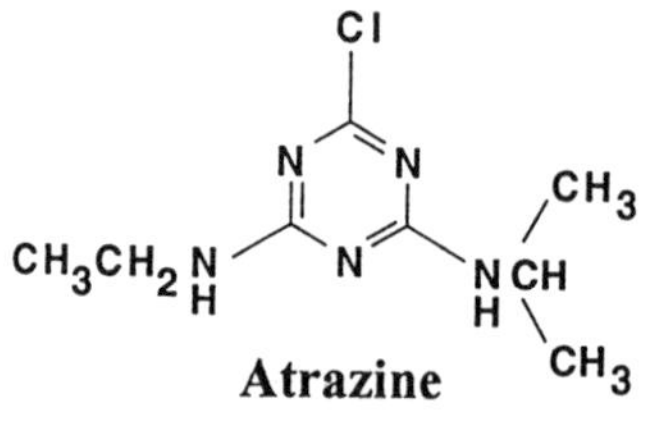
Cl
N
N
N
CH3CH2 N H
N CH H
CH3
CH3

Atrazine

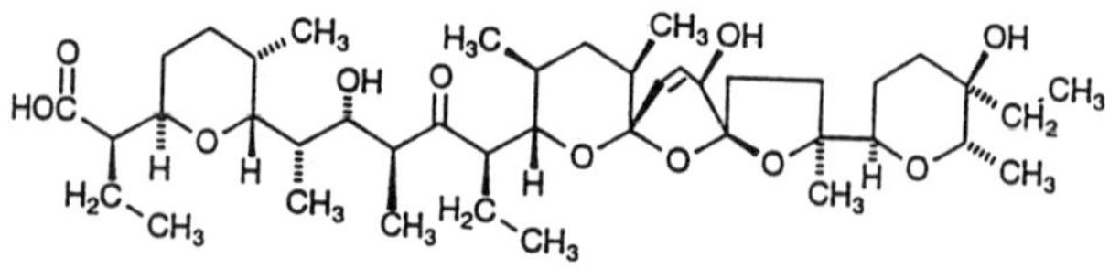
HOC
O
CH3
H3C
CH3 OH
OH
OH
O
H
H
H
CH3
CH3
H2C
CH3
H2C
CH3
CH2
CH3
CH3
CH3

Salinomycin

Figure 3. Structures of atrazine and salinomycin.

nature as those which occur during the initial steps of polymer synthesis with some exceptions (*15*). Multiple contact points or interactions between the print molecule and the MIP are optimal for the production of a useful MIP (*2*). However, these interactions must be reversible if the MIP is to be used for chromatographic applications. Reversible covalent bond interactions typically utilized in MIPs include boronic acid ester linkages and Schiff base formation (*7, 25*). The most important non-covalent interaction utilized for MIP recognition is the electrostatic bond (*11*). Hydrogen bonding also is important and, in combination with other interactions, probably accounts for the high resolving power of some non-covalent binding MIPs (*26*). Non-covalent interactions are more useful for rapid chromatographic applications since they are more readily reversible than covalent ones.

   Polymer synthesis is typically performed in an inert non-aqueous organic solvent (e.g., chloroform, hexane, acetonitrile). This condition facilitates the association and binding of the functionalized polymer to the print molecule. Interestingly, rebinding and specific recognition of the print molecule by the MIP is usually optimal when carried out in the same solvent as originally used for polymer synthesis (*27*). We have synthesized MIPs against several agrochemicals and have observed this effect using both an MIP against the herbicide atrazine and an MIP against the coccidiostat salinomycin (Muldoon, M.T. and Stanker, L.H., unpublished results). The structures of atrazine and salinomycin are shown in Figure 3.   Both of these MIPs used methacrylic acid as the functional monomer, ethylene glycol dimethacrylate as the co-monomer, and chloroform as the solvent. Figure 4 shows the results from binding studies using both the anti-atrazine and the anti-salinomycin MIPs and control polymer (synthesized in the absence of print molecule). Of the solvents tested, specific binding of print molecule to the respective MIP was greatest in chloroform, the solvent used for synthesis. Specific binding diminished as the polarity of the solvent increased. The decrease in specific binding that occured in polar organic solvents (acetonitrile and dimethylformamide) was probably due to the disruption of hydrogen bonding between the print molecule and the MIP by the solvent. For atrazine, this was not surprising since it has been shown that this molecule can hydrogen bond to organic acids as well as amides (*6*). In water, we observed a large increase in non-specific binding to the control polymer. This may be caused by a hydrophobic effect, that is, at high amounts of water the print molecule may non-specifically partition into the organic-based polymer. This effect was also reported for other anti-atrazine MIPs in ligand binding experiments (*28*) and using the MIP in HPLC (*29*).

Applications of MIPs for the Analysis of Low Molecular Weight Compounds

Since the initial studies on the development of MIPs for sugar and amino acid derivatives, MIPs for over 20 classes of compounds have been made. Table I is a listing of some of the compounds to which MIPs have been made and their applications. A more comprehensive review has recently been published (*2*). As shown in Table I, most of the work has involved the development of MIPs for sugars, amino acids, and their derivatives.   This research has primarily focused on the separation of racemic mixtures of these compounds by HPLC, which otherwise are cumbersome to separate using conventional solid-phase adsorbents.   Racemic

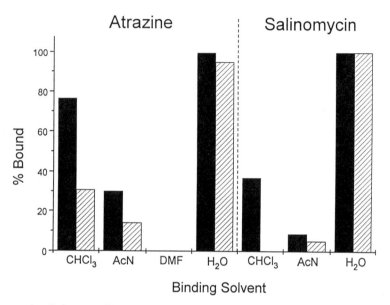

Figure 4.   Solvents effects on analyte binding to specific MIPs and control polymers.   Analytes (atrazine or salinomycin) (10 μg) in 1 mL of either chloroform ($CHCl_3$), acetonitrile (AcN), dimethylformamide (DMF) (atrazine only), or water (H_2O) were incubated with the specific MIP (solid bar) or control polymer (made without print molecule present) (hatched bar) at room temperatrue for 1 hr with agitation. Following centrifugation, supernatants were analyzed for unbound analyte by reversed-phase HPLC. The percent bound to the polymer was calculated as follows: % Bound = [(Amount added - Amount measured in supernatant) / Amount added] x 100.   Atrazine data is from Muldoon, M.T. and Stanker, L.H. *Anal. Chem.* (submitted).   Salinomycin data is from Muldoon, M.T. and Stanker, L.H. (unpublished results).

**Table I. Examples of Various Classes of Compounds to Which MIPs Have Been
Made and Their Applications**

Compound Class	Application	Reference
sugars and derivatives	ligand binding studies, HPLC	7, 8, 17, 23, 25
amino acids and derivatives	ligand binding studies, HPLC, TLC	9, 12, 16, 18, 20, 22, 26, 30-35
peptides and proteins	HPLC	13, 35
therapeutic drugs	ligand binding studies, MIA, sensors	36-38
steroids	ligand binding studies	15
metal ions	sensors, catalysts	39, 40
aromatic hydrocarbons and derivatives	ligand binding studies, HPLC	14, 21, 24
dyes	HPLC	41
phosphonate esters	catalysts	42-44
pesticides	ligand binding studies, MIA, HPLC, sensors	5, 28, 29, 45

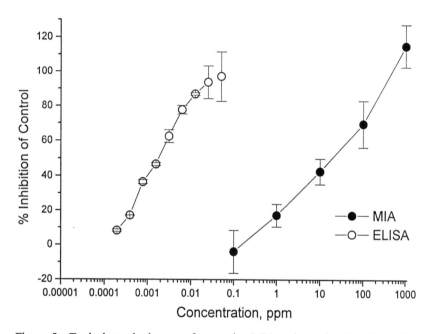

Figure 5. Typical standard curves for atrazine MIA and atrazine ELISA. MIA data is from reference 5.

separations were also accomplished using MIPs as the solid-phase adsorbent for thin-layer chromatography (TLC) (*34*).

The variety of compounds to which MIPs have been made has dramatically increased within the last few years. In addition, the applications in which they have been used has also greatly expanded. Anti-theophylline and -diazepam MIPs were developed and used as the specific adsorbent for a competitive binding assay called a Molecularly-Imprinted Sorbent Assay, or MIA (*36*). This technique is formatted in a similiar manner as a radioimmunoassay (RIA), however MIPs are substituted for antibodies. The sensitivities of the MIAs were in the low micromolar range, which are not quite as sensitive as typical modern immunoassay techniques. A "pseudobiosensor" specific for morphine was prepared using a potentiometric electrode coated with anti-morphine MIP (*37*). Thus, the MIP replaced the biological receptor which is used in a "true" biosensor (such as antibodies and enzymes). A molecularly-imprinted membrane was developed which was specific for theophylline (*38*). The inclusion of analyte-specific binding within a membrane should have wide applications for the development of rapid analytical techniques. A major advantage of using MIPs over biological receptors is their resistance toward environmental and matrix effects. Analyte binding to MIPs can be performed in pure organic solvent, an environment in which most biological receptors do not function.

Molecularly-imprinted polymers have been produced using polymerizable ionophore (metal-binding) complexes that specifically bind particular metal ions (such as calcium and magnesium) (*39*). These were used for the fabrication of ion-selective electrodes. Polymerizable metal ion complexes also were used for the synthesis of polymers with catalytic activities (*40*).

Using "transition state analogues", which are compounds that resemble the structure of high-energy enzyme substrate intermediates, as the print molecules, MIPs with predictable catalytic activities have been made (*42*). These "plastic enzymes", preferentially catalyzed particular reactions depending on the structure of the print molecule used in MIP synthesis (*42-44*). This approach is analogous to using transition state analogues as the immunizing haptens for the generation of antibodies with catalytic activities (*46*).

Development of MIPs for Agrochemicals

The application of MIP technology for the analysis of agrochemicals has recently been described by several groups. Most of this work has focused on the development of MIPs for the herbicide atrazine (*5, 28, 29, 45*). These anti-atrazine MIPs were used as solid-phase adsorbents in MIAs, HPLC, sensors, and solid-phase extraction (SPE) columns (Muldoon, M.T. and Stanker, L.H. *Anal. Chem.*, submitted).

Molecularly-Imprinted Sorbent Assay (MIA). We recently reported on the development of an anti-atrazine MIP and used it in a MIA technique for atrazine (*5*). The binding specificity of the MIP was characterized and shown to be similiar to an immunoassay for the *s*-triazine herbicides (*47*). Analogous to the antibody studies, the anti-atrazine MIP recognized other chlorinated *s*-triazine herbicides such as simazine and propazine. Recognition was diminished for dealkylated metabolites. Figure 5 shows typical standard curves obtained for atrazine using both the MIA and the ELISA. The MIA was sensitive toward atrazine for over two log units (1.0 to 100.0

Table II. Structures of the Various s-Triazines and Their Relative Retention Factors (K) as Measured by Molecularly-Imprinted Sorbent HPLC

Compound	R_1	R_2	R_3	K^a
atrazine	Cl	$NHCH_2CH_3$	$NHCH(CH_3)_2$	1.00
propazine	Cl	$NHCH(CH_3)_2$	$NHCH(CH_3)_2$	1.04
simazine	Cl	$NHCH_2CH_3$	$NHCH_2CH_3$	0.81
cyanazine	Cl	$NHCH_2CH_3$	$NHC(CH_3)_2CN$	0.46
ametryne	SCH_3	$NHCH_2CH_3$	$NHCH(CH_3)_2$	0.36
deethylatrazine	Cl	NH_2	$NHCH(CH_3)_2$	0.60
deisopropyl-atrazine	Cl	$NHCH_2CH_3$	NH_2	0.52
chlorodiamino-s-triazine	Cl	NH_2	NH_2	0.50
hydroxyatrazine	OH	$NHCH_2CH_3$	$NHCH(CH_3)_2$	0.00
ammeline	OH	NH_2	NH_2	0.06
ammelide	OH	OH	NH_2	0.20
cyanuric acid	OH	OH	OH	0.20

[a] The relative retention factors (K) of the atrazine MIP were determined using the polymer as the solid phase for HPLC (Muldoon, M.T. and Stanker, L.H. *Anal. Chem.*, submitted). Analyte capacity factors (k') were calculated according to the equation: $k' = t_R - t_0 / t_0$, where t_R is the retention time of the analyte and t_0 is the retention time of the unretained solute (acetone). The relative retention factors (K) were calculated according to the equation: $K = (k'_{s, imp}/k'_{s, ctl})/(k'_{atr, imp}/k'_{atr, ctl})$, where k'_s is the capacity factor for the test substrate and k'_{atr} is the capacity factor for the imprint molecule (atrazine) on imprinted (imp) and non-imprinted (ctl) polymers.

ppm). The IC_{50}'s for the atrazine MIA and the atrazine ELISA were approximately 10 ppm and 2 ppb, respectively. The MIA was not as sensitive as the ELISA, however the sensitivity of the MIA may be greatly improved with the use of a higher specific activity tracer. Furthermore, the MIA was conducted in pure acetonitrile, whereas the ELISA was carried out in phosphate buffer. The ability to function in concentrated organic solvents has important implications for the analysis of highly lipophilic compounds or crude organic extracts using the MIA method. The MIA technique may alleviate the need for solvent exchange which may facilitate its field adaptability.

Molecularly-Imprinted Sorbent HPLC. Anti-atrazine MIPs have been used as solid-phases for HPLC (*28, 29*). Separation of related *s*-triazine herbicides and metabolites was accomplished using MIPs. Table II lists the relative capacity factors (K) we obtained for the various *s*-triazine herbicides and metabolites shown using an anti-atrazine MIP in HPLC (Muldoon, M.T. and Stanker, L.H. *Anal. Chem.,* submitted). Specific retention on the column was dependent on the presence of a chlorine as well as side-chain alkylation. Specific retention of the hydroxy metabolites was minimal. However, in this application, non-specific binding of the dealkylated metabolites appeared to be greater than in the MIA technique. Similiar results were previously reported by Siemann et al. (*28*) using their anti-atrazine MIP in both HPLC and MIA.

Sensor Techniques. Anti-atrazine MIP membranes which specifically bind atrazine were synthesized (*45, 48*). Atrazine binding was measured as a change in electroconductivity. The use of diethyl aminoethyl methacrylate as the functional monomer gave better results than when methacrylic acid was used in polymer synthesis. In addition, these MIPs appeared to function well in aqueous-based systems. The response and recovery times of the sensors were rather long, 30 min and 12 hr, respectively. However, this may be improved by optimization of the polymerization process and improvements in sensor washing techniques.

Molecularly-Imprinted Sorbent Solid Phase Extraction (MISPE). Most applications of MIPs have been limited to the analysis of standard mixtures and other relatively pure samples. We have recently used an anti-atrazine MIP as the solid-phase for solid-phase extraction of complex biological samples (Muldoon, M.T. and Stanker, L.H. *Anal. Chem.,* submitted). In this application, atrazine was recovered from organic extracts of beef liver tissue. A relatively large volume of dilute crude extract (20 mL) was passed through the column, and then after rinsing the column, atrazine was eluted. This had the effect of removing 93% of the organic-extractable solids. These extracts were analyzed by reversed-phase HPLC and ELISA. Figure 6 shows reversed-phase (C_{18}) HPLC chromatograms of organic extracts from atrazine-spiked beef livers (0.1 ppm) before (top panel) and after (bottom panel) molecularly-imprinted sorbent solid-phase extraction (MISPE). From these chromatograms it is clear that MISPE can be used to recover the analyte from complex biological extracts. By significantly reducing the amount of non-specific material in these extracts, baseline resolution of the atrazine peak was accomplished. This improved atrazine recovery, as measured by HPLC, from approximately 60% prior to MISPE to greater than 70% following MISPE.

We also observed improvements in recovery using an ELISA for atrazine as a result of MISPE cleanup. Figure 7 shows atrazine recovery from beef liver tissues before and after MISPE cleanup of the extracts as determined by the ELISA method.

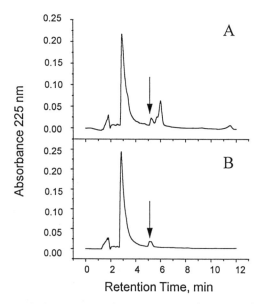

Figure 6.   Reversed-phase HPLC chromatograms of extracts from a 0.1 ppm atrazine-spiked beef liver homogenate before (A) and after MISPE (B).   Extract from a 0.1 ppm atrazine-spiked liver homogenate (in 20 mL chloroform) was applied to the column and rinsed with chloroform.   Atrazine was eluted with 10% (v/v) acetic acid/acetonitrile.   The arrow indicates the atrazine peak.   Data is from Muldoon, M.T. and Stanker, L.H. *Anal. Chem.* (submitted).

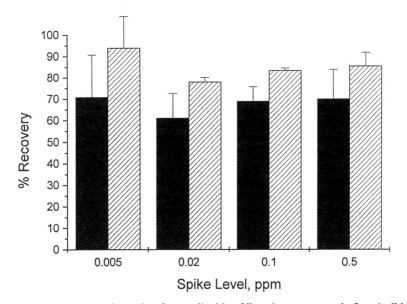

Figure 7. Recovery of atrazine from spiked beef liver homogenates before (solid bars) and after (hatched bars) MISPE as determined by ELISA. Extracts from atrazine-spiked liver homogenates (in 20 mL chloroform) were applied to the columns and rinsed with chloroform. Atrazine was eluted with 10% (v/v) acetic acid/acetonitrile. The error bars represent the standard deviation of the mean from the determination of duplicate extracts. Data is from Muldoon, M.T. and Stanker, L.H. *Anal. Chem.* (submitted).

At spike levels above 0.005 ppm, the MISPE procedure gave improvements in both the accuracy and precision of the ELISA determination. After MISPE, atrazine recoveries were greater than 70% and the average coefficient of variation was 6.7%. These improvements in the determination were probably due to the removal of fats and lipids from the extract by MISPE. Otherwise, the presence of these insoluble substances may sequester the lipophilic analyte in the buffer-reconstituted extract, making the analyte unavailable for antibody binding. The use of MIPs in this application, verified by two different determinative methods, suggested that this technology is robust and may be generally applicable to many other environmentally-important compounds.

Conclusions

Molecular imprinting technology has many applications in residue analysis. Molecularly-imprinted polymers can be used as binding matrices in "stand alone" techniques such as the MIA and in sensors, and as solid-phases for chromatography, such as in HPLC and MISPE. The use of MIPs in the analysis of residues of environmental and agricultural importance is new. This arena offers several analytical challenges that are unique to this field and molecular imprinting technology should find many useful applications. Many of the compounds that are particularly important are present at extremely low levels and are highly lipophilic, such as the dioxins, PCBs, PAHs, and many pesticides. Therefore, analytical procedures often require large sample volumes in order for these compounds to be detected, even by ultra-sensitive techniques such as mass spectrometry or immunoassay. However, since MIPs function in organic solvents, they may be particularly suited for use in conjuction with conventional types of extraction and detection methods. As new analysis situations arise, new alternative approaches to MIP synthesis, such as new functional reagents and polymerization techniques, should further increase the application of MIPs to residue analysis.

Literature Cited

1. Mosbach, K. *Trends Biochem. Sci.* **1994**, *19*, 9-14.
2. Wulff, G. *Agnew. Chem. Ind. Ed. Engl.* **1995**, *34*, 1812-1832.
3. Sherry, J.P. *Crit. Rev. Anal. Chem.* **1992**, *23*, 217-300.
4. Stanker, L.H.; Beier, R.C. In: *Immunoassays for Residue Analysis: Food Safety;* Beier, R.C.; Stanker, L.H., Eds. ACS Symposium Series 621; American Chemical Society: Washington, D.C., 1996; pp 2-16.
5. Muldoon, M.T.; Stanker, L.H. *J. Agric. Food Chem.* **1995**, *43*, 1424-1427.
6. Welhouse, G.J.; Bleam, W.F. *Environ. Sci. Technol.* **1990**, *27*, 500-505.
7. Wulff, G.; Sarhan, A.; Zabrocki, K. *Tetrahedron Lett.* **1973**, *44*, 4329-4332.
8. Wulff, G.; Vesper, W.; Grobe-Einsler, R.; Sarhan, A. *Makromol. Chem.* **1977**, *178*, 2799-2816.
9. Andersson, L.; Sellergren, B.; Mosbach, K. *Tetrahedron Lett.* **1984**, *25*, 5211-5214.
10. Andersson, L.; Sellergren, B.; Mosbach, K. *Tetrahedron Lett.* **1985**, *26*, 3623-3624.

11. Sellergren, B.; Lepisto, M.; Mosbach, K. *J. Am. Chem. Soc.* **1988,** *110,* 5853-5860.

12. O'Shannessy, D.J.; Ekberg, B.; Andersson, L.I.; Mosbach, K., *J. Chromatogr.* **1989,** *470,* 391-399.

13. Andersson, L.I.; O'Shannessy, D.J.; Mosbach, K. *J. Chromatogr.* **1990,** *513,* 167-179.

14. Matsui, J.; Kato, T.; Takeuchi, T.; Suzuki, M.; Yokoyama, K.; Tamiya, E.; Karube, I. *Anal. Chem.* **1993,** *65,* 2223-2224.

15. Whitcombe, M.J.; Rodriquez, M.E.; Villar, P.; & Vulfson, E.N. *J. Am. Chem. Soc.* **1995,** *117,* 7105-7111.

16. Dabulis, K.; Klibanov, A.M. *Biotechnol. Bioeng.* **1992,** *39,* 176-185.

17. Wulff, G.; Kemmerer, R.; Vietmeier, J.; Poll, H. *Nouv. J. Chim.* **1982,** *6,* 681-687.

18. Sarhan, A.; Wulff, G. *Makromol. Chem.* **1982,** *183,* 1603-1614.

19. Wulff, G. In *Polymeric Reagents and Catalysts;* Ford, W.T., Ed.; ACS Symposium Series 308; American Chemical Society: Washington, D.C., 1986, pp 186-230.

20. O'Shannessey, D.J.; Ekberg, B.; Mosbach, K. *Anal. Biochem.* **1989,** *177,* 144-149.

21. Dunkin, I.R.; Lenfield, J.; Sherrington, D.C. *Polymer* **1993,** *34,* 77-84.

22. Ramstrom, O.; Andersson, L.I.; Mosbach, K. *J. Org. Chem.* **1993,** *58,* 7562-7564.

23. Wulff, G.; Schauhoff, S. *J. Org. Chem.* **1991,** *56,* 395-400.

24. Shea, K.J.; Sasaki, D.Y. *J. Am. Chem. Soc.* **1991,** *113,* 4109-4120.

25. Kugimiya, A.; Matsui, J.; Takeuchi, T.; Yano, K.; Muguruma, H.; Elgersma, A.V.; Karube, I. *Anal. Lett.* **1995,** *28,* 2317-2323.

26. Nicholls, I.A.; Ramstrom, O.; Mosbach, K. *J. Chromatogr. A* **1995,** *691,* 349-353.

27. Kempe, M.; Mosbach, K. *J. Chromatogr. A* **1995,** *694,* 3-13.

28. Siemann, M.; Andersson, L.I.; Mosbach, K. *J. Agric. Food Chem.* **1996,** *44,* 141-145.

29. Matsui, J.; Miyoshi, Y.; Doblhoff-Dier, O.; Takeuchi, T. *Anal. Chem.* **1995,** *67,* 4404-4408.

30. Wulff, G.; Gimpel, J. *Makromol. Chem.* **1982,** *183,* 2469-2477.

31. Sellergren, B.; Ekberg, B.; Mosbach, K. *J. Chromatogr.* **1985,** *347,* 1-10.

32. Sellergren, B.; Andersson, L.I. *J. Org. Chem.* **1990,** *55,* 3381-3383.

33. Kempe, M.; Mosbach, K. *Anal. Lett.* **1991,** *24,* 1137-1145.

34. Kriz, D.; Kriz, C.B.; Andersson, L.I.; Mosbach, K. *Anal. Chem.* **1994,** *66,* 2636-2639.

35. Kempe, M.; Mosbach, K. *J. Chromatogr. A* **1995,** *691,* 317-323.

36. Vlatakis, G.; Andersson, L.I.; Muller, R.; Mosbach, K. *Nature* **1993,** *361,* 645-647.

37. Kriz, D.; Mosbach, K. *Anal. Chim. Acta* **1995,** *300,* 71-75.

38. Kobayashi, T.; Wang, H.Y.; Fujii, N. *Chem. Lett.* **1995,** *10,* 927-928.

39. Rosatzin, T.; Andersson, L.I.; Simon, W.; Mosbach, K. *J. Chem. Soc. Perkin Trans. II.* **1991,** *8,* 1261-1265.

40. Gamez, P.; Dunjic, B.; Pinel, C.; Lemaire, M. *Tetrahedron Lett.* **1995,** *36,* 8779-8782.
41. Glad, M.; Norrlow, O.; Sellergren, B.; Siegbahn, N.; Mosbach, K. *J. Chromatogr.* **1985,** *347,* 11-23.
42. Robinson, D.K.; Mosbach, K. *J. Chem. Soc., Chem. Comm.* **1989,** 969-967.
43. Sellergren, B; Shea, K.J. *Tetrahedron: Assym.* **1994,** *5,* 1403-1406.
44. Ohkubo, K.; Urata, Y.; Hirota, S.; Funakoshi, Y.; Sagawa, T.; Usui, S.; Yoshinaga, K. *J. Mol. Catal.* **1995,** *101,* L111-L114.
45. Piletsky, S.A.; Piletskaya, E.V.; Elgersma, A.V.; Yano, K.; Karube, I. *Biosensors and Bioelectronics* **1995,** *10,* 959-965.
46. Schultz, P.G.; Lerner, R.A. *Science* **1995,** *269,* 1835-1842.
47. Karu, A.E.; Harrison, R.O.; Schmidt, D.J.; Clarkson, C.E.; Grassman, J.; Goodrow, M.H.; Lucas, A; Hammock, B.D.; White, R.J.; Van Emon, J.M. In: *Immunoassays for Trace Chemical Analysis: Monitoring Toxic Chemicals in Humans, Foods, and Environment*; Vanderlaan, M., Stanker, L.H., Watkins, B.E., Roberts, D.W., Eds.; ACS Symposium Series 451; American Chemical Society: Washington, D.C., 1991; pp 59-77.
48. Piletsky, S.A.; Parhometz, Yu. P.; Lavryk, N.V.; Panasyuk, T.L.; El'skaya, A.V. *Sensors and Actuators B.* **1994,** *18-19,* 629-631.

Chapter 27

Immunoassay of Cross-Reacting Analytes

G. Jones[1], M. Wortberg[2], D. M. Rocke[1], and Bruce D. Hammock[3]

[1]Center for Statistics in Science and Technology, Graduate School
of Management, University of California, Davis, CA 95616
[2]BASF AG, Agricultural Research Station, APS/US, D–67114
Limburgerhof, Germany
[3]Departments of Entomology and Environmental Toxicology,
University of California, Davis, CA 95616

Cross-reactivity is often seen as a drawback in the use of immunoassay
for environmental analysis. We consider here the problem of
identifying and quantitating single compounds and mixtures from
within a large class of cross-reacting analytes, combining work by
earlier authors on pattern recognition and mixture analysis. Careful
choice of antibodies together with appropriate statistical analysis has
the potential not only to overcome the cross-reactivity problem but to
turn it into an advantage.

A great deal of effort goes into the development of antibodies which are monospecific
for a particular target analyte. Often there is still some cross-reactivity with other
compounds of similar structure, although this may be slight and may be discounted in
the analysis of field samples if the cross-reactants can reasonably be assumed to be
absent. There are many classes of environmental pollutants, however, which can
sometimes occur together in samples and which are so similar in structure that 100%
monospecificty is difficult to obtain. Furthermore, the number of members in the class
may be so large that the development of specific antibodies to each one is extremely
costly, and the subsequent monitoring of each member of the group by a different
assay may be inefficient and slow.

One such class of compounds is the triazine herbicides. Triazines are used in
large quantities throughout wide areas of the world. In 1991, eight different triazines
were in use in the State of California alone; the total amount being nearly 1.4 million
pounds (*1*). Some triazines have been found in groundwater samples from California,
Iowa, Maryland, Nebraska, Pennsylvania and Wisconsin at concentrations above
health advisory levels (*2*). There is a need therefore for a general groundwater
monitoring program for pesticide contamination. This will involve analysis of a large

number of samples for many different pesticides, and will therefore require a simple, inexpensive and rapid method that can identify and quantitate a range of analytes. Enzyme-linked immunosorbent assay (ELISA) has been shown to be a sensitive analytical tool for such pesticide analysis (3), although it is generally thought to be a single-analyte method. By generalizing the format to the use of an array of antibodies with different patterns of affinity for the target class of compounds, it is possible to retain many of the advantages of low cost and rapid process time of single-antibody immunoassay, without the need for complete monospecificity. Moreover the number of antibodies required will typically be less than the number of distinct analytes in the class, so that the procedure is potentially more efficient than the use of many mono-specific antibodies.

Investigation of multiple immunoanalysis using an immunoarray, a panel of less selective antibodies with differing affinity patterns, has been proceeding in two slightly different directions. One is the identification and quantitation of single-analyte samples, where the sample submitted for analysis is assumed to contain one unknown analyte from a large class of possible candidates. The reponses to the immunoarray are used first to identify the analyte and then to estimate the concentration. The general approach is discussed by Kauvar (4). Cheung et al. (5) give some experimental results, using a number of different multivariate statistical techniques to select the identity of test samples. Karu et al. (6) present an overview and evaluation of various alternative methods of analysis. Wortberg et al. (7) describe the construction and application of an immunoarray for the triazine herbicides, concentrating on the problem of selecting a suitable small set of antibodies and giving an approximate statistical criterion on which to base decisions.

An approach with a slightly different emphasis has been that of mixture analysis. Here, samples are assumed to contain possible mixtures of analytes, with the components of the mixture coming from a known small set of cross-reacting analytes (usually no more than four). Mixture analysis requires at least as many antibodies as there are components in the mixture. The approach was proposed by Rocke (8) and implemented by Muldoon et al. (9) for mixtures of three triazines at the high levels typical of pesticide waste rinsate. A model for mixture analysis was given by Jones et al. (10), and implemented successfully for mixtures of up to four analytes at low ppb levels (11).

Either one of these approaches may prove useful in a particular setting, depending on the assumptions one is able to make about the possible identity of the sample and the likely complexity of a mixture. It is helpful, however, to have a unified approach to the modeling and analysis of immunoarray responses, so that appropriate decisions can be made and tested. The situation is of sufficient complexity as to require the close cooperation of immunologists and statisticians to obtain the best possible assay design and a reliable method of data analysis.

The assays used in our description are competitive ELISAs with a coating hapten format conducted on a 96-well microtiter plate, with the response, an optical density, being read by an automatic photometric plate-reader. Details of equipment, reagents and experimental procedures may be found in the literature cited (7). The general ideas we present for modeling and analyzing multiple immunoassay response are not necessarily limited to this format, but may be extended to other assay types with appropriate modification.

Methods

Quantitative immunoassay is a controlled calibration experiment in which standard laboratory-prepared samples of known concentration are assayed together with unknown samples. The responses from the standards are used to estimate a relationship between sample concentration and assay response. This is then inverted to estimate the concentrations of the unknowns from their responses. In the case of multiple immunoassay, the response from each sample is a vector $Y = (Y_1, Y_2, ..., Y_n)$ in which each component is the optical density when the sample is assayed with a different antibody.

Taking the simplest case of two antibodies, the responses (Y_1, Y_2) for each sample can be represented on a two-dimensional plot as in Figure 1. Suppose we have standards for two analytes. Provided the antibodies used have different affinity patterns for the analytes, the responses will tend to occur in different regions of the two-dimensional space. If we now add the response from an unknown sample, its position in this space should inform us as to its identity: it will either "look like" Analyte 1, or like Analyte 2. If it looks like neither, we might conclude that it is a mixture (assuming no other cross-reactants). We need a reasonable method or algorithm for deciding among these alternatives.

This is both a calibration and a discrimination problem, and a number of methods have been suggested for tackling it (6). Some, such as discriminant analysis, have been taken from standard statistical practice, and inevitably ignore some aspects of the problem. In fact much is known already about the nature of immunoassay response. Our approach is to use this knowledge to form an explicit statistical model, with explicitly-stated assumptions, so that coherent rules can be derived regarding the decision of identity of the unknown sample and the estimation of its concentration. More "automatic" methods can also achieve good results, and are attractive because of the apparent lack of need for assumptions or an algorithm. In fact, there is always an algorithm, and there are always assumptions, although these may be hidden from the user and may, if made explicit, be found to be inappropriate.

Response Paths. One obvious fact concerning the multivariate response Y from a single analyte is that, as the concentration increases, the expected response will trace out a path in n-dimensional space, starting from a point representing maximal response at zero concentration and finishing at minimal response (corresponding to non-specific binding). It seems sensible to use these response paths, rather than the points corresponding to a few particular concentrations, to make our decisions. The paths are clearly nonlinear, so that the pattern of responses to the immunoarray changes with analyte concentration; any attempt to use pattern-matching techniques borrowed from the analysis of spectra, as for example in spectroscopy, must take account of this. One could employ some simple smoothing technique, such as smoothing splines, to estimate the paths; again, the use of this automatic method does not necessarily treat errors or departures from the true path caused by experimental variation in an appropriate way. We know that the errors in the individual responses $Y_1,...Y_n$ are independent since they are derived from independent assays, and it is reasonable to assume that the likely size of the error depends on the response. It is often assumed that enzyme immunoassay errors have constant coefficient of variation (cv). The expected response for each individual assay is S-shaped, and can usually be represented reasonably well by a parametric curve (12). We next consider how to build this prior knowledge into our analysis.

Statistical Model. As with any fully-specified statistical model, two components need to be considered in modeling immunoassay response. The first is the deterministic part, which specifies the expected or typical response level Y for a given concentration of the analyte; it is this part which is used to produce the characteristic response path of the analyte. It is usually given as a mathematical function of analyte concentration and some other parameters which depend on the affinity of the antibody for the analyte and other immunochemical conditions. Some practitioners rely on a linear function, restricting the range to "the linear part of the curve". This is not appropriate for multiple immnoassay since even a null response for one component may be informative about the identity of the analyte. A common choice of function for modeling the whole curve is the four-parameter log-logistic model (12):

$$Y = \frac{A - D}{1 + (x / C)^B} + D$$

where x is the analyte concentration, A and D are the maximum and minimum responses (corresponding to zero and infinite concentration respectively), C is the concentration which gives 50% inhibition of the signal (commonly the IC_{50}) and B is a slope parameter. The parameters A, B, C and D have to be estimated from the responses of the standard concentrations, to give the standard or calibration curve. It is common practice to re-estimate the curve for each experiment, although some preliminary work has been done on "borrowing" parameters from another plate (13).

The second component of the model is the stochastic part which represents the part played by experimental error in producing an observed response, and is important in constructing an appropriate estimation procedure and valid decision criteria. For definiteness we will assume a constant cv model in which errors are multiplicative in effect, and use a logarithmic transformation to transform to additive errors with constant variance. Specifically, we model the response Y_i of the ith antibody to a concentration x of analyte j as:

$$\log Y_i = \log \left(\frac{A_i - D_i}{1 + (x / C_{ij})^{B_{ij}}} + D_i \right) + \varepsilon_i$$

and assume that the error term ε_i follows a zero-mean gaussian distribution with constant variance σ_i^2. The parameters A_i, B_{ij}, C_{ij}, D_i and σ_j^2 are estimated from the standards of analyte j assayed with antibody i. We assume that the maximum and minimum binding constants A and D will be the same for each analyte, but will vary with different antibodies.

Given the response vector Y of an unknown sample, we want to determine its identity and then estimate the concentration of that particular analyte. Figure 1 suggests an intuitive approach based on the distance of the sample point from the response paths, namely to choose the analyte whose response path comes closest to the observed sample response, and take the concentration corresponding to the point on that response path where that distance is a minimum. It can be shown that, with a suitably chosen scaling, this intuitive approach corresponds to a standard statistical procedure. In the case of our model above, we would calculate the distance for each component on a log scale (because of the constant cv assumption) and divide by the estimated σ_i so that the difference in precision of different component assays is allowed for. It may also be necessary to adjust for uncertainty in the estimated standard curves. Details of the calculation, along with statistical arguments and simulation studies, are given by Jones and Rocke (Jones, G.; Rocke, D.M. *J. Am. Stat. Ass.*, submitted). With these adjustments, the squared distance (d^2) from the sample point to the correct response path will follow a known statistical distribution, whereas the distance to an incorrect path will be too large. Tabulated values of the appropriate chi-squared distribution can thus be used to decide whether a particular analyte identity is reasonable or not. Either the distance to the response path will be acceptably small or it will be implausibly large.

It is clear from Figure 1 that the method will not reliably differentiate between analytes when the response paths are close together. This occurs inevitably at large and small concentrations, but can also be caused by having similar patterns of cross-reactivity to the antibodies used, so that the response paths of the analytes are always close together. Thus, there will be a range of concentrations, as in single immunoassay, within which analysis is feasible. For this range to be useful in practice, a good combination of antibodies in the immunoarray which can look at the analytes in different ways is required.

Mixture Analysis. If all the single-analyte identities for a given sample are found to be inadequate (i.e. the sample response is too far away from all the individual response paths) the obvious conclusion would be that the sample contains a mixture of analytes. One possible next step would be to try a separation technique such as liquid chromatography (*14*). However, it is also feasible to investigate the possibility of simple mixtures using the immunoassay responses already obtained. The extended four-parameter log-logistic model (*10*) enables us to estimate the response paths (now response surfaces) for mixtures using only the standard curves for the individual analytes. Thus, using the existing data we can now examine possible mixtures to see if they give a plausible identity for our sample. For any given mixture we find the concentrations of the components in the mixture which minimize the distance from the sample point to the response surface for that mixture, and refer the minimum distance to the appropriate statistical distribution.

In practice there will be a very large number of possible mixtures (e.g. with eight individual analytes there will be $2^8-9=247$ different mixtures), so it seems

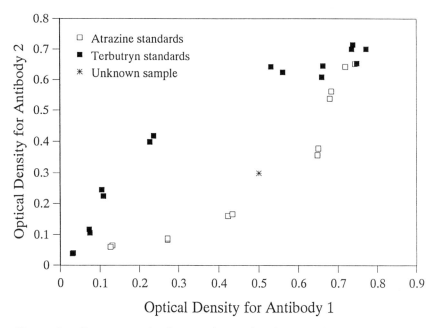

Figure 1. Response paths for atrazine and terbutryn with two antibodies (standard concentrations were a dilution series of each analyte in duplicates, with zeros and blanks).

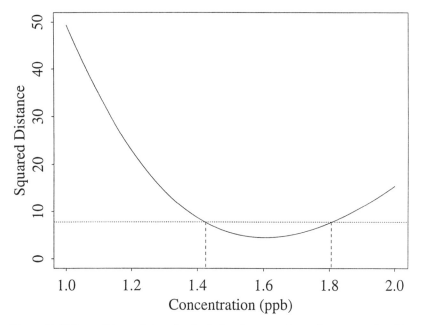

Figure 2. 90% confidence interval calculation for an estimated concentration (the dotted line represents the cutoff value, which is the upper 10% point of a chi-squared distribution with four degrees of freedom).

sensible to order them in some way. If it is believed that simple mixtures, with few components, are more likely than complex ones, the following iterative procedure could be followed:

(1)   Assay unknown samples together with standards for all analytes under consideration using each antibody in turn, with separate microplates being used for each antibody assay.

(2)   Estimate the standard curve parameters A_i, B_{ij}, C_{ij}, D_i and σ_i^2 for each antibody.

(3)   Calculate the minimum distance from the sample point to each single-analyte response path, together with the analyte concentration which gives this minimum. This requires the use of a nonlinear minimization computer routine, many of which are available in standard statistical packages and subroutine libraries.

(4)   By comparing d^2 with the appropriate statistical distribution, find which, if any, of these distances is plausible (see examples below). Report all plausible analyte identities together with the estimated concentration.

(5)   If no single analyte is found to be plausible in (4), return to (3) but substituting binary mixtures for single analytes.

(6)   Repeat until the number of components in the mixture equals the number of antibodies in the immunoarray or the total number of analytes, whichever is smaller.

Results

We have applied the above methodology to single-analyte samples and binary mixtures in the class of triazine herbicides and their derivatives, using immunoarrays of four or five mono- and polyclonal antibodies and up to eight candidate analytes with concentrations in the range 0.5-5.0 ppb. Experimental details are given in a previous work (7). To test our procedure we selected from our library of assays those which were more general in their cross-reactivity for a variety of triazines. In the case of single-analyte samples identification was usually successful and the resulting concentration quite well-estimated, although there was sometimes confusion of identity within subgroups, particularly between prometon, prometryne and terbutryn or atrazine and simazine. This occurs when subgroups of analytes exhibit similar patterns of affinity for the antibodies in the immunoarray, and is perhaps indicative of the need for a better choice of antibodies. Based on our results improved assays have been developed to distinguish among some of these subgroups. For illustrative purposes we show in Table I the complete results of single-analyte analysis (i.e. up to step (4)) for two "unknowns".

For the first sample, containing 1.5 ppb prometryne, the assay procedure is unable to decide between prometon, prometryne and terbutryn: all three have response paths which come reasonably close to the sample response. We can see however that the estimated concentration of prometryne is quite accurate. The second sample has a higher concentration of the analyte, and now the identification is unambiguous, with again a reasonably accurate estimate of concentration. The statistical analysis can be carried a stage further by evaluating the distance over a range of possible concentrations and using a 90% cutoff point to give a confidence interval for the estimate. The procedure is illustrated graphically in Figure 2, the cutoff value (the

upper 10% point of a chi-squared distribution with four degrees of freedom) being
shown as a dotted line.

Table I. Multiple Immunoassay Results for Two Single-Analyte Samples

Analyte:	Sample 1 (1.5 ppb Prometryne)			Sample 2 (5.0 ppb Prometryne)		
	Conc.	d^2	P-value	Conc.	d^2	P-value
Prometon	1.05	0.94	0.816	2.69	68.52	0.000
Atrazine	1.07	214.35	0.000	2.30	414.81	0.000
Simazine	4.10	161.50	0.000	1.91	830.34	0.000
Cyanazine	0.56	415.84	0.000	1.60	848.93	0.000
OH-atrazine	0.01	396.26	0.000	0.02	1197.84	0.000
Prometryne	1.60	4.49	0.213	4.12	2.05	0.562
Terbutryn	1.31	1.41	0.703	3.29	12.75	0.005
DEatrazine	0.01	395.74	0.000	0.05	1197.40	0.000

Samples containing binary mixtures of analytes are, as expected, more
difficult to identify. Often there were a number of possible identities for the samples,
with the problem of confusion within subgroups being compounded. Thus for
example a mixture of atrazine and prometon might look like simazine and terbutryn,
or atrazine and prometryne. In most cases, mixtures were clearly identified as such,
i.e. not as single analytes, although mixtures of prometon and terbutryn, or
prometryne and terbutryn, were incorrectly classified as containing terbutryn only.
Table II shows the acceptable results for a mixture of 1 ppb atrazine with 1 ppb
cynazine. There were eight candidate analytes, so 28 possible binary mixtures. The
computer searches through all 28, looking for acceptable solutions. There were three
acceptable solutions found, one being the correct identity.

Table II. Multiple Immunoassay Results for 1 ppb Atrazine + 1 ppb Cyanazine

Analyte 1:	Analyte 2:	Conc. 1	Conc. 2	d^2	P-value
Atrazine	Cyanazine	0.35	1.61	1.41	0.494
Simazine	Cyanazine	0.38	1.84	1.28	0.527
Cyanazine	DIatrazine	2.03	2.46	1.16	0.560

As with single-analyte solutions, we can evaluate the distance over a grid of
values to produce a confidence region for the concentration estimates. If we add
contour lines corresponding to different distances, we can see how the estimation of
one analyte concentration affects the other (see Figure 3). In the case of atrazine and
cyanazine the estimates are correlated so that over-estimation of one component
causes under-estimation of the other component. We can see that the point estimates

of the atrazine and cyanazine concentrations are not very accurate, but that the total amount is well-estimated.

Discussion

We have demonstrated an approach to the immunoanalysis of samples containing one or more members of a group of cross-reacting analytes, using a parametric model for the assay responses and making some explicit assumptions about the way in which experimental error affects them. Other approaches could also be taken. For example, neural networks have also been applied to the analysis, with the inputs being either the untransformed responses (6) or an estimated concentration of a chosen reference analyte (15). The results so far have been similar to ours in terms of correct identification and concentration accuracy, although with neural networks the algorithm and assumptions are hidden from the user so that there is no way of telling whether they are appropriate, and no way of applying statistical tests or calculating confidence intervals. It seems that the most important limiting factor on the reliability of these methods at present is the discriminatory power of the immunoarray used. We now consider some aspects of this.

Since mixture analysis requires at least as many antibodies in the immunoarray as there are components in the mixture, it might be supposed that our antibody array should be as large as the class of candidate analytes. If this were the case, we would clearly be better off using the same number of monospecific antibodies, and this multiple immunoanalysis would be useful only for those cases where specific antibodies were unavailable. However, there may be situations in which one might want to perform the kind of analysis described above, in which a small immunoarray is capable of differentiating between subgroups of the analyte class. If necessary more specific antibodies could be used at a second stage for any samples not clearly identified at the first stage, or else the ambiguous cases could be submitted to a different form of analysis. Alternatively, if one is prepared to believe that simple mixtures, with only two or three components at the most, are more likely than complex mixtures, then a small immunoarray would again suffice, with any unresolved samples being submitted to a second stage. Such a procedure could be much more efficient than many single immunoassays while still providing an effective analysis for the majority of samples.

One might suppose that more antibodies would always give better resolution, but this is not necessarily the case. Each additional antibody adds both signal and noise to the system, and if the cross-reactivity pattern of the new antibody is not sufficiently different from those already present, the additional information it provides will be swamped by the increase in noise, leading to a deterioration in performance. The most important consideration then is to choose antibodies which contribute independent or near-independent information; geometrically this means that their cross-reactivity patterns, measured perhaps by their IC_{50} "spectra", should be as close to orthogonal as possible. It would be relatively straightforward to construct a mathematical optimality criterion along these lines, and by searching through all possible choices to construct an "optimal" immunoarray, but the nonlinearity complicates this. In practice, each assay has a limited workable range, and the

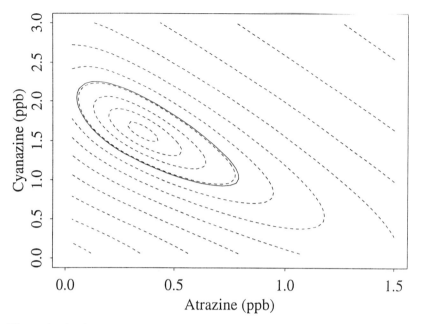

Figure 3. 90% confidence region for a binary concentration estimate (the true value of (1,1) is inside the 99% region).

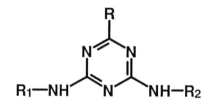

triazine	R	R$_1$	R$_2$
atrazine	Cl	isopropyl-	ethyl-
propazine	Cl	isopropyl-	isopropyl-
simazine	Cl	ethyl-	ethyl-
cyanazine	Cl	ethyl-	NHCCN(CH$_3$)$_2$
prometryn	S-CH$_3$	isopropyl-	isopropyl-
prometon	O-CH$_3$	isopropyl-	isopropyl-
terbutryn	S-CH$_3$	tert. butyl-	ethyl-

Figure 4. Molecular structure of triazine herbicides.

overlapping or intersection of these ranges is also important. It would probably be advisable to weaken the sensitivity of some of the assays in order to have a suitable range of accurate identification and quantitation for each analyte.

The performance of an immunoarray itself provides information on where the resolution is weakest, for example in the existence of indistinguishable subgroups, thus helping to direct research toward the development of reagents most valuable for multianalyte determination. In practice the immunoarray might be tailored to fit the requirements of a particular geographical location based on known triazine usage. Basing our procedure on a fully-specified statistical model enables decisions to be made concerning the performance of a particular hypothetical immunoarray, including whether it would be desirable to add more antibodies to an existing array.

A complementary approach to the development of immunoarrays is to consider the molecular structures of the analytes; antibodies can perhaps be chosen or developed to bind to particular molecular moieties or substituents, so that classification uses recognition of details of the molecular structure of the analytes. This allows the possibility of designing an immunoarray with deliberately-created less-specific antibodies so that the cross-reactivities are used positively in an efficient assay system. The concept can be well illustrated by using the triazine example (see Figure 4): antibodies capable of distinguishing among the common substituents R at C-2 on the triazine heterocycle will define several major classes while other antibodies selective for R1NH or R2NH groups at C-4 and C-6 will define subclasses. Such work, requiring the close collaboration of immunologists, synthetic chemists and statisticians, could provide another useful immunochemical tool to assist in the efficient and cost-effective analysis of complex environmental samples.

Acknowledgments

We wish to thank B. Hock, T. Giersch, R.O. Harrison and A. E. Karu for donating antibodies, and M. Goodrow for providing triazine derivatives. This research was supported by NIEHS Superfund 2P42-ES04699, NSF DMS 93-01344, NSF DMS 94-06193, NSF DMS 95-10511, US EPA CR 819047, USDA Forest Service NAPIAP R8-27, Center for Ecological Health Research CR 819658, NIEHS Center for Environmental Health Sciences IP30-ES05707 and Water Resources Center grant No. W-840.

Literature Cited

1.   California Environmental Protection Agency, Department of Pesticide Regulation, Pesticide Use Report, Annual Manual, Sacramento, CA, 1992.
2.   Cohen, S.Z.; Eiden, C.; Lorber, M.N. In *Evaluation of Pesticides in Ground Water*; Garner, W.J.; Honeycutt, R.C.; Nigg, H.N., Eds.; ACS Symposium Series 315; American Chemical Society: Washington, DC, 1986, pp 170-196.
3.   Vanderlaan, M.; Watkins, B.E.; Stanker, L.H. *Environ. Sci. Technol.* **1988**, 22, 247-254.
4.   Kauvar, L.M. In *New Frontiers in Agrochemical Immunoassay*; Kurtz, D.A.; Sherritt, J.H., Stanker, L., Eds.; AOAC International: Arlington, VA, 1995, pp 305-312.

5.   Cheung, P.Y.K.; Kauvar, L.M.; Engqvist-Goldstein, A.E.; Ambler, S.M.; Karu, A.E.; Ramos, S. *Anal. Chim. Acta* **1993**, *282*, 181-191.

6.   Karu, A.E.; Lin, T.H.; Breiman, L.; Muldoon, M.Y.; Hsu, J. *Food & Agric. Immunol.* **1994**, *6*, 371-384.

7.   Wortberg, M.; Jones, G.; Kreissig, S.B.; Rocke, D.M.; Gee, S.J.; Hammock, B.D. *Anal. Chim. Acta* **1996**, *319*, 291.

8.   Rocke D.M. Statistical design of immunoassay protocols, presented at the American Chemical Society Symposium on Emerging Pesticide Residue Analysis Techniques, San Francisco, April 1992.

9.   Muldoon, M.T.; Friese, G.F.; Nelson, J.O. *J. Agric. Food Chem.* **1993**, *41*, 322-328.

10. Jones, G.; Wortberg, M.; Kreissig, S.B.; Gee, S.J.; Hammock, B.D.; Rocke, D.M. *J. Immunol. Meth.* **1994**, *177*, 1-7.

11. Wortberg, M.; Kreissig, S.B.; Jones, G.; Rocke, D.M.; Hammock, B.D. *Anal. Chim. Acta* **1995**, *304*, 339-352.

12. Rodbard, D. In *Ligand Assay: Analysis of International Developments on Isotopic and Nonisotopic Immunoassay*, Masson: New York, 1981, pp 45-99.

13   Jones, G.; Wortberg, M.; Kreissig, S.B.; Hammock, B.D.; Rocke, D.M. *Anal. Chim Acta* **1995**, *313*, 197-207.

14. Kramer, P.M.; Li, Q.X.; Hammock, B.D. *J. AOAC Int.* **1994**, *77*, 1275-1287.

15. Wittman, C. Application of neural networks for pattern recognition of pesticides in water samples by different immunochemical techniques, presented at the American Chemical Society Symposium on Development and Applications of Immunoassays for Environmental Analysis, New Orleans, March 1996.

Chapter 28

Application of a Neural Network for Pattern Recognition of Pesticides in Water Samples by Different Immunochemical Techniques

C. Wittmann[1], R. D. Schmid[2], S. Löffler[3], and A. Zell[3]

[1]Department of Food Technology, Technical College of Neubrandenburg, 17033 Neubrandenburg, Germany
[2]Institute for Technical Biochemistry, University of Stuttgart, 70569 Stuttgart, Germany
[3]Department of Rechnerarchitektur, Wilhelm-Schickard Institut für Informatik, University of Tübingen, 72074 Tübingen, Germany

The application of a neural network was started using the enzyme immunoassay format (ELISA) followed by a dipstick immunoassay (IA). Training was performed with the Stuttgart Neural Net Simulator (SNNS). A simulation model was used to increase the training data set. Training was carried out with the following learning procedures: Backpropagation, Resilient Propagation and Cascade Correlation. The best configuration was a net with 2 hidden layers with 15 neurons in total. The best learning method was Standard Backpropagation (learning parameter: 0.2). All environmental water samples containing none of the three s-triazines were correctly determined as zero samples. The correct s-triazine classification of environmental samples containing different analyte mixtures was possible in 80% of all cases. The dipstick IA format could be successfully used in combination with the neural network data to correctly identify terbuthylazine in 100% of the cases as the only s-triazine contaminant in environmental water samples.

During the last 10 years university research groups and several companies came up with a series of highly sensitive enzyme immunoassays for the analysis of pesticides in drinking water (*1, 2, 3, 4, 5*). Despite the obvious success of IAs this approach is not yet considered an established methodology in the field of environmental analytics, comparable to the classic gas chromatography (GC) or high pressure liquid chromatography (HPLC) methods. This argument especially holds true for Europe. One reason is the availability of only about 100 existing pesticide IAs compared with the number of applied active ingredients which amounts to ca. 300, excluding the relevant metabolites. This bottleneck can only be opened by more efficient methods of antibody production using recombinant techniques. The second restraint pertains to the habit of erroneously classifying IAs among biological methods including their assessment as imprecise and/or unreliable tools. In addition, comparability of results is required for long-term or long-range monitoring programs, for instance if the quality change of rivers or lakes is to be judged. It is obvious that for practical purposes group-specific assays providing a sum parameter would be most attractive. However, this goal appears to be the most difficult one for reasons that most often the given antibody cross-reacts to different extents with compounds of similar structure as the analyte. In this case no classification of the analyte present in the sample is possible. The cross-reactivity can be calculated starting from the normalized % B/B_0 values obtained by the following formula:

$$\%B / B_0 = \frac{A - A_E}{A_0 - A_E} (100) \tag{1}$$

where:
%B/B_0: ratio (in %) of enzyme tracer binding to the pesticide-specific antibodies in the presence of a defined pesticide concentration (B) in relation to the absence of pesticide (B_0).
A: absorption of the sample or standard
A_E: absorption at the excess concentration
A_0: absorption of the zero concentration
and the cross-reactivities were then determined comparing the mid-points of the assay for different compounds:

$$\%cross - reactivity = \frac{analyte\,conc.\ at\,50\%B / B_0}{conc.\ of\,cross - reacting\,compounds\,at\,50\%B / B_0} (100) \tag{2}$$

One possibility was to use the cross-reactivities to identify and correctly analyze the different compounds from analyte mixtures in the samples. This would, at least in theory, only be possible under the prerequisite that the standard curves for cross-reacting compounds show a parallel curvature to the one for the analyte, i.e. that the slope is identical. If one uses monoclonal antibodies, the sensitivities of the standard curves for different compounds is in many instances identical.

Another approach was tested by Wortberg et al. (6) in applying an extended 4-parameter logistic function. The four-parameter logistic model is often used to calculate data from the sigmoidal standard curves and has the following equation:

$$y = \frac{a - d}{1 + \left(\dfrac{x}{c}\right)^b} + d \tag{3}$$

where:
y: ELISA response (e.g. absorption)
x: analyte concentration
a, b, c, d: four constants; a, d: upper and lower asymptote of the curve; c: analyte concentration at the mid-point of the assay; b: the slope at the mid-point of the assay.
Derived from this model an extended 4-parameter logistic model for two different analytes would follow the extended equation:

$$y = \frac{a - d}{1 + \left[\left(\frac{x_1}{c_1}\right)^{\frac{b_1}{b^*}} + \left(\frac{x_2}{c_2}\right)^{\frac{b_2}{b^*}}\right]^{b^*}} + d \qquad (4)$$

where:
x_1, x_2: analytes 1 and 2 in 2 different concentrations
b_1, b_2: different slopes of the curves for analytes 1 and 2
b^*: "average" slope, i. e. geometric mean of the slope parameters of the individual analytes.

If there is no possibility to use a mathematical model to describe and/or to predict the standard curve for an analyte mixture, then this principal drawback can be overcome by the application of a neural network (7, 8) for pattern recognition enabling a multiresidue screening.

Experimental

Materials.

Chemicals. Triazine standards were kindly provided by Riedel de Haen (Seelze, Germany). The monoclonal antibodies K4E7, K1F4, and P6A7 were a generous gift from Dr. Thomas Giersch, Technical University München (Weihenstephan), Department of Botany, Freising, Germany. The enzyme tracer was prepared using, as the haptens, the two s-triazine derivatives, 4-chloro-6-(isopropylamino)-1,3,5-triazine-2-(6-aminohexanecarboxylic acid) for the K4E7 and the P6A7 assay and 4-chloro-6-(tert-butylamino)-1,3,5-triazine-2-(6-aminohexanecarboxylic acid) for the K1F4. In addition, the following reagents were used: N,N`-dicyclohexylcarbodiimide (DCC, Sigma, Deisenhofen, Germany), dioctylsulfosuccinate, sodium salt (DSS) (Sigma, Deisenhofen, Germany), absolute ethanol, p. a. (Merck, Darmstadt, Germany), goat anti-mouse immunoglobulin G (IgG) (Sigma), horseradish peroxidase (HRP, 1350 U/mg = 22,505 nkat; Serva, Heidelberg, Germany), hydrogen peroxide, 30% (Merck), N-hydroxysuccinimide (NHS; Aldrich, Gillingham, Dorset, UK), poly(oxyethylenesorbitan)monolaurate (Tween 20; Merck), and 3,3`,5,5`-tetramethylbenzidine (TMB) (Sigma). All other reagents were of the highest purity grade available.

Buffers and solutions. The following were used: (1) carbonate buffer, 50 mmol/L, pH 9.6, for coating; (2) phosphate-buffered saline (PBS), 40 mmol/L, pH 7.2 (containing 8.5 g/L NaCl), for the dilution of the peroxidase tracer; (3) PBS washing buffer, 4 mmol/L, pH 7.2 (containing 0.85 g/L NaCl and 0.5 mL/L Tween 20), for washing the microtiter plates and the dipsticks; (4) TMB substrate, (a) for the microtiter plate assay: 400 µL of TMB stock solution [6 mg of TMB were dissolved in 1 mL of dimethyl sulfoxide (DMSO)] + 100 µL of 1% (v/v) hydrogen peroxide solution in 25 mL of 0.1 mol/L acetate solution, pH 5.5 (pH adjusted by addition of citric acid); (b) for dipstick assay: 400 µL of TMB stock solution [6 mg of TMB were dissolved in 1 mL of dimethyl sulfoxide (DMSO)] + 1 mL of DSS solution (8 mg of DSS per mL of ethanol) + 3.6 mL of 0.1 mol/L acetate solution, pH 5.5 (pH adjusted by addition of citric acid) + 50 µL of 1% hydrogen peroxide solution.

Preparation of standards. Ten milligrams of the s-triazine compounds were dissolved in 10 mL of absolute ethanol with the aid of an ultrasonic bath (1 min). Starting with this solution, a standard series was prepared in distilled water containing the following analyte concentrations: 0.01, 0.03, 0.1, 0.3, 1, 10, 100, 1000, and 10,000 µg/L.

Equipment. The laboratory equipment used consisted of a photometer for 96-well microtiter plates (ICN, Munich, Germany), a microtiter plate washer with 96 channels (ICN), an ultrasonic bath (Sonorex Bandelin, Bandelin Electronic, Berlin Germany), and an RQflex reflectometer (Merck) for dipstick IA measurement including the barcode "testroutine" to measure the output of the absorption values in percent, in the transmission mode, at 657 nm.

Support materials. These included the following: 96-well microtiter plates, type F-form, high binding capacity (Nunc, Roskilde, Denmark), and the Biodyne B membrane (PALL, Dreieich, Germany).

Methods.

(a) ELISA performance. The ELISAs were performed as described by Giersch (9). The microtiter plates were precoated overnight with 300 µL/well of goat-anti mouse IgG (5 µg/mL carbonate buffer) at 4°C. After a washing step with PBS washing buffer, 300 µL of the monoclonal antibodies (K4E7, dilution 1:20,000; K1F4, dilution 1:5000; P6A7, dilution 1:20,000) were added per well to the microtiter plate. After incubation at room temperature for 2 h, unbound antibodies were washed off with PBS washing buffer. 250 µL aliquots of standard or sample were added together with 50 µL aliquots of the respective enzyme tracer (for the K4E7 and the P6A7 test in a dilution of 1:20,000; for the K1F4 test in a dilution of 1:10,000). After a 60-min incubation, the plates were washed and 200 µL/well TMB substrate was added. The substrate reaction was stopped after 5 min in the case of the K4E7 assay and after 25 min in the case of the K1F4 and the P6A7 assay with 100 µL of 2 mol/L H$_2$SO$_4$, and the absorption was measured at 450 nm with the ELISA reader.

(b) Dipstick IA performance. The same protocol as described by Wittmann et al. (10) was investigated. All incubation steps were performed in 2 mL plastic tubes. For the immunoreaction, the test strips were incubated with a mixture of 800 µL of the standard or sample and 200 µL of the enzyme tracer (dilution 1:20,000) for 10 min. After washing three times with PBS washing buffer, 800 µL of TMB enzyme substrate were added and incubation was continued for a further 10 min. The total assay time was about 25 min. The dip-sticks were taken out of the tubes and the absorption of the coloured product was immediately measured with the RQflex reflectometer at 657 nm (red LED).

SNNS (Stuttgart Neural Network Simulator)

SNNS is a simulator for neural networks. The SNNS was installed on a DEC station 3100 with Ultrix V4.2 as the operating system (SNNS needs an ANSI-C compiler to run). Most current neural network architectures do not try to closely imitate their biological model but rather can be regarded simply as a class of parallel algorithms (7, 8). In these models, knowledge is usually distributed throughout the net and is stored in the structure of the topology and the weights of the links. The networks are organized by automated training methods which greatly simplify the development of specific applications. The inherent fault tolerance of connectionist models is another advantage. Furthermore, neural nets can be made tolerant against noise in the input: with increased noise, the quality of the output usually degrades only slowly.

Neural Network Terminology. A network consists of units and directed, weighted links between them. One can distinguish three types of units: input units, output units and hidden units describing the topology of the net. The actual information processing within the units is modeled in the SNNS simulator with the activation function and the output function. The activation function first computes the net input of the unit from the weighted output values of prior units. It then computes the new activation from this net input and possibly its previous activation. The general formula of the activation function f_{act} (cf. Figure 1) is:

$$a_j(t) = f_{act}(net_j(t) - \theta_j) \tag{5}$$

where:
$a_j(t)$: activation of unit j in step t
$net_j(t)$: net input in unit j in step t
θ_j: threshold (bias) of unit j.
In many cases a sigmoidal activation function is used, e. g.:

$$f_{act}(x) = \frac{1}{1 + e^{-x}} \tag{6}$$

This yields the logistic activation function.
The output function f_{out} takes the result of the activation function to generate the output of the unit:

$$o_j(t) = f_{out}(a_j(t)) \tag{7}$$

where:
$a_j(t)$: activation of unit j in step t
$o_j(t)$: output of unit j in step t
j: index for all units of the net

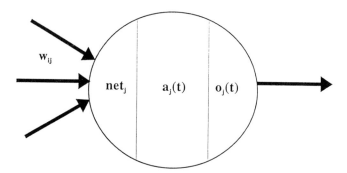

Figure 1. Model of a neuron. It consists of an input value netj(t), has an activation aj(t) and an output value oj(t). Weighted links (wij) connect the neuron with following neurons.

Feedforward neural network

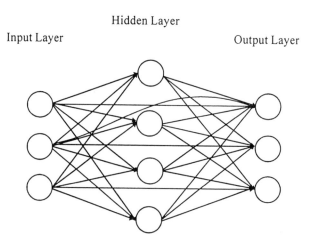

Figure 2. Feedforward net with an input layer, an output layer, and the hidden layer. All units of the input layer are connected with every unit of the following hidden layers which are then fully connected with the neurons of the subsequent output layer. Additional shortcut connections are added linking the input with the output layer directly.

If the activation function already incorporates a nonlinearity, as with the logistic activation function, the output function is usually set to be the identity function, giving:

$$o_j(t) = a_j(t) = f_{act}(net_j(t) - \theta_j) \tag{8}$$

The direction of a connection shows the direction of the transfer of activation. Each connection has a weight (or strength) assigned to it. Weights are represented as floats with nine decimal digits of precision.

Feedforward nets (cf. Figure 2):
The most frequently used network architecture is built hierarchically bottom-up. The input into a unit comes only from the units of preceding layers. Because of the unidirectional flow of information within the net they are also called feed-forward nets. The kernel sorts the units by their topology. This order corresponds to the natural propagation of activity from input to output.

Learning in Neural Nets. One of the major advantages of neural nets is their ability to generalize. This means that a trained net could classify data from the same class as the learning data that it has never seen before. In real world applications developers normally have only a small part of all possible patterns for the generation of a neural net. To reach the best generalization, the data set should be split into three parts:
1. The training set is used to train a neural net. The error of this data set is minimized during training.
2. The validation set is used to determine the performance of a neural network on patterns that are not trained during learning.
3. A test set for finally checking the overall performance of a neural net.
The learning should be stopped in the minimum of the validation set error. At this point the net generalizes best. When learning is not stopped, overtraining occurs and the performance of the net on the whole data decreases, despite the fact that the error on the training data still gets smaller.

An important focus of neural network research is the question of how to adjust the weights of the links to get the desired system behavior. Training of a feed-forward neural network with supervised learning consists of the following procedure: An input pattern is presented to the network. The input is then propagated forward in the net until activation reaches the output layer. This constitutes the so called forward propagation phase. The output of the output layer is then compared with the teaching input. The error, i.e. the difference (δ_j) between the output (o_j) and the teaching input (t_j) of a target output unit (j) is then used together with the output (o_i) of the source unit (i) to compute the necessary changes of the link (w_{ij}). To compare the differences of inner units for which no teaching input is available (units of hidden layers) the differences of the following layer, which are already computed, are used in a formula given below. In this way the errors (differences) are propagated backward, so this phase is called backward propagation. The general formula for standard backpropagation (online version) is:

$$\Delta_p w_{ij} = \eta\, o_{pi}\, \delta_{pj} \tag{9}$$

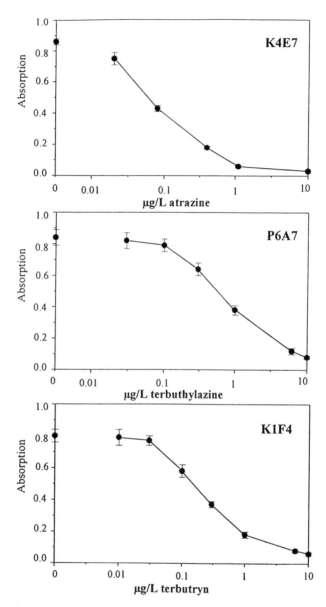

Figure 3. Standard curves for atrazine (mab K4E7), terbutryn (mab K1F4), and terbuthylazine (mab P6A7). The ELISA performance is described in the Materials and Methods section.

where:

δ_{pj}: error (difference between the real output and the teaching input) of unit j for pattern p

o_{pi}: output of the preceding unit i for pattern p

η: learning factor eta (a constant)

There are several variations of standard backpropagation as backpropagation with momentum term and backpropagation with weight decay.

Backpropagation with momentum term uses a momentum term (α) introducing the old weight change as a parameter for the computation of the new weight change. This avoids problems common with the regular backpropagation algorithm when the error surface has a very narrow or extremely flat minimum area. The new weight change is computed by a formula where these enhancements effect that flat spots of the error surface are traversed relatively rapidly with a few big steps, while the step size is decreased as the surface gets rougher. This adaptation of the step size may increase learning speed significantly.

Backpropagation with weight decay decreases the weights of the links while training them with backpropagation. In addition to each update of a weight by backpropagation, the weight is decreased by a part d of its old value.

Rprop (resilient propagation) is a local adaptive learning scheme, performing supervised batch learning in multi-layer perceptrons. The basic principle of Rprop is to eliminate the harmful influence of the size of the partial derivative on the weight step. As a consequence, only the sign of the derivative is considered to indicate the direction of the weight update. The size of the weight change is exclusively determined by a weight-specific so called "update value" $\Delta_{ij}(t)$. Every time the partial derivative of the corresponding weight w_{ij} changes its sign, which indicates that the last update was too big and the algorithm has jumped over a local minimum, the update value is decreased by the factor η^-. If the derivative retains its sign, the update value is slightly increased in order to accelerate convergence in shallow regions. Additionally, in case of a change in sign, there is no adaptation in the succeeding learning step.

Cascade correlation is characterized as a constructive learning rule. It starts with a minimal network, consisting only of an input and an output layer. Minimizing the overall error of a net, it adds step by step new hidden units to the hidden layer. Cascade correlation (CC) is a supervised learning architecture which builds a near minimal multi-layer network topology. CC combines two ideas: The first is the cascade architecture, in which hidden units are added only one at a time and whose input links do not change after they have been added. The second is the learning algorithm, which creates and installs the new hidden units. For each new hidden unit, the algorithm tries to maximize the magnitude of the correlation between the new unit's output and the residual error signal of the net. As embedded learning algorithms to train the new hidden unit backpropagation, quickprop or Rprop are available in SNNS. The two advantages of this architecture are that there is no need for a user to worry about the topology of the network, and that frequently CC learns much faster than the other learning algorithm.

Generalization.   For neural network application at first the ELISA format was applied. The three ELISAs for atrazine, terbutryn and terbuthylazine determination were studied to distinguish between the 3 s-triazines. The standard curves obtained with the three different monoclonal antibody based ELISAs and the cross-reactivities are shown in Figures 3 and 4.

1. For the training set the 6 analyte concentrations (0, 0.01, 0.1, 0.5, 1, 5 µg/L) were prepared using tap water from Stuttgart (s-triazines < 0.025 µg/L as checked by GC/MS analysis) for all possible mixtures of atrazine, terbutryn and terbuthylazine (i. e. $6^3 = 216$ training patterns).

2. For the validation set on the basis of the 216 patterns, 162 were taken as the training patterns and the remaining 54 patterns were used for the internal validation set.

3. As the test set 52 real environmental water samples (from different origins in Germany, and 11 samples from the Veneto area, Italy) and 64 spiked water samples (40 surface water and 22 tap water samples) were applied. The latter samples were spiked in concentrations between 0.01 and 5 µg/L with the 3 analytes and thus, were complete mixtures.

The input values for the neural net were modified starting with the logarithmic results of the ELISA test. The values are transformed due to the logistic activation function used to lie between the interval [0, 1]. The transformation was performed:

$$i_{new} = \frac{\log_{old} - \log_{min}}{\log_{max} - \log_{min}} \tag{10}$$

where:

i_{new}: input value for the neural network

$\log_{min}$ and $\log_{max}$: lower and upper limit of range to be transformed

$\log_{old}$: value calculated from ELISA measurement

As output values, the concentrations of the training patterns and the GC/MS data for the real environmental water samples were used.

Simulation of training data. Because of the time needed to prepare and measure the 216 training patterns with the three ELISAs, a simulation model modified from Muldoon et al. (11) was developed. A linear system was used calculating the expected concentrations with the help of the cross-reactivities. The calculated mean cross-reactivities were transferred to the reactivity coefficients used in the following equation by dividing the percent values by a factor of 100. The concentrations measured were composed of the weighted sum of the single concentrations:

$$i_A = o_A + 0.003 \, o_T + 0.26 \, o_{TBA}$$
$$i_T = 0.14 \, o_A + o_T + 0.43 \, o_{TBA} \tag{11}$$
$$i_{TBA} = 0.07 \, o_A + o_{TBA}$$

where:

i: input data, o: output data, A: atrazine, T: terbutryn, TBA: terbuthylazine

The same concentrations as for the measured training patterns were selected. Transformation was performed equally as for the measured data.

Parameters tested. The first step is to find a suitable network topology. With this topology different learning methods (backpropagation, variations of backpropagation, Rprop and Cascade Correlation; cf. paragraph: Learning in Neural Nets) were investigated. For each learning method different learning parameters were compared. For training with standard

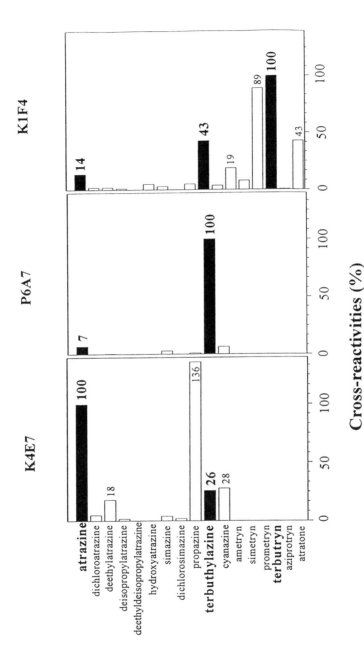

Figure 4. Cross-reactivities of the 3 different monoclonal antibody based ELISAs studying 17 different s-traizine compounds.

backpropagation, the learning factor η was varied from 0.1 to 0.4. Using backpropagation with momentum term an α-value of 0, 0.2, 0.4, and 0.6 was investigated. For training with backpropagation with flat-spot elimination, the constant was set to 0.1. Using backpropagation with weight decay the part d was set to 0.001 and 0.0001 and its influence was studied. For the training with Rprop, the weight change Δ_0 was varied between 0.1 and 0.3 and α-values between 0.001 and 0.00001 were investigated. As embedded learning methods for the training with Cascade correlation, backpropagation and Rprop were studied. In addition, a comparison of the network with the simulated data was performed with the one for the measured training patterns. For evaluation of training runs SNNS contains a tool called "analyze" which calculates the total error of the net. The purpose of this tool is to analyze the result files that have been created by SNNS. The result file has to contain the teaching output and the output of the network. In our case this tool was modified in a way that the mean deviation of the output neurons was calculated. In addition, the mean and maximum deviation of the backtransformed output values were estimated. Therefore, the nets could be compared and the optimized network configuration could be determined as the one exhibiting the lowest mean deviation (prerequisite: always the same number of training patterns is used).

In order to work with various learning algorithms, different initialization functions that initialize the components of a net are required. The connections between the neurons were initialized with random weights from the interval [-1, 1] before training. As different initializations may give different qualities in the results, for each topology first a suitable initialization was searched. The best of 5 initializations was then chosen for training. The criterium was the summed square error after 5000 learning cycles. The net is trained for 5000 cycles with the learning method backpropagation and the learning factor η=0.2.

1. Topology. A feedforward net with 3 input and 3 output units was the basis to find out the ideal topology. Tested were one and more hidden layers, and additional shortcut connections in subsequence. The mean and maximum deviation for the backward propagated output values were compared.

2. Learning methods. The following learning methods were studied:
- standard backpropagation
- backpropagation with momentum term, with flat-spot elimination, and with weight decay
- Resilient Propagation (Rprop)
- Cascade Correlation: learning with backpropagation, Rprop.

Results and Discussion

The best topology for the measured data was a net with two hidden layers with 5 units in the first and 10 in the second layer. The net was completely connected and did not contain any shortcut connections (cf. Figure 5). Comparing different learning methods, the best results were obtained with standard backpropagation. This was surprising as it was expected that variations of standard backpropagation, Rprop or at least Cascade correlation would decrease the mean and maximum deviation of the output neurons for the training and the test data. An interesting point was that we observed an effect on the monoclonal antibodies K4E7 and especially P6A7 when several analyte mixtures were measured in the ELISA. With increasing amounts of an s-triazine (in most cases mainly terbuthylazine) in the analyte mixture first an overestimation of the s-triazine was observed. If the s-triazine concentration was further increased then a dramatic decrease of the measured concentration, thus a distinct underestimation, could be observed. Therefore, standard backpropagation turned out ot be the learning method superior to all the other learning methods investigated. Modifications of backpropagation and Rprop did not result in any improvement but in a deterioration. For training with Cascade correlation the embedded learning method Rprop was better than backpropagation but for the test patterns (which are of greater concern) backpropagation was superior. An ideal learning factor of 0.2 was determined. The best results using the simulated data were obtained from the training with a net with one hidden layer consisting of 10 neurons. The variances were lower than with the net with the measured data although the relative error decreased even here with increasing s-triazine concentration. Table I shows the optimized neural network protocol for the measured and the simulated data.

As the Tables II, III and IV show, the mean deviation increases with increasing concentration of the s-triazines although the relative error decreases for increasing concentrations. For the concentration 0.01 µg/L the relative error lies between 100 and 1000%, whereas for the concentration of 5 µg/L a relative error between 12.6 and 48.4% was observed. The main reason for this poor accuracy is that for smaller concentrations of the s-triazines the measured data are distributed over the whole measuring range, whereas for the higher concentrations the measured data are highly concentrated in a small area of the measuring range. Therefore, the higher the concentration of an s-triazine in the sample, the more exact its concentration can be determined. The same situation occurred with the simulated data, especially at very low concentrations where higher distributions of the values are observed. The higher the s-triazine concentrations are the better the predictions. The relative errors were about 40% on average for the training patterns and about 50% on average for the test patterns. In comparison, the relative errors for the measured data were over 100% on average for the training patterns and over 200% for the test patterns. As the data were only simulated, the system has to prove its suitability in measuring real environmental samples.

As Table V shows, the correct analyte classification of environmental water samples containing different analyte mixtures was possible in 80% of all cases. All environmental water samples containing none of the three s-triazines were correctly determined as zero samples. However, the concentration values could be determined more precisely by the neural net than with the ELISA only where overestimations occurred with analyte mixtures, as it was obvious from Table VI. A main factor for the high overestimations occurring are matrix effects. This can be derived mainly from the antibodies but also from the other assay components, e.g. enzyme tracer. This kind of overestimations can only be prevented by using more robust antibodies.

Measurements of natural water samples, from the Veneto region, Italy, which were contaminated with terbuthylazine only, confirmed the potential of the neural network in discriminating between the three analytes. In 100% of the samples only terbuthylazine was detected as the values for atrazine and terbutryn were smaller than the respective detection limit. On the basis of the neural network results, the dipstick IA for terbuthylazine analysis (using the monoclonal antibody K4E7 and a terbuthylazine standard series) was used to analyze 11 water samples collected in the Veneto area, Italy, from April-June 1995. The results for the 11 water samples by the dipstick IA compared favorably to the GC/MS analyses (cf. Table VII). Only slight overestimations were observed (results on average were

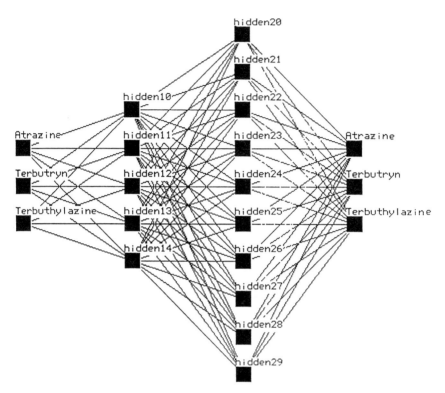

Figure 5. Topology of the optimized feedforward net for identification of 3 s-triazines applying 3 different monoclonal antibody based ELISA tests.

Table I. Optimized protocol of the neural network

Parameter	Measured data	Simulated data
Topology	3-5-10-3	3-10-3
Learning method	standard backpropagation	standard backpropagation
mean deviation in µg/L:		
1. Training	0.29	0.25
2. Test	0.48	0.32
maximum deviation in µg/L:		
1. Training	3.59	2.82
2. Test	4.98	1.74

Table II. Validation Set for Atrazine

Measured data					Simulated data				
concentration in µg/L	mean deviation in µg/L		relative error in %		concentration in µg/L	mean deviation in µg/L		relative error in %	
	Training	Test	Training	Test		Training	Test	Training	Test
0	0.01	0.03	-	-	0	0.005	0.009	-	-
0.01	0.01	0.06	100	600	0.01	0.005	0.007	50	70
0.1	0.08	0.08	80	80	0.1	0.05	0.05	50	50
0.5	0.33	0.32	66	64	0.5	0.21	0.27	42	54
1	0.38	0.32	38	32	1	0.43	0.52	43	52
5	0.63	1.33	12.6	26.6	5	0.44	0.48	8.8	9.6

Table III. Validation Set for Terbuthylazine

Measured data					Simulated data				
concentration in µg/L	mean deviation in µg/L		relative error in %		concentration in µg/L	mean deviation in µg/L		relative error in %	
	Training	Test	Training	Test		Training	Test	Training	Test
0	0.03	-	-	-	0	0.003	-	-	-
0.01	0.05	0.05	500	500	0.01	0.004	0.005	40	50
0.1	0.08	-	80	-	0.1	0.05	-	50	-
0.5	0.60	0.41	120	82	0.5	0.22	0.23	44	46
1	0.63	-	63	-	1	0.36	-	36	-
5	0.93	0.91	18.6	18.2	5	1.11	1.06	22.2	21.2

Table IV. Validation Set for Terbutryn

Measured data					Simulated data				
concentration in µg/L	mean deviation in µg/L		relative error in %		concentration in µg/L	mean deviation in µg/L		relative error in %	
	Training	Test	Training	Test		Training	Test	Training	Test
0	0.02	0.09	-	-	0	0.006	0.003	-	-
0.01	0.03	0.10	300	1000	0.01	0.005	0.005	50	50
0.1	0.08	0.10	80	100	0.1	0.07	0.08	70	80
0.5	0.47	0.39	94	78	0.5	0.24	0.35	48	70
1	0.93	0.54	93	54	1	0.52	0.66	52	66
5	1.79	2.42	35.8	48.8	5	1.22	0.59	24.4	11.8

Table V. Test Set: Measurement of 52 Environmental Water Samples

Measured data			
s-triazine present in the sample	samples correctly detected in %	mean deviation in μg/L	maximum deviation in μg/L
atrazine	71.4	1.32	3.31
terbutryn	85.7	1.81	6.34
terbuthylazine	87.5	5.55	7.02
none	100	-	-
Simulated data			
s-triazine present in the sample	samples correctly detected in %	mean deviation in μg/L	maximum deviation in μg/L
atrazine	57.1	1.85	4.56
terbutryn	19.1	2.83	3.44
terbuthylazine	87.5	2.82	4.25
none	0	-	-

Table VI. Test Set: Measurement of 64 Spiked Water Samples

	Measured data			
s-triazine	mean deviation in μg/L		maximum deviation in μg/L	
	neural net	ELISA	neural net	ELISA
atrazine	1.52	2.03	5.72	4.11
terbutryn	2.10	3.05	5.06	8.79
terbuthylazine	1.92	1.26	5.24	4.95
	Simulated data			
s-triazine	mean deviation in μg/L		maximum deviation in μg/L	
	neural net	ELISA	neural net	ELISA
atrazine	0.78	2.03	3.62	4.11
terbutryn	1.32	3.05	4.49	8.79
terbuthylazine	1.08	1.26	3.19	4.95

Table VII. Measurement of 11 Environmental Water Samples from the Veneto Area, Italy, Using the K4E7 Dipstick IA with a Terbuthylazine Standard Series

Sample No.	Terbuthylazine concentration analyzed by GC/MS (μg/L)	Terbuthylazine concentration determined with the K4E7 dipstick IA (μg/L)
1000	3.3	3.8
1001	2.3	2.6
1150	5.3	5.1
1194	1.65	2.0
1195	could not be determined	1.9
1196	1.15	1.5
1197	1.35	1.9
1198	0.72	1.1
1199	2.98	3.4
1185	< 0.5	< 0.5
1189	< 0.5	< 0.5

10% higher). In the case where no terbuthylazine was present in the sample, no false positive results were obtained with the dipstick IA. The detection limit of the atrazine dipstick IA is still quite high (0.3 µg/L, as shown in Figure VI) whereas legislation in Europe requires detection limits lower than 0.1 µg/L, as this is the maximum permissible concentration for a single pesticide in drinking water (cf. European Drinking Water Guidelines).

Conclusions and Outlook

It was shown that the neural networks can be successfully applied to solve the problem of overestimations deriving from cross-reactivities of antibodies directed against small molecular weight compounds such as pesticides. In addition, at least for the 3 monoclonal antibody based assays used in this study, it was not possible to use a mathematical model to describe and predict the influence of analyte mixtures to the assay. In only using simulated data as a basis in 100% of all cases, even in the confirmed absence of s-triazines, one or more of the s-triazines was indicated in the water samples by SNNS and by the ELISAs. For comparison of our results, Karu et al. (*12*) used 14 different antibody based assays for the determination of 9 analytes and yielded similar values of about 80% correct pesticide determination using a neural network. In our case, we only investigated the absolute minimum of 3 assays to distinguish between 3 different analytes. If we consider all the samples below the detection limits as negative samples then we would even reach 100% correct identification of the 3 analytes. Therefore, one possibility is to start with these 3 monoclonal antibody based ELISAs to set 0.5 µg/L as the threshold value, which is of interest in Europe because the maximum permissible concentration is 0.5 µg/L for the sum of all pesticides as prescribed by

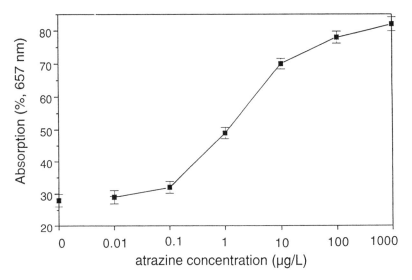

Figure 6. Representative atrazine calibration graph obtained with the dipstick IA based on the monoclonal antibodies K4E7. The tests were run six times. The two-fold standard deviations (± 2s, representing 95% confidence intervals) are indicated as error bars.

the EC Drinking Water Guidelines. It is obvious that the data set should be increased using additional antibodies directed against other pesticide compounds to enable a class-specific pesticide multiresidue screening.

The dipstick IA with multiple antibodies immobilized on one test strip as field test device is a promising tool for multiresidue screening on site.

Acknowledgements

We thank Dr. Thomas Giersch, Department of Botany, Technical University München (Weihenstephan), 85350 Freising, Germany, for the generous gift of the s-triazine-specific monoclonal antibodies K4E7, K1F4 and P6A7. We gratefully acknowledge Dr. Placido Bertin, Ente di Sviluppo Agricolo del Veneto, Centro Agrochimico, 31033 Castelfranco Veneto, Italy, who kindly provided us with environmental water samples collected from the Veneto area from April - June 1995 together with the respective GC/MS data for s-triazines. We thank Sigrid Unseld and Helmut Herzog for their excellent technical assistance in ELISA measurements to provide the ELISA data set.

Literature cited

1. Hammock, B. D.; Gee, S. J.; Cheung, P. Y. K.; Miyamoto, T.; Goodrow, M. H.; van Emon, J.; Seiber, J. N. In: *Pesticide Science and Biotechnology*, Greenhalgh, R.; Roberts, T., Blackwell, Eds.; Oxford, London, Edinburgh, 1987, pp. 309-316.
2. Bushway, R. J.; Perkins, B.; Savage, S. A.; Lekousi, S. J.; Ferguson, B. S. *Bulletin Environ. Contam. Toxicol.* **1988**, *40*, 647-654.
3. Vanderlaan, M.; Watkins, B. E.; Stanker, L. *Environ. Sci. Technol.* **1988**, *22*, 247-254.
4. Thurman, E. M.; Meyer, M.; Pomes, M.; Perry, C. A.; Schwab, A. P. *Anal. Chem.* **1990**, *62*, 2043-2048.
5. Wittmann, C.; Hock, B. *Nachr. Chem. Tech. Lab.* **1991**, *39*, M1-M .
6. Wortberg, M.; Kreissig, S.B.; Jones, G.; Rocke, D. M.; Hammock, B. D. *Analytica Chimica Acta* **1995**, *304*, 339-352.
7. Jansson, P.A. *Anal. Chem.* **1991**, *63*, 357A-362A.
8. Zell, A. *Simulation neuronaler Netze*. Addison Wesley Publishing Company, Bonn, Paris, 1994, 1st edition, 624 pages.
9. Giersch, T. *J. Agric. Food Chem.* **1993**, *41*, 1006-1011.
10. Wittmann, C.; Bilitewski, U.; Giersch, T.; Kettling, U.; Schmid, R. D.. *Analyst* **1996**, *121* (6), 863-869.
11. Muldoon, M. T.; Fries, G. F.; Nelson, J. O. *J. Agric. Food Chem.* **1993**, *41*, 322-328.
12. Karu, A. E.; Lin, T. H.; Breiman, L.; Muldoon, M. T.; Hsu, J. *Food & Agricultural Immunology* **1994**, *6*, 371-384.

Chapter 29

Role of Quality Assurance in Immunoassay Methods Used for Field Screening

Robert O. Harrison

Cape Technologies, Inc., 4 Washington Avenue, Scarborough, ME 04074

The expansion of environmental immunoassays into the industrial waste market over the last several years has brought these simple and inexpensive methods into many new situations, often lacking adequate analytical oversight. Many of these tests are now being used in kit form by environmental field workers with limited analytical training or experience. Problems resulting from such situations generally are not publicized because of possible negative reflections on kit users or manufacturers. This paper attempts to highlight some of the problems in environmental immunochemistry by examination of the current market structure and by presentation of several case studies of troubleshooting by the manufacturer. Clinical immunochemistry is identified as a possible "role model" for development of a quality assurance infrastructure. However, numerous unique problems in environmental immunochemistry will demand unique solutions. Responsibility for maintaining analytical quality ultimately falls to the analyst, but this must be supported by kit manufacturers for the industry to mature.

Since 1990, environmental immunoassays have expanded rapidly into the industrial waste analysis market. These simple and inexpensive methods have been thrust into many new situations, often without adequate analytical oversight. Most of the environmental immunoassays available are commercial kits and are often used by environmental field workers with limited analytical training or experience. Such analysts frequently expect immunoassay kits to perform flawlessly without quality assurance (QA), often comparing kit results to lab results produced under stringent QA protocols. Site managers often make decisions with inadequate understanding of the quality of their data, regardless of source. Analytically inexperienced users of immunoassay kits must be encouraged to evaluate the performance of all analytical methods relative to some QA baseline before making decisions about their analytical results. Immunoassay kit users and site managers may benefit from an explicit description of the questions they should be asking about their analytical methods and the resulting data. This paper attempts to provide that description, with guidance on specific QA procedures capable of answering many of the questions posed. The structure of the current environmental immunoassay market will be described as background. Several instructive comparisons between environmental and clinical immunochemistry will be used to illuminate some important QA issues in the environmental field.   Selected troubleshooting case studies will be presented to illustrate real-world errors and QA actions which users may find helpful. Recommendations to improve overall performance are offered for the benefit of kit users, kit manufacturers, and analytical standards manufacturers.

Quality Assurance as an Ongoing Thought Process

In the context of current environmental immunochemistry, it is appropriate to introduce basic quality assurance concepts as though they are being encountered for the first time (as will be seen from the case studies presented here).   Quality assurance will be broadly defined here as any activity directed toward maintaining overall analytical quality.   Quality assurance should be viewed not as a specific product or method, but rather a process of careful examination of analytical methods, data, and decisions.   A basic framework for this process is given in Table I. The goals of this process are 1) to anticipate and avoid analytical problems, 2) to verify that the entire sample preparation and analytical system is working properly, and 3) to determine the reason(s) for failure if it is not working. These tasks can be greatly simplified and much time, money, and effort can be saved if this thought process is engaged before generating results in the field.   Once in the field, the analyst also must provide documented results supporting either positive or negative conclusions about the quality of the analysis. The scale of QA for any individual project will often be determined by the scope and budget of that project.   However, there are many levels of QA which are appropriate for field immunoassay screening. Table II presents a list of questions the analyst should be asking at various points in the analytical process. Following each question are QA actions which should be taken in the process of answering that question.

Current Environmental Immunoassay Market Structure: The QA Problem

To understand the different QA problems and needs of environmental immunoassay users, one must recognize the segmentation of the environmental immunoassay market. This market can be divided into two quite different components. The industrial waste segment is newer and is dominated by field use of test kits during remediation actions and secondary site assessments. The pesticide segment of the market is older and more diverse, is largely based on fixed analytical laboratories, and focuses on environmental residue monitoring and product testing. The differences between these two market segments are critical to understanding the roles of each type of kit and their respective QA issues. The industrial waste market has grown faster recently and has been driven primarily by the desire of environmental engineers for tools which enable decision making in the field. This capability has fostered a new level of competition among engineering firms bidding on site remediation contracts, with economic analyses now based on short turnaround times and dramatically reduced analytical costs.   In this market, rapid expansion of field kit use has often bypassed the conventional QA world and has led frequently to preventable technical difficulties. In contrast, fixed labs doing pesticide analyses have been slower to assimilate immunoassay kit technology, largely because of the less forceful economics and the different nature of their sample streams. Their routine investment in QA for conventional methods has both helped and hindered the process of adoption of this new technology. The frequent view that kits are not amenable to existing QA methods has reduced kit penetration into this market. However, among some labs, the ability to treat immunoassay kits as just another method requiring sound QA has produced significant progress toward routine kit use. The remainder of this paper will focus on the field use market segment because of the combination of severity of problems and lack of analytical/QA experience evident in that area.

Table I. Framework of a typical quality assurance process

1. Preparatory work: conducted prior to field work
   Choice of analytical method matched against analytical needs
   Identification of QA methods appropriate for the planned field work
   Verification of analytical method performance using fortified samples

2. Field work: conducted on site
   Record keeping of sample collection, site delineation, and field analysis
   Recovery analyses using spiked samples and reference materials
   Precision analyses using field duplicates and other replication

3. Data interpretation: conducted on or off site
   Interpretation of raw data based on sample source, appearance. and history
   Interpretation of raw data based on recovery and precision results

4. Decision making: conducted on or off site
   Choice of actions based on sound interpretation of raw data
   Documentation of support for actions based on earlier QA process

Table II. Questions to be asked and quality assurance actions to be considered during field use of immunoassay kits

Phase of Project	Question to be asked	QA Action
1. Before choosing a field screening kit-	A. Is the kit appropriate for the site and analyte?	Examine site history for potential problems and conflicts.
		If analyte is a mixture, evaluate whether cross-reactivity profile of kit adequately matches composition of mixture.
		Look for likely interferences based on both site knowledge and kit information.
		Evaluate site potential for analyte weathering, especially including differential volatilization, leaching, and degradation of mixtures.
	B. Is the existing kit calibration strategy valid for the analyte and the site, or would it require modification?	If significant weathering is expected or is demonstrated in the initial assessment, evaluate how the expected change in analyte composition should affect the kit performance. Consult the kit manufacturer if necessary.
		If a change in kit performance is expected or demonstrated, compensate for this change, if possible, through consultation with

Continued on next page

Table II. *Continued*

Phase of Project	Question to be asked	QA Action
		the manufacturer and modification of the kit calibration.
	C. Is the kit appropriate for the sample matrix being tested? Or conversely, is the target sample type within the manufacturer's recommendations for that method?	Verify the appropriateness of the manufacturer's protocol for the intended matrix, including drying protocol, if any. Assess potential problems due to soil types, extremely wet or high organic sediments, etc.
2. Before taking the chosen screening method to the field-	A. Has an adequate QA plan been prepared, including confirmatory analysis with adequate lab QA?	Refer to regulatory requirements, data quality objectives, economic considerations, and client preferences. Review with QA adviser.
	B. Will the various required QA samples be available to the field analyst in a timely fashion?	Define, prepare, or obtain matrix blanks, matrix spikes, controls, spiking solutions, standard reference materials (SRMs), or other characterized samples specifically intended for QA. Review with QA adviser.
	C. Are the QA methods to be used amenable to semiquantitative screening analysis if that is the intended use of the kit? If not, have the required changes been made?	Verify compatibility of QA samples with the planned use of the kit. If changes are required, check their validity with the manufacturer of the kit or QA sample.
3. While performing the screening method in the field-	A. Has the analyst read and followed all of the precautions and instructions in the kit insert?	Conform to manufacturer's instructions for kit storage conditions, warming time from cold storage, water content of soil, maximum batch size, timing, etc.
	B. Is the sample being prepared properly?	Verify acceptability of sample appearance, especially regarding homogeneity and wetness. Be certain that treatment does not cause analyte loss by volatilization (use matrix spike, SRM, or other defined QA sample). Conform to manufacturer's instructions for sample handling and extraction.

Table II. *Continued*

Phase of Project	Question to be asked	QA Action
	C. Is the kit working properly?	Track calibrators and control samples (either spiked method blank or other control sample) to be sure they are within specified ranges.
	D. Is the test protocol being executed properly?	Track calibrators and control samples (either spiked method blank or other control sample) to be sure they are within specified ranges. Verify that precision is acceptable using field duplicates or similar QA samples.
4. When evaluating confirmatory data from lab-	A. Has the proper confirmatory method been used?	Compare to QA plan and regulatory requirements if appropriate. Note that some lab methods may be controversial (such as 418.1 for hydrocarbons). Consult QA adviser as necessary. Note also that multiple methods for one analyte may be available (e.g. 8270 vs. 8310 for PAHs), though these may be very different in performance.
	B. Is the lab performing this method correctly, including extraction and QA?	Check for proper sample holding time, correct quantitation method from chromatograms, acceptable data from QA samples, etc. Consult QA adviser as necessary.
5. When resolving conflicts between lab and screening methods-	A. Have the QA actions described above been used correctly to assure the quality of the process to this point?	Review QA plans and results for both field and lab analyses.
	B. Does the sample matrix interfere with the extraction?	Analyze spiked field samples by quantitative immunoassay to check for recovery.
	C. Does extracted material cause a positive or negative interference in the screening method?	Analyze spiked extracts of field samples by quantitative immunoassay to check for over- or under-recovery.
	D. Is the observed difference between methods due to differences in extraction or analysis?	Use double crossover- extract samples by both extraction methods, analyze each extract by both analytical methods (using proper solvent exchange for immunoassay).

Clinical Immunochemistry as QA Role Model: Key Differences

The conventional environmental analysis industry has both strong regulatory input regarding QA (1,2) and a healthy QA infrastructure. However, because this industry is based on instrumental analysis rather than immunochemistry, it is helpful to look elsewhere for a model of how QA can be applied to immunochemistry. An obvious and vigorously healthy example is the clinical immunochemistry field. Many independent companies focus solely on the QA market represented by analytical labs. The intention of this paper is not to minutely examine how QA functions in clinical immunochemistry, but rather to point out that there is no fundamental reason why environmental immunochemistry cannot develop the same vigorous QA infrastructure as that which exists in the clinical field.

Parallels and differences between environmental and clinical immunochemistry can be instructive in assessing the potential for analytical problems and the prospects for development of improved QA in the environmental immunoassay kit market. Clinical immunochemistry has been a well established core medical technology for decades and constitutes a multi-billion dollar per year industry in North America. Many clinical immunoassays are quantitative methods for small molecules, conceptually identical to some current environmental immunoassays. The systematic approach to QA that has served clinical immunochemistry so effectively could greatly benefit the environmental field. However, there are numerous differences between the two fields which prevent a wholesale adoption of clinical QA methods and infrastructure in the environmental field. The critical differences between these two fields can be classified into several areas. These are summarized below, along with discussion of the significance of those differences.

Market Size. The small size of the environmental market has prevented the development of the same economies of scale that have been crucial in the success of clinical immunoassays. In turn, there has been little pressure to develop the same type of QA infrastructure that both the clinical market and the conventional environmental market support. The overall quality in the clinical analysis industry is very high because of the intersection of several factors. In addition to the QA infrastructure, these factors include strong centralized oversight by both regulatory agencies and trade groups, profound understanding of test chemistries by the manufacturers, use of simple test formats with limited opportunity for user error, consistent and defined sample matrices, and predictable environmental conditions. The high level of overall quality exemplified by these tests is the model to which the field environmental immunoassay market aspires. However, some of the potential barriers to achieving this model may be intrinsic to the environmental field and therefore impossible to surmount.

Methods. In part because of the low market volume noted above, the sophistication of environmental immunochemistry methods has lagged significantly behind the clinical market. This has meant a nearly total lack of automation and its accompanying refinement in laboratory methods. However, the impact on field analysis has been greater. The absence of simple one-step disposable devices has required field analysts to perform relatively complicated assays with multiple pipetting operations and two or more manually timed incubation steps. In addition, the level of QA readily available to inexperienced users is minimal. This level of complexity and lack of user-friendliness is in stark contrast to the simple home pregnancy tests which are now familiar. QA is sometimes built into these simple devices, but user execution of specific QA protocols is usually not required. For the devices which do not include integral QA, test reliability still is very high, for the reasons noted in the preceding paragraph.

Matrices. Sample matrices present much greater problems for the environmental market than for the clinical market. In general, environmental matrices, such as soil, sediment, or water, are much more complex and less well defined than typical clinical matrices such as serum and urine. In addition, environmental matrices exhibit more variation within one matrix and sometimes show significant cyclical variation. For example, the range of matrix variability represented by soils from clays to high organic mucks to sediments is unheard of among major clinical sample matrices. In addition, other factors such as microbial flora, pH, inorganic content, chemical reactivity, and water content of solid samples may vary widely. All of these factors have the potential for significantly affecting the performance of an immunoassay.

Analytes. The differences between clinical and environmental analytes constitute a potentially large hurdle for environmental immunochemistry. A prime concern is that typical environmental analytes are less polar than clinical analytes. But a major complicating factor and one of the most difficult problems in environmental immunochemistry is that many common "analytes" are actually complex mixtures. In some cases an "analyte" in one sample can actually be quite different than the same "analyte" in another sample, though they may be widely presumed to be identical. An example of this is the different compositions of gasolines by season, manufacturer, and crude oil source. These may contain very different levels of the alkyl benzenes which are typically detected by an immunoassay and can therefore require different calibration procedures. Also, difficulties will arise in interpretation of analytical results from any method whenever there are changes from the starting composition due to selective leaching, volatilization, or degradation. For example, the technical polychlorinated biphenyl (PCB) products responsible for most of the original PCB contamination in the environment consist of mixtures of 15 to 20 of 209 discrete PCB congeners. Each of these congeners is a distinct compound with slightly different chemical and biological properties, including environmental mobility, bioconcentration factor, toxicity, and susceptibility to chemical and microbial degradation. Consequently, even if PCB contamination in the environment always started with the same congener mixture, which it often does not, there is clear potential for divergent sample compositions over time and under different weathering conditions. This problem is much more severe for petroleum fuels immunoassays due to their detection of a variety of alkyl benzenes. The higher polarity and volatility of these compounds compared to PCBs greatly exacerbate the differential weathering problem.

Toxaphene, polycyclic aromatic hydrocarbons (PAHs), and polychlorinated dibenzo-*p*-dioxins/furans are also complex groups of related compounds which present the kit user with some of these same difficulties. Tests for each of these analytes presents its own unique QA problems. Certain analyte groups may contain many structurally related compounds which are not normally measured by the conventional method, but are detected by the immunoassay. This can lead to irreconcilable discrepancies between analytical methods even if both methods were working correctly. For example, though PAH methods by GC or LC measure non-alkylated PAHs well, the methods are not designed for the alkylated PAHs common in crude oil. In contrast, immunoassays designed to be class specific may cross-react to some extent with such compounds even though they are not intended targets. Likewise, dioxin GC-MS methods do not report congeners with less than 4 chlorines because of their lower toxicity. However, two different dioxin immunoassays both have significant cross-reactivity for at least one trichlorodioxin (3,4). In this case, proper validation studies would need to develop modifications of standard methods to adequately address the issue of cross-reaction by related congeners which are commonly present, but at unknown concentrations. Avoidance of such pitfalls requires careful planning built on a base of understanding of both analytical methods. In all of these cases good QA is critical to maintaining confidence in the procedure, but flexibility is required in the design of appropriate QA procedures.

Analysts. One surprising aspect of the field analysis segment of the environmental immunoassay market is that it has been dominated by professionals who, upon entry, have neither immunoassay experience nor broad analytical experience of any kind. No counterpart of this analyst group exists in the clinical field. These analysts often are field engineers from environmental engineering firms who work with analytical data often, but have little or no formal analytical or QA training. Their own inexperience and/or the lack of QA infrastructure noted above often prevents them from obtaining guidance when needed. Because of these deficiencies, they are often unable or unwilling to invest in adequate QA in the field, where they need it most. This shortcoming is the unifying foundation of the four case studies presented below and is the biggest single reason for this paper. However, this bleak picture has begun to change through the efforts of some kit manufacturers and third parties in the environmental QA industry (5). Continued improvement will require commitment by each kit manufacturer to develop QA samples and procedures compatible with their own kits. As the environmental immunochemistry matures and analyst QA awareness increases, manufacturers will need to remain attentive to customer QA needs to maintain market growth.

Testing Environment. The last area of comparison to clinical immunochemistry is one which was expected by some early industry participants to cause most of the technical service problems. The controlled operating environment which is essential to high level performance in clinical laboratory settings is currently impossible in most environmental field situations. However, the use of immunoassays in uncontrolled environments has caused problems only in the most severe cases. The lack of control of ambient temperature has generally not been a significant primary cause of technical problems. However, timing and other factors related to analyst training and experience have been directly responsible for significant customer problems in many cases. In many other cases, such as the case studies below, these factors are secondary to the main problem, but they complicate the troubleshooting process sufficiently to make it difficult for anyone not highly trained in the area. As the field matures, simplification of test protocols and increased sophistication of test formats should reduce the significance of these factors further.

Technical Service Case Studies

Many of the common problems encountered at lower levels of technical service have been attributable to procedural problems related to user inexperience and test complexity. However, a number of problems defied the usual simple fixes and required higher level troubleshooting by the manufacturer (Millipore). As shown below, these cases would never have required manufacturer assistance if thorough QA planning had occurred prior to field screening and adequate QA procedures had been in place during the field use of the immunoassay kits.

Case Study 1. An engineering firm remediating an industrial PCB site reported high variability of results for calibrators and samples both within and among immunoassay runs. Routine technical service assistance did not alleviate the problem. Testing of the same kit lot by the manufacturer and use of new kits by the customer indicated that the kit was not at fault. A visit was made to the site, during which the analyst demonstrated her technique alone, side by side with the manufacturer representative. The comparison runs revealed that the customer was causing compound errors through a combination of simple and avoidable procedural errors which were undetected by the normal technical service phone work-up because of their inconsistent occurrence. These included poor pipetting technique, inadequate mixing after sample addition, and differential mixing within a run (by treating different sub-groups differently within a run). Correction of these errors brought subsequent results into acceptable ranges. The recommendations to the customer focused largely on immunoassay technique and analyst training. The manufacturer's internal recommendations were to improve user training, refine the question list used in support of telephone technical service, improve the kit instructions, provide QA control samples, and provide the customer with response ranges for the kit calibrators, to be used by the customer to verify acceptable kit performance.

Case Study 2. An engineering firm testing and disposing of soil from an industrial PCB site reported many borderline false negatives, primarily at the 1 ppm decision level. The customer was using the kit quantitatively and comparing numerical results from immunoassay and gas chromatography-electron capture detection (GC-ECD). A secondary complaint was variability of results, especially for wet soils. The initial troubleshooting, as for case study 1, was unproductive, due in part to the combination of slow reporting of lab results and pressure for rapid resolution of the field analytical problems. A site visit provided opportunity for observation and discussion of customer immunoassay technique, lab analytical results, and the limited QA data available. Field observations were initially keyed to patterns of errors in the customer's immunoassay technique, but patterns were difficult to establish because three different analysts had each performed a significant fraction of the tests, with minimal use of QA samples. However, the variability in their heat lamp drying method, which was caused by multiple analysts, contributed significantly to the variability in the final results. Changing to sodium sulfate drying reduced this variability and avoided the loss by volatilization that was apparently occurring in extreme cases. A new round of soil coring by the customer allowed observation of both the sampling technique and sample variability in the field, the latter being crucial to the ultimate resolution of the problem. The sampling area consisted of both well drained graded loam and a swampy zone of about one foot of high organic muck over a clay pan. The site map indicated a disproportionate frequency of false negative samples from the swampy zone, but only at shallower levels, corresponding to the muck layer over the clay pan (no PCB could penetrate the clay so all the deeper clay samples were true negatives).

Several samples in this new round of sampling were obtained within inches of the previous cores and were taken off site for more complete analysis. Subsamples of these soils were dried with sodium sulfate, extracted and analyzed quantitatively by immunoassay, confirming the previous immunoassay results. Other subsamples were spiked, dried, extracted, and analyzed quantitatively by immunoassay, confirming adequate recovery through the sample drying, methanol extraction, and analysis. Methanol extracts from unspiked subsamples were then split, spiked or mock spiked, and analyzed quantitatively by immunoassay to test for extractable interferences (none were observed). Several soils were then submitted for independent analysis by GC-ECD.

At about this time, it was discovered that the field staff had incorrectly reported the GC-ECD results to be on a wet weight basis. In addition, the project manager had been unaware that the record of decision for the site expressed concentration limits based on wet weight rather than dry weight. The comparative data were evaluated again in light of these problems and found to be in general agreement. Because the record of decision specified limits on a wet weight basis, the immunoassay results were not adjusted upward, but rather the GC-ECD results were adjusted downward to account for the change from dry to wet weight basis. Nearly all the samples which had appeared initially to be immunoassay false negatives at the 1 ppm level were reclassified as below 1 ppm by both methods. The independent GC-ECD results confirmed these results and the soils in question were disposed of as uncontaminated.

Further review of the accumulated data showed that some of these samples were extremely wet and should have been rejected by the lab as too wet to analyze properly (>50% moisture). However, when the lab inappropriately reported PCB results for these wet samples, the customer was unaware of this rejection criterion and accepted the results. These samples were responsible for a disproportionate share of the initial complaints. Correction of all the errors of procedure, communication, and interpretation allowed the customer to use the immunoassay kit at the site for the remainder of that field season and all of the next season. Subsequent performance data and the resulting user confidence were good enough to persuade state regulators to accept a reduction in the rate of GC-ECD confirmation to 10% for the remainder of the project.

Case Study 3. An engineering firm performing an initial assessment of an industrial site prior to sale discovered sporadic PCB contamination during routine screening (Soxhlet extraction with hexane followed by Florisil clean-up and GC-ECD analysis). During secondary assessment using semiquantitative immunoassay to determine the extent of the contamination, they observed high immunoassay false negative rates at all decision levels, relative to their ongoing GC-ECD confirmation program. Routine troubleshooting did not reveal any obvious problems, showing instead that the customer was performing the test properly and attaining acceptable precision. Samples were obtained from the customer for independent immunoassay analysis. These were extracted and analyzed quantitatively by immunoassay, confirming the prior immunoassay results. These same samples were then spiked, extracted, and analyzed quantitatively by immunoassay, confirming adequate spike recovery through the extraction and analysis. Methanol extracts from these unspiked soils were then split, spiked or mock spiked, and analyzed quantitatively by immunoassay to test for extractable interferences (none were observed).

Because the conflict between GC-ECD and immunoassay results was supported by both analytical methods and their respective extraction methods, extracts of several samples were prepared by substituting hexane for methanol in the immunoassay 2 minute shake extraction. When these extracts were evaporated for exchange into methanol and subsequent immunoassay analysis, a greasy residue remained which was methanol insoluble. For some samples, this residue was more than 1% of the original sample by weight. Several of these soils were then submitted for independent analysis by GC-ECD for PCBs and by GC-FID for hydrocarbons. The PCB analysis confirmed the previous lab results and the GC-FID analysis indicated the presence of high levels of 20 to 40 carbon aliphatic hydrocarbons. Examination of the site history revealed that through at least 2 decades, ending in the early 1960's, large quantities of oil were recycled at the site for the purpose of manufacturing petroleum jelly. The residues from this process became contaminated by PCB at some point and coated the soil particles. The chemical and physical stability of the residual hydrocarbons protected the PCB over the intervening years, while the insolubility of this residue in methanol prevented the analysis by immunoassay. The customer discovered that they were unable to use the immunoassay for its intended purpose only after significant and unnecessary expenditure of time, money, and effort.

Case Study 4. An engineering firm was testing soil by semiquantitative PCB immunoassay for classification into different disposal categories. The customer reported a high immunoassay false positive rate relative to their analytical lab GC-ECD results. A second lab had performed GC-ECD analysis on two of the samples sent to the first lab, obtaining significantly higher results for one sample. However, the second sample was only slightly higher and was interpreted by the customer as confirming the prior GC-ECD result. The customer was concerned that their disposal would become uneconomical if the many immunoassay positive samples were classified into the most contaminated category. They clearly hoped to find that the GC-ECD results were correct and that the immunoassay results were false positives. This hope, coupled with lack of experience with the immunoassay, led them to question the correctness of the immunoassay results first. However, their field QA plan was well designed, well executed and generally supported the quality of the immunoassay results. The customer was running duplicates for selected samples in each immunoassay batch and was also running frequent field duplicates. Additionally, the customer was tracking and accumulating immunoassay calibrator results for routine performance monitoring. All of the data from these samples indicated competent execution of the immunoassay, acceptable kit performance, and good precision.

Several soil samples were selected for independent testing by quantitative immunoassay, the results of which confirmed the customer's results closely. In order to separate problems in the extraction method from problems in the analytical method, selected samples were subjected to a "double crossover" analysis. Soil samples were mixed thoroughly and split into two subsamples. One subsample was extracted with methanol according to standard immunoassay procedure. This extract was analyzed quantitatively by immunoassay, but also was submitted for GC-ECD analysis (by direct injection) as a separate sample. The other subsample was submitted for routine GC-ECD analysis (Soxhlet extraction with hexane followed by Florisil clean-up and GC-ECD). In addition, a portion of the extract from this subsample was exchanged to methanol for quantitative immunoassay analysis. These results indicated that both extraction methods successfully recovered the PCB from the samples and that both analytical methods accurately measured the PCB in the extracts. In light of this determination, QA records from the first analytical lab were investigated further, revealing a history of QA problems. This lab had used a less efficient non-standard

Table III. Recommendations for improved field immunoassay performance

Recommendations for Kit Users

1. Read background information on QA and understand basic concepts.
2. Before going to the site, know site history and possible restrictions it may introduce.
3. Understand basic QA for the laboratory method as well as the field screening method.
4. Maintain access to analytical QA expertise, either on staff or as an outside adviser. Consult frequently, especially during the planning phase, to avoid unnecessary analytical costs.
5. Know the quality of the supporting lab. Demand performance data when soliciting proposals and consider quality as well as price when choosing a lab.
6. Require adequate QA from both lab and field methods.
7. Return performance comments and QA data to kit manufacturers.

Recommendations for Kit Manufacturers

1. Provide QA information to customers to increase awareness and understanding.
2. Develop and provide performance data for calibrators to allow initial performance evaluation by the user in the field.
3. Design QA samples appropriate for specific kits and make them available to customers.
4. Provide descriptions of specific QA protocols for troubleshooting.
5. For troubleshooting purposes, support quantitative use of kits. This should include both technical capability and user guidance.
6. Compile field data from kit users on an ongoing basis and make it available to all users.
7. Publicize QA pitfalls in both conventional and immunoassay methods.

Recommendations for Analytical Standards Manufacturers

1. Work with kit manufacturers and users toward developing a QA infrastructure.
2. Design QA samples appropriate for specific kits and make them available to customers and kit manufacturers.
3. Educate immunoassay developers and manufacturers about conventional QA methods.
4. Publicize QA pitfalls in both conventional and immunoassay methods.
5. Support continuing QA education for immunoassay method users.

version of a sonication method for soil extraction, without adequately defining the substitution for their client. This method had recently demonstrated a low bias due to low recovery, but because of poor QA documentation and the client's analytical inexperience, this problem was not realized until the conflict arose between independent analytical methods. A key factor was that the small number of samples (2) sent to the second lab was inadequate basis for the conclusion made about the quality of data from the first lab, especially considering the incomplete agreement of results. After a time consuming and expensive troubleshooting process, the customer was considering legal action against the first analytical lab. Subsequently, the client negotiated a settlement with the first lab and chose a different lab for further analytical support of their field immunoassay program.

Conclusions

The creation of the current market in field immunoassay kits has placed analytical capabilities in the hands of many analytically untrained and inexperienced users who often have a considerable financial stake in the correctness of the results obtained with such kits. It is in the best interests of all parties involved to maintain the integrity of the process by using adequate QA to track the quality of the analytical procedures in both field and lab. Table III lists several specific recommendations toward this end for both kit users and manufacturers, as well as for standards manufacturers who support QA for conventional environmental analyses. Commitment to QA by kit manufacturers will be necessary for maturation of the environmental immunoassay industry, especially in the field analysis segment of the market. Development of a modest QA infrastructure would provide a much needed boost in overall quality. This could be driven by either kit manufacturers or by third parties, but it is in the self-interest of the kit manufacturers to at least facilitate this process as much as possible. In the end, users of immunoassays and other field analytical kits must realize that while manufacturers can guarantee kit quality up to arrival at the site, it is the responsibility of the analyst to assure the quality of the results obtained thereafter. The best way for the kit user and site manager to accomplish this is to prepare and execute a QA plan which allows a level of confidence in the field screening results that is acceptable to all involved parties.

Acknowledgments

Thanks to Bob Robertson for enthusiastic and competent assistance with Case Study 4 and to all of the customers who cooperated in the troubleshooting process.

Literature Cited

1. U.S. Environmental Protection Agency, Office of Emergency and Remedial Response and Office of Waste Programs Enforcement **1987**. Document number EPA/540/G-97/003.
2. U.S. Environmental Protection Agency, Office of Solid Waste and Emergency Response **1993**. Document number EPA/540/R-93/071.
3. Harrison, R.O.; Carlson, R.E. *Chemosphere*, in press.
4. EnSys Environmental Products Inc. **1995**, Research Triangle Park, NC.
5. Keith, L.H.; Patton, G.L.; Lewis, D.L.; Edwards, P.G. in *Principles of Environmental Sampling*, 2nd. Ed., ACS Books, American Chemical Society: Washington DC, 1995, Chapter 1.

Chapter 30

Organic Solvent Modified Enzyme-Linked Immunoassay for the Detection of Triazine Herbicides

Walter F. W. Stöcklein, A. Warsinke, and Frieder W. Scheller

University of Potsdam, c/o Max-Delbrück Center, Robert-Roessle-Strasse 10, 13122 Berlin, Germany

The specificity (cross-reactivity) of monoclonal anti-atrazine antibody K4E7 was influenced by the addition of 10 to 20% water-miscible organic solvent to the assay buffer for the competitive binding step of enzyme-labelled atrazine and free triazines. The different cross-reactivities of the antibody in buffer and buffer containing 20% ethanol were used for the calculation of atrazine and propazine concentrations from mixtures. The activity of the enzyme label horseradish peroxidase was not inhibited by the solvent. The benefits and limitations of solvent modified enzyme immunoassays are discussed.

Organic solvents have been shown to influence the specificity of enzymes. A relationship was found between the solvent-to-water partition coefficients of substrates and subtilisin activity: the specificity of the enzyme was changed in favor of the more hydrophilic substrates, when water was replaced by anhydrous organic solvents as the reaction medium (*1*). Also antibodies against aminobiphenyl (*2*) and atrazine (*3*) exhibited different specificity in water and anhydrous organic solvents. Evidently, aqueous mixtures of water miscible organic solvents influence antibody specificities too (*4*).

The use of organic solvents in immunochemical methods promises several benefits: Hydrophobic analytes in nonaqueous environments (e.g. oil) or in solvent extracts (e.g. from soil samples) can be directly captured by antibodies. Elution of antigen from immobilized antibodies for purification, preconcentration or antibody regeneration can be improved with respect to antigen or antibody stability, elution time and yield. Lastly, influences of solvents on antibody specificity can result in more specific and sensitive immunoassays. The aim of this paper was to demonstrate the possibility of distinguishing and quantifying two analytes in mixtures by changing the specificity of a monoclonal antibody via solvent modification.

Materials

Monoclonal atrazine-antibody K4E7 was obtained from Dr. Thomas Giersch (Technical University of München, Freising, Germany). 2-Aminohexanoic-acid-4-chloro-6-isopropylamino-1,3,5-triazine was a gift from and triazine standards were obtained from Riedel de Haen AG, Seelze, Germany. Conjugation of the carboxylated atrazine with horseradish peroxidase (HRP, Boehringer Mannheim, Germany) was performed according to the EDC/NHS procedure (5). Excess reagents were removed by passage over a Sephadex G25 column (PD 10, Pharmacia, Uppsala, Sweden). The solvents dimethylsulfoxide (DMSO), dimethylformamide (DMF), methanol, acetonitrile, ethanol, acetone, n-propanol and all other reagents were of analytical grade from Merck (Darmstadt, Germany). Phosphate buffered saline (PBS, 50 mM phosphate, 150 mM NaCl pH 7.2) was used as the standard buffer for immunoassays. PBS supplemented with 0.05% Tween 20 (Sigma, Deisenhofen, Germany) was used for the washing steps. Microtiter plates (MaxiSorp) were obtained from Nunc (Roskilde, Denmark).

Enzyme Linked Immunosorbent Assay (ELISA)

The competitive assay was based on the coating antibody format which requires simultaneous incubation of antigen and antigen-enzyme conjugate on microtiter plates coated with 80 μg/L antibody. The atrazine and propazine dilutions (from 0.2 mg/L) were prepared in PBS containing 80 μg/L atrazine-HRP conjugate, with or without 20 % (v/v) solvent. The dilutions were incubated on the plates for 3 hours. The last step was the enzymatic oxidation of the substrate ABTS. The green product was measured at 405 nm. All incubations were performed at 26°C. Mean values were obtained from four wells per antigen concentration. The triazine stock solutions were adjusted with reference to their absorbance at 222 nm. The molar extinction coefficient ε was determined to be 5,360 $M^{-1}cm^{-1}$ for atrazine and 4,430 $M^{-1}cm^{-1}$ for propazine.

The following ELISA procedure was used for the determination of solvent effects on the desorption of antibodies: the coated wells were incubated with 10% or 20% solvent in PBS for 2 hours. After a washing step, atrazine-enzyme conjugate in PBS was incubated, as described above. The antibody concentration for coating and the conjugate concentration was 2 mg/L. The overall effects of solvents on the ELISA, including antibody desorption and conjugate binding to antibodies were determined by incubation of conjugate, dissolved in the respective medium, in the coated wells for 3 hours, as described for the competitive assay.

Effect of Solvents on Single Binding Steps in ELISA

Effects of Solvents on Antibody Desorption. The water miscible solvents used had only moderate effects on the interaction of antibody with the surface of the wells, at a concentration of 10%. The observed activities were between 80% and 120% of the control, in which the solvent incubation step was omitted. With 20% solvent, activity decreased from DMSO to n-propanol, which exhibited a remarkably high antibody desorption (Table I). The results were based on the absorbance changes obtained in the substrate incubation step within 10 min, which are a measure of the amount of antibodies adsorbed to the well surface (via bound conjugate). The background signals obtained in the absence of antibodies were negligible. The relative absorbance signals (100 % without solvent) were compared with the log P values, which are a measure of the hydrophobicity of the solvents. The log P value is the logarithm of the partition coefficient of a substance in a two-liquid phase octanol/water system (6). A high log P value indicates high hydrophobicity. The absorbance data show approximate correlation with the log P value, which is in agreement with the expected desorption of antibodies, adsorbed to the wells by hydrophobic interaction, by organic solvents.

Effects of Organic Solvents on Conjugate-Antibody Binding. The effects of organic solvents were also examined with respect to binding of the conjugate to antibody in the presence of solvent (Table I, right column). In fact the results of this experiment also include the antibody desorbing effects of solvents described above. Comparing the results of both experiments it can be concluded that, among the solvents tested, only DMSO and DMF decrease the antibody affinity for antigen-enzyme conjugate. The other solvents seem to increase the affinity, as the values in the right column of Table I are higher than in the left column. However, the solvents also have an influence on the enzyme stability: 20% ethanol and 20% DMF were tested, and it was found that the activity of the diluted enzyme-atrazine conjugate (80 µg/L) was unchanged after 1 hour storage at 26°C in the presence of the solvent, whereas in PBS the activity decreased to 63%. A stabilization effect of 10-30% ethanol has already been described for peroxidase and laccase (4). Therefore, an increased stability of an antigen-enzyme conjugate in the presence of a solvent may result in higher absorbance changes in the substrate incubation step, and may thus

Table I. Effects of Water Miscible Solvents (20% v/v in PBS) on Atrazine-ELISA

Solvent	log P	Antibody adsorption (% activity)[a]	Conjugate binding (% activity)[a]
Buffer	-2.7	100	100
DMSO	-1.3	91	56
DMF	-1.0	59	23 (6[b])
Methanol	-0.76	82	120
Acetonitrile	-0.33	65	100
Ethanol	-0.23	74	92 (61[b])
Acetone	-0.23	67	109
n-Propanol	+0.28	11	29

[a] activity of bound conjugate after substrate addition; the activity obtained in the assay without the use of organic solvent, was defined as 100%

[b] the value in brackets were obtained in the competitive ELISA (with 80 µg/L concentrations of antibody and conjugate instead of 2 mg/L)

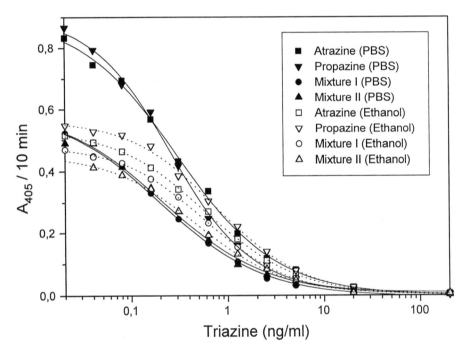

Figure 1. Semilogarithmic plots of standard curves for atrazine, propazine and the mixtures I and II , containing both triazines (concentrations see Table III). Conjugate and triazines were incubated in PBS with and without 20% ethanol.

compensate for a lower antibody affinity. Furthermore, antigen-enzyme conjugate and antibody may change their behaviour in organic solvents, when they bind to another.

The solvents (20% ethanol and DMF were tested) also exhibited a more negative effect on the atrazine-ELISA at higher working dilutions of antibody and conjugate (Table I, values in brackets). That means that the highest possible dilution of antibody and conjugate in PBS, resulting in the highest sensitivity of the assay, cannot be applied in the presence of the solvents.

Effect of 20% Ethanol on Standard Curves for Atrazine and Propazine

As mentioned above, changes of the substrate specificity of enzymes and the cross-reactivity of antibodies by substitution of water with organic solvents have been published. We looked for a solvent which fulfilled the following criteria: it should change the cross-reactivity of antibody K4E7 for atrazine and propazine; not severely impair the adsorption of antibodies and the binding of conjugate (resulting in low signals); be water-miscible; be preferentially non-toxic; have low vapour pressure. 20% (v/v) solvent was chosen as the upper limit, as higher solvent compositions were expected to desorb antibodies from the wells. Water immiscible solvents were not taken into consideration, as their use implies complications for the analysis of water samples (e.g. affords extraction). 20% ethanol seemed to fit best to the abovementioned criteria, although the solvents were not screened for their effects on antibody cross-reactivity, and was selected for further studies.

Standard curves for atrazine, propazine and mixtures I and II (see below) were obtained with buffer or buffer containing 20% ethanol. Semilogarithmic plots were prepared from the resultant data. Figure 1 shows the plots. The curves were fitted by a four-parameter log-logistic model (Microcal Origin, sigmoidal fit). The parameters are summarized in Table II.

The maximum absorbance values for the triazine standards in different media can be directly compared, as they were determined in the same experiment. The values for the mixtures I and II were obtained in another experiment. The maximum absorbance (in the absence of antigen) was lower compared with the first experiment, presumably due to the lower activity of the enzyme conjugate after freezing and thawing.

Two effects of ethanol can be seen. At first, the apparent affinities of the antibody for both triazines and also for the conjugate decrease, as compared with buffer. At second, the cross-reactivity of the antibody is changed in favor of atrazine. The first effect was also quantified for other solvents: The IC_{50} for atrazine was 1.4 (20% DMSO) and 5.9 (20% DMF), compared with 0.3 (PBS). The main reason for the decreased apparent affinities is that the solvents impair hydrophobic interactions between the triazines and the antibody. However, other factors may also contribute: partial desorption of antibodies leads to a mixed equilibrium; solvents may affect enzyme-conjugate complexes, free and adsorbed antibodies and free conjugate differently.

The solvent effect on cross-reactivity can be interpreted with respect to antigen hydrophobicities. Solubility data (7) and the effect of detergents on triazine ELISA (8) show that atrazine is more hydrophilic than propazine. Therefore, the cross-reactivity is shifted by the solvents in favor of the more hydrophilic substrate atrazine. This result is in accordance with the effect of nearly anhydrous toluene and hexane on the cross-reactivity of anti-atrazine antibodies (*3*) and also with published results on solvent effects on enzyme specificity (*1*).

Table II. Parameters for Atrazine and Propazine Standard Curves Obtained by Fitting with a four-parameter log-logistic Model[a]

	Atrazine	Propazine	I	II
	PBS	*PBS*	*PBS*	*PBS*
Substrate incub. time (min)	10	10	10	10
A1	0.9006	0.9059	0.594	0.594
A2	0.0012	0.01	-0.003	0.0031
IC50	**0.302**	**0.2613**	**0.207**	**0.2204**
p	0.856	1.0450	0.834	0.8254
CR %	87%	100%		
	Ethanol	*Ethanol*	*Ethanol*	*Ethanol*
Substrate incub. time (min)	10	10	12	12
A1	0.5342	0.5657	0.4822	0.4538
A2	0.0021	0.0038	0.0023	0.0035
IC50	**0.6012**	**0.758**	**0.5852**	**0.4788**
p	0.9294	0.9787	1.0023	0.9677
CR %	126%	100%		

[a]Abbreviations used: I and II = mixtures of atrazine and propazine. A1 = calculated maximum absorbance in the absence of antigen, A2 = calculated minimum absorbance in the presence of excess antigen, IC_{50} = test midpoint (center), p = slope parameter (rate), CR = cross-reactivity. Each point represents the mean of 4 wells.

Calculations of Atrazine and Propazine Concentration from Mixtures.

The different cross-reactivity of antibody K4E7 in PBS in the presence and absence of ethanol should allow the calculation of the individual triazine concentrations in mixtures. Therefore mixtures of atrazine and propazine were prepared and standard curves produced. For calculation purposes, the mixtures were assumed to be solutions of 2 mg/L atrazine. The concentrations of atrazine and propazine were calculated from the IC_{50} values. The contribution of atrazine and propazine to the measured absorbance can be formulated for mixture I with buffer as the incubation medium as follows:

$$c_{atrazine\ in\ I} = IC_{50\ (atrazine)} / IC_{50\ (I)} * p * 0.2\ mg/L \qquad [1]$$
$$c_{propazine\ in\ I} = IC_{50\ (propazine)} / IC_{50\ (I)} * (1-p) * 0.2\ mg/L \ [2]$$

The IC_{50} values from buffer experiments were used with the fraction parameter p, the IC_{50} values from the ethanol experiments with the fraction parameter z, as the contribution of atrazine and propazine to the absorbance signal depends on the medium. Therfore it can be written:

$$IC_{50} \text{ (atr., PBS)} / IC_{50} \text{ (I, PBS)}^{*p} = IC_{50} \text{ (atr., solvent)} / IC_{50} \text{ (I, solvent)}^{*z} \quad [3]$$
$$IC_{50} \text{ (prop., PBS)} / IC_{50} \text{ (I, PBS)}^{*(1-p)} = IC_{50} \text{ (prop., solv.)} / IC_{50} \text{ (I, solv.)}^{*(1-z)} \quad [4]$$

These equations can be solved for p and z and the concentrations of atrazine and propazine calculated according to equations [1] and [2] (Table III).

Table III. Theoretical and Experimentally Determined Concentrations of Atrazine and Propazine Concentrations in Mixture I and II

Concentration (ng/mL)	I	II
Atrazine theoretical	0.1	0.2
Propazine theoretical	0.2	0.1
Atrazine calculated	0.017	0.201
Propazine calculated	0.238	0.063

The calculated concentrations of atrazine and propazine in the mixed samples show moderate correlation with the theoretical values. The compound with the lower concentration was under-estimated in both mixtures. It was also considered to take into account the different slope factors of the standard curves. Indeed, the calculated slope factors qualitatively reflect the relative concentrations of atrazine and propazine in the mixtures, as the curves for propazine and mixture I are steeper than for atrazine and mixture II. However, the slope factors were not consistent throughout the experiments, and no significant solvent effect on the slope factor was observed.

Conclusion

It could be shown that the addition of 20% ethanol to the analyte/conjugate incubation buffer caused an inversion of cross-reactivity of the monoclonal antibody K4E7 for atrazine and propazine. A similar effect was observed with polyclonal anti-atrazine-antibodies in nonpolar organic solvents (*3*), where toluene shifted the cross-reactivity 5.6 fold (from 80 to 450%) in favor of atrazine, compared with propazine. However, quantitation of results may become difficult with polyclonal antibodies, unless the solvent effects are high enough and similar for the individual antibody populations.

The results show that, in principle, multi-analyte detection by solvent modification of ELISA with a single antibody is possible. However, the measured solvent effects were not high enough to allow an accurate differentiation between single analytes by comparison of the IC_{50} values in buffer and solvent. Minor changes in the IC_{50} have a large effect on the calculated concentrations. For example, the A1 value (maximum absorbance in the absence of antigen) is sometimes ambigous due to the appearance of a low dose ʹhookʹ effect, and therefore the determination of the IC_{50} becomes less precise. However, if one analyte is present in excess, the proposed method allows its identification and approximate quantitation.

It has been proposed by G. Jones et al. (*9*) to include the slope factors of standard curves with cross-reacting analytes for calculations in multianalyte analysis. This approach is necessary if undiluted (or properly diluted) samples are analyzed and if high differences in cross-reactivities can be achieved by different experimental conditions. This was not the case here. The triazine concentrations in the mixtures were chosen high enough to allow the determination of IC_{50} from a wide range of dilutions in analogy to the standard curves.

Within the specified range, it is worthwile to examine the influence of solvent composition on cross-reactivities. It has been shown recently that 10% (v/v) ethanol improved binding of antibodies to 2,4-dichlorophenoxy acetic acid immobilized on the grating coupler surface of an optical sensor whereas 20% ethanol had no effect (10). In another example, anti-testosterone antibodies exhibited a maximum affinity towards testosterone at low concentrations of several water-miscible organic solvents (11).

In principle, multianalyte detection may be possible by the selection of a correspondingly large number of solvent/water systems causing different cross-reactivity patterns. Standardization of experimental parameters can be achieved by automated immunoassays, like flow injection analysis in combination with immunosensors and automated signal evaluation. A novel approach to achieve multianalyte detection of triazines is based on multisensor arrays (12). In contrast to the method presented here this methods depends on the availability of different antibodies with related, but different specificities.

Recently, non-ionic surfactants were shown to influence the cross-reactivities of three antibodies against triazine herbicides (8). Due to solubilization by detergents the more hydrophobic triazines become shielded from binding to the antibody to a higher degree than the hydrophilic. Therefore the situation seems to be analogous to the solvation of triazines with organic solvent molecules, with a comparable effect on cross-reactivities. Depending on the individual binding mechanism, the affinity of antigens can also be influenced more or less by the ionic strength, pH and temperature. Further work will include these parameters. Future experiments will also involve the use of activated microtiter plates for covalent antibody immobilization, in combination with higher concentrations of organic solvents than used here.

Acknowledgments

This work was supported by the Deutsche Forschungsgemeinschaft, Innovationskolleg INK 16 A1-1

Literature Cited

1. Wescott, C.R.; Klibanov, A.M. *Biochim. Biophys. Acta* **1994,** *1206,* 1-9.
2. Russell, A.J.; Klibanov, A.M. *J. Biol. Chem.* **1988,** *263,* 11624-11626.
3. Stöcklein, W.; Gebbert, A.; Schmid, R.D. *Anal. Lett.* **1990,** *23,* 1465-1476.
4. Stöcklein, W.M.F.; Scheller, F.W. *Sensors Actuators B* **1995,** *24-25,* 80-84.
5. Tijssen, P. In *Practice and Theory of Enzyme Immunoassays*; Burdon, R.H.; van Knippenberg, P.H., Eds.; Laboratory Techniques in Biochemistry and Molecular Biology; Elsevier: Amsterdam, Netherlands 1989, Vol. 15; p. 285.
6. Laane, C.; Boeren, S.; Vos, K.; Veeger, C. *Biotechnol. Bioeng.* **1987,** *30,* 81-87
7. Perkow, W.; Ploss, H. *Wirksubstanzen der Pflanzenschutz- und Schädlings-bekämpfungsmittel*; 3rd edition; Verlag Paul Parey: Berlin, Germany, 1993; Vol. 2; data sheets

8. Stangl, G.; Weller, M.G.; Niessner, R. *Fresenius J. Anal. Chem.* **1995**, *351*, 301-304.

9. Jones, G.; Wortberg, M.; Kreissig, S.B.; Bunch, D.S.; Gee, S.J.; Hammock, B.D.; Rocke, D.M. *J. Immunol. Meth.* **1994**, *177*, 1-7

10. Bier, F.F.; Stöcklein, W.F.M.; Ehrentreich-Förster, E.; Scheller, F.W. In *Environmental Monitoring and Hazardous Waste Site Remediation;* Vo-Dinh, T.; Niessner, R., Eds.; SPIE: Bellingham, U.S.A.,1995, Vol. 2504; pp 153-158.

11. Giraudi, G.; Baggiano, C. *Biochim. Biophys. Acta* **1993**, *1157,* 211-216.

12. Piehler, J; Brecht, A.; Kramer, K.; Hock, B., Gauglitz, G. In *Environmental Monitoring and Hazardous Waste Site Remediation;* Vo-Dinh, T.; Niessner, R., Eds.; SPIE: Bellingham, U.S.A., 1995, Vol. 2504; pp 185-194.

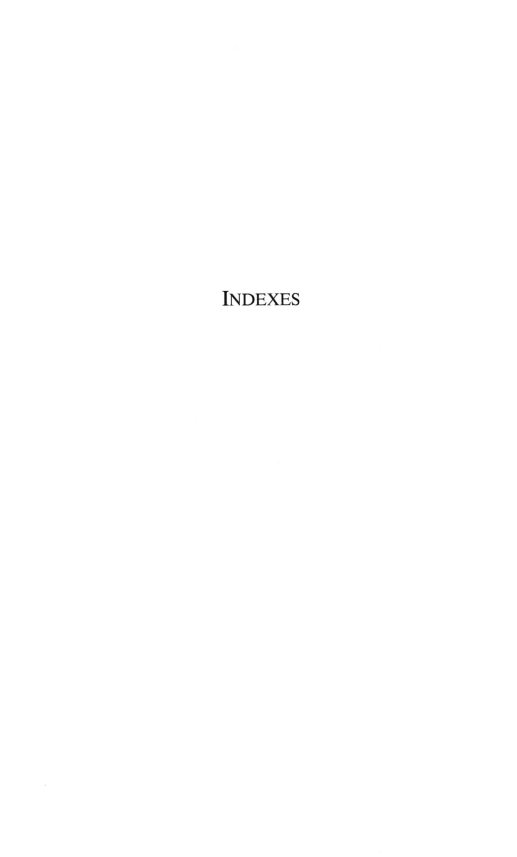

INDEXES

Author Index

Affiliation Index

Subject Index

Highlights from ACS Books

Chemical Research Faculties, An International Directory
1,300 pp; clothbound ISBN 0–8412–3301–2

College Chemistry Faculties 1996, Tenth Edition
300 pp; paperback ISBN 0–8412–3300–4

Visualizing Chemistry: Investigations for Teachers
By Julie B. Ealy and James L. Ealy
456 pp; paperback ISBN 0–8412–2919–8

Principles of Environmental Sampling, Second Edition
Edited by Lawrence H. Keith
700 pp; clothbound ISBN 0–8412–3152–4

Enough for One Lifetime: Wallace Carothers, Inventor of Nylon
By Matthew E. Hermes
364 pp; clothbound ISBN 0–8412–3331–4

Peptide-Based Drug Design
Edited by Michael D. Taylor and Gordon Amidon
650 pp; clothbound ISBN 0–8412–3058–7

Attenuated Total Reflectance Spectroscopy of Polymers: Theory and Practice
By Marek W. Urban
232 pp; clothbound ISBN 0–8412–3348–9

Teaching General Chemistry: A Materials Science Companion
By Arthur B. Ellis, Margaret J. Geselbracht, Brian J. Johnson, George C. Lisensky,
and William R. Robinson
576 pp; paperback ISBN 0–8412–2725–X

Understanding Medications: What the Label Doesn't Tell You
By Alfred Burger
220 pp; clothbound ISBN 0–8412–3210–5; paperback ISBN 0–8412–3246–6

For further information contact:

American Chemical Society
Customer Service and Sales
1155 Sixteenth Street, NW
Washington, DC 20036

Telephone 800–227–9919
202–776–8100 (outside U.S.)

The ACS Publications Catalog is available on the Internet at
http://pubs.acs.org/books

Bestsellers from ACS Books

The ACS Style Guide: A Manual for Authors and Editors
Edited by Janet S. Dodd
264 pp; clothbound ISBN 0–8412–0917–0; paperback ISBN 0–8412–0943–X

Writing the Laboratory Notebook
By Howard M. Kanare
145 pp; clothbound ISBN 0–8412–0906–5; paperback ISBN 0–8412–0933–2

Career Transitions for Chemists
By Dorothy P. Rodmann, Donald D. Bly, Frederick H. Owens, and Anne-Claire Anderson
240 pp; clothbound ISBN 0–8412–3052–8; paperback ISBN 0–8412–3038–2

Chemical Activities (student and teacher editions)
By Christie L. Borgford and Lee R. Summerlin
330 pp; spiralbound ISBN 0–8412–1417–4; teacher edition, ISBN 0–8412–1416–6

Chemical Demonstrations: A Sourcebook for Teachers, Volumes 1 and 2, Second Edition
Volume 1 by Lee R. Summerlin and James L. Ealy, Jr.
198 pp; spiralbound ISBN 0–8412–1481–6
Volume 2 by Lee R. Summerlin, Christie L. Borgford, and Julie B. Ealy
234 pp; spiralbound ISBN 0–8412–1535–9

From Caveman to Chemist
By Hugh W. Salzberg
300 pp; clothbound ISBN 0–8412–1786–6; paperback ISBN 0–8412–1787–4

The Internet: A Guide for Chemists
Edited by Steven M. Bachrach
360 pp; clothbound ISBN 0–8412–3223–7; paperback ISBN 0–8412–3224–5

Laboratory Waste Management: A Guidebook
ACS Task Force on Laboratory Waste Management
250 pp; clothbound ISBN 0–8412–2735–7; paperback ISBN 0–8412–2849–3

Reagent Chemicals, Eighth Edition
700 pp; clothbound ISBN 0–8412–2502–8

Good Laboratory Practice Standards: Applications for Field and Laboratory Studies
Edited by Willa Y. Garner, Maureen S. Barge, and James P. Ussary
571 pp; clothbound ISBN 0–8412–2192–8

For further information contact:

American Chemical Society
1155 Sixteenth Street, NW ◆ Washington, DC 20036
Telephone 800–227–9919 ◆ 202–776–8100 (outside U.S.)
The ACS Publications Catalog is available on the Internet at
http://pubs.acs.org/books